BASIC METHODS OF SOLITON THEORY

ADVANCED SERIES IN MATHEMATICAL PHYSICS

Published

Vol. 10: Yang-Baxter Equations in Integrable Systems
edited by M Jimbo

Vol. 11: New Developments in the Theory of Knots
edited by T Kohno

Vol. 12: Soliton Equations and Hamiltonian Systems
by L A Dickey

Vol. 13: The Variational Principles of Dynamics
by B A Kupershmidt

Vol. 14: Form Factors in Completely Integrable Models of Quantum Field Theory
by F A Smirnov

Vol. 15: Non-Perturbative Quantum Field Theory – Mathematical Aspects and Applications
Selected Papers of Jürg Fröhlich

Vol. 16: Infinite Analysis – Proceedings of the RIMS Research Project 1991
edited by A Tsuchiya, T Eguchi and M Jimbo

Vol. 17: Braid Group, Knot Theory and Statistical Mechanics (II)
edited by C N Yang and M L Ge

Vol. 18: Exactly Solvable Models and Strongly Correlated Electrons
by V Korepin and F H L Ebler

Vol. 19: Under the Spell of the Gauge Principle
by G. 't Hooft

Vol. 20: The State of Matter
edited by M Aizenman and H Araki

Vol. 21: Multidimensional Hypergeometric Functions and Representation Theory of Lie Algebras and Quantum Groups
by Alexander Varchenko

Vol. 22: *W*-Symmetry
by P. Bouwknegt

Advanced Series in Mathematical Physics
Vol. 25

BASIC METHODS OF SOLITON THEORY

Ivan Cherednik
University of North Carolina, USA

World Scientific
Singapore • New Jersey • London • Hong Kong

Published by

World Scientific Publishing Co. Pte. Ltd.
P O Box 128, Farrer Road, Singapore 912805
USA office: Suite 1B, 1060 Main Street, River Edge, NJ 07661
UK office: 57 Shelton Street, Covent Garden, London WC2H 9HE

British Library Cataloguing-in-Publication Data
A catalogue record for this book is available from the British Library.

Translated by Takashi Takebe, University of Tokyo.

BASIC METHODS OF SOLITON THEORY

ISBN 981-02-2643-8

Printed in Singapore.

CONTENTS

Dedicated to my teacher
Yurii I. Manin

AGE, QUOD AGIS,
AT RESPICE FINEM

PREFACE

This book is an introduction to the soliton theory from a mathematical point of view. It is a study of general integrable two-dimensional systems of differential equations and related (mostly algebraic) structures. The very first papers on the integration of the celebrated Korteweg - de Vries equation demonstrated the algebraic character of the remarkable new phenomenon. Later the algebraic (in a broad sense) methods of soliton theory became a relatively independent direction close to differential algebra, infinite dimensional representaion theory, and the theory of algebraic curves. The book contains a systematic theory of local conservations laws, Bäcklund transformations, algebraic-geometric solutions, and the inverse scattereing technique. We constantly use the zero curvature representations and the necessary elements of the theory of loop groups, though the latter are not discussed here. They will play the main role in the next book on the so-called adele approach in the soliton theory. On the other hand, we always try to adapt general methods utilizing special symmetries of concrete equations as much as we can. *All non-integrable equations are non-integrable the same way, all integrable ones are integrable in their own way.*

The book is oriented towards mathematicians and specialists in "one-dimensional" mathematical physics. The reader is not required to be acquainted with nonlinear equations, solitons, chiral fields. We use only basic mathematical knowledge of undergraduate/graduate calculus and algebra. Each section is rather autonomous to make it possible to study a particular question without reading the whole book. The same holds true (with some reservations) for the concrete equations (the equation of chiral fields, the Heisenberg magnet, Sin-Gordon, the nonlinear Schrödinger equation and others).

One can use the book to master the modern soliton technique. However it is not quite a text book. We mostly tried to present a compact updated introduction interesting to the specialists. We hope that it can draw attention to open questions and stimulate research in the soliton theory.

I am very grateful to my teacher Yurii Manin, who inspired me to take up the soliton theory. I thank I.M. Gelfand, L.D. Faddeev, S.P. Novikov, V.E. Zakharov, who influenced my interests very much. My thanks also go to H. Flashka, V.G. Kac, I.M. Krichever, L.A. Takhtadjan, V.A. Vysloukh, to the translator Takashi Takebe, and Chawne Kimber for help in editing the book. I should like to acknowledge my special indebtedness to my wife Tatiana.

The author was supported in part by the National Science Foundation and the University of North Carolina at Chapel Hill.

Notations

$\mathbb{C}, \mathbb{R}(\mathbb{C}^*, \mathbb{R}^*)$	—	the (nonzero) complex, real numbers
$\mathbb{Z}, (\mathbb{Z}_+, \mathbb{Z}_n)$	—	the (non-negative, modulo n) integers,
$\operatorname{Re}\alpha, \operatorname{Im}\alpha (\alpha \in \mathbb{C})$	—	the real, imaginary part of α
$GL_n, \mathfrak{gl}_n$	—	the complex linear group, its Lie algebra
$GL_n(\mathbb{R}), \mathfrak{gl}_n(\mathbb{R})$	—	the real linear group, its Lie algebra
$U_n, \mathfrak{u}_n$	—	the unitary group, its Lie algebra
$O_n, \mathfrak{o}_n$	—	the orthogonal group, its Lie algebra
$SL_n, \mathfrak{sl}_n$	—	the special linear group, its Lie algebra
$\otimes, \bigwedge^p$	—	tensor product, exterior product
$\operatorname{Sp}, \det$	—	matrix trace, determinant
$(p, q), \quad p, q \in \mathbb{C}^n, \mathbb{R}^n$	—	euclidean or hermitian scalar product
${}^t(\cdot), \overline{(\cdot)}, (\cdot)^\dagger$	—	transposition, complex, hermitian conjugation
$(p \wedge q)z = (z, p)q - (z, q)p$	—	the $\mathfrak{u}_n$-valued "vector" product
$\|z\|^2 = z^\dagger z = (z, z)$	—	the hermitian norm of $z \in \mathbb{C}^n$
$\delta_i^j, \delta(x)$	—	Kronecker's symbol, Dirac's delta
$A = (a_i^j) = \sum_{i,j} a_i^j I_i^j$	—	a_i^j is the (i, j) element of a matrix A
$\boldsymbol{a} = {}^t(a_i) = \sum_i a_i e^i$	—	a_i is the i-th component of a vector $\boldsymbol{a}$
$\operatorname{diag} A = \sum_i a_i^i I_i^i$	—	the diagonal part of A
$e^l = {}^t(\delta_i^l)$	—	the standard bases of $\mathbb{C}^n$
$I_k^l = (\delta_i^k \delta_l^j)$	—	the standard bases of $\mathfrak{gl}_n$
$\mathbf{I}$	—	the matrix unit
$[A]_p = \sum_{i,j=1}^p a_i^j I_i^j$	—	the cross of the first p rows and columns
$l_p(A) = \operatorname{Sp}[A]_p$	—	the trace of the matrix $[A]_p$
$m_p(A) = \det[A]_p$	—	the p-th principal minor of A
$\boldsymbol{a}^p = {}^t(a_i^p)$	—	the p-th column of A
$g_x = \dfrac{\partial g}{\partial x}, g_t = \dfrac{\partial g}{\partial t}$	—	the derivatives of g
$\mathbb{C}[\lambda], \mathbb{C}[[\lambda]]$	—	the ring of polynomials of λ, formal power series
$\mathbb{C}(\lambda), \mathbb{C}((\lambda))$	—	their quotient fields
Φ, Ψ, ϕ, ψ	—	the λ-analitical solutions of spectral problem
E_+, E_-	—	the solutions normalized at $x \to \pm\infty$
$T, t_p = m_p(T)$	—	the monodromy matrix and its minors
(1.2), §1.2	—	the second formula, item of §1
Theorem 1.2	—	the second theorem of §1

INTRODUCTION

After the fundamental work of Gardner, Green, Kruskal and Miura [GGKM] on the integrability of the *Korteweg-de Vries equation* (KdV) $u_t = 6uu_x - u_{xxx}$, a whole series of two-dimensional nonlinear KdV-like differential equations have been discovered. All these equations are integrated by the inverse scattering technique, possess infinitely many local conservation laws, parametric Bäcklund transformations, and solutions in terms of theta functions of algebraic curves as well as KdV does. The theory of these equations was named the soliton theory.

The main tool in the theory of the KdV equation is the ***Lax pair*** [Lax1], a representation of KdV in the form of the commutation relation $L_t = [A, L]$, where

$$L = -\frac{\partial^2}{\partial x^2} + u, \qquad A = -4\frac{\partial^3}{\partial x^3} + 6u\frac{\partial}{\partial x} + 3u_x.$$

The integrability and other remarkable properties of KdV result from this interpretation.

Lax pairs became rather popular. However, it was quickly realized that for many equations and for quite a few problems they must be modified. Zakharov, Shabat for the *nonlinear Schrödinger equation*, Ablowitz, Kaup, Newell and Segur for the *Sin-Gordon equation*, and Novikov for KdV suggested the so-called *zero curvature representations* which appeared to be very important for the systematic development of the theory. The new idea was to write differential equations as the relation

$$M_t - N_x = [N, M]$$

where M, N are matrix functions depending rationally on the auxiliary parameter. For example, the nonlinear Schrödinger equation (NS) $ir_t = r_{xx} + 2r^2\bar{r}$ can be written in this form for

$$M = \begin{pmatrix} -i\lambda & \bar{r} \\ r & i\lambda \end{pmatrix}, \qquad N = 2\lambda M + i\begin{pmatrix} r\bar{r} & -\bar{r}_x \\ -r_x & -r\bar{r} \end{pmatrix}.$$

The equations KdV, NS, and Sin-Gordon are still the most famous two-dimensional soliton equations, with numerous applications in both physics and mathematics. For instance, the NS equation describes the evolution of the complex envelope of ultra-short impulses in one-mode optical fiber. The stability of one-soliton impulses (which have constant velocity and form) in optical fibers is 4-5 orders greater than the stability of the "solitary waves", observed by J.Scott-Russel in a narrow canal in 1834. An impulse of length $\sim 10^{-12}$ sec can survive in an optical fiber over 100km.

An outline of the main directions. The paper [GGKM] and the work by Lax [Lax1] were the first. Then the paper [ZF] by Zakharov and Faddeev and the results by Gardner gave rise to the Hamiltonian approach in the theory of solitons and were developed algebraically in the works by Gel'fand and Dickey. Novikov [N1] and Lax [Lax2] introduced the finite zoned solutions of the KdV equation, which was a starting point of the algebraic-geometric direction.

The next achievement was a unification of all these results by means of the zero curvature representations and a generalization to the multi-component (matrix) soliton equations. Paper [ZMa1] about the *N-wave equation* and then the papers by Pohlmeyer [Poh] and Zakharov, Mikhailov [ZMi2] (in which zero curvature representations for the chiral fields were found) became important steps. The Riemann-Hilbert problem based on the theory by M. Krein was adapted to the soliton theory by Shabat. It eventually replaced the Gel'fand-Levitan-Marchenko equations (which were used to integrate KdV) in the multi-component case.

The equations of N-waves, of the *principal chiral fields* (PCF), and the *generalized Heisenberg magnet* (GHM) appeared to be the most universal soliton multi-component equations with zero curvature representations. Many others (including KdV, NS and the Sin-Gordon equation) are their reductions.

A qualitatively new type of zero curvature representations with an elliptic dependence on the spectral parameter was proposed by Borovik [Bor] and Sklyanin [Sk] for the *Landau-Lifshitz equation* of magnetism. This equation is an asymmetric variant of the equation of the Heisenberg magnet. Analogous representation of an asymmetric analogue of the PCF equation (in 2×2 matrices) was found in paper [Ch12]. The problem was to prove that the elliptic case is the most general and somehow list all soliton equations.

The quantum inverse scattering method originated by Faddeev and others gave birth to the *classical r-matrices*, which in their turn changed the classical theory a lot. Applied for the first time for the description of the Poisson brackets of integrable equations, r-matrices also appeared to be directly related to the zero curvature representations.

The problem of the classification of fundamental types of hamiltonian soliton equations was successfully formulated with the help of r-matrices and then partially solved in the paper [BD]. The zero curvature representations were subjected to the analogous algebraization in [Ch13]. It was proved that under natural assumptions the genus of the algebraic curve on which an r-matrix is defined does not exceed one. Thus the elliptic representations are indeed the most universal in the soliton theory.

Here we remark that considering more complicated coset spaces of loop groups, one can get certain zero curvature representations over an arbitrary algebraic curve. However the corresponding differential equations have essentially more complicated nature and no one has succeeded in describing and studying them so far.

Another important notion which allowed to systematize the soliton theory and involve the representation theory was the so-called *τ-function*, introduced by Hirota and Sato and then developed in the papers by Date, Jimbo, Kashiwara, Miwa and other (mainly Japanese) researchers. First elaborated for the *Kadomtsev-Petviashvili equation*, the technique of the τ-functions was transfered to many soliton equations (including matrix ones). In particular, invariant methods of obtaining formulas of determinant type for N-soliton solutions and the so called Hirota identities (closely related to the local conservation laws) were suggested. We note also the applications of τ-functions to algebraic-geometric solutions and discrete equations.

Nowadays, a large part of the mathematical theory of solitons can be included in the theory of infinite dimensional groups and their representations. What are these groups? The *loop groups* over a projective complex line and other one-dimensional complex or real curves should be mentioned first. They are closely connected with the others that are the corresponding diffeomorphism groups and the groups of matrices of infinite order.

The process of development and algebraization of the theory of solitons turned out to be and still remains fruitful not only for mathematical physics but also for modern mathematics. The theory has gone far beyond the initial range of ideas. At the present time, the quantum theory with its original methods and impressive applications plays a proponderant role. On the other hand, the possibilities of the development of the classical theory of solitons are far from being exhausted. The aim of this book is to demonstrate it.

0.1. Plan of this book.

We are trying to unify and present systematically the soliton theory on the basis of a limited number of algebraic notions. The first step in this direction was made in the work by Manin [Mani1]. Since then almost all tools of the soliton theory have radically advanced. Describing this development is another aim of this book.

Among the books written before, this one is close mostly to the books [ZMa2] and [TF2]. We would like to mention [PS] as well. Supplementing the theory presented in the monograph of Takhtajan and Faddeev, we study mainly the algebraic aspects and the matrix theory(GL_n and other Lie groups). The book can be read irrespective of whether one has read these or other books and papers, though a certain preliminary acquaintance with the soliton theory would be desirable. The readers can find proper recommendations in the introductions to each paragraph.

We give complete proofs for all statements apart from those included as exercises. The exercises which require more complicated considerations (additional research) are marked with an asterisk. In order to get familiar with the material of a chapter or a paragraph for the first time, it is recommended to read the corresponding introduction, successively looking at the theorems and corollaries (containing the main statements). Perhaps after that it is worth turning to the commentaries placed at the ends of paragraphs. Almost all paragraphs are independent units supplied with introductions and the necessary comments. Results of other paragraphs and general mathematical information are reminded.

Though a large part of this book is understandable to mathematics and physics students, it is not a textbook in a regular sense of the word. As already mentioned our main purpose is to expose modern methods of the soliton theory in the form convenient for independent study and reading special literature. We tried to compensate certain (almost inevitable) concentration, dividing the material into relatively small logically self-contained sections("solitons"). We hope that the brevity will not be an obstruction and, perhaps, make it easier to use this book as a guide in the soliton theory and its recent applications.

Issues that have been left out. In this book general mathematical theory is investigated. As for applications in theoretical physics, it is worth noting that the soliton theory plays an important role in the study of "one-dimensional" classical and quantum physical models. Multi-dimensional counterparts of the soliton techniques are also remarkably intresting. The soliton theory (especially the inverse problem technique) penetrates deeper and deeper into the numerical methods for integrable equations.

The concrete equations considered in this book are mainly selected for the following reason. The general theory of the zero curvature representations is needed for them with relatively simple modifications. Of course, there are still quite a few equations interesting from various viewpoints which are not even mentioned in this book. Some are barely discussed. The general constructions of this book can be adapted to many concrete equations. The same is true for the proofs of the basic statements, which are almost always simpler for general equations than for their reductions.

We note that the references to literature (and also the comments at the end of each paragraph) are somewhat fragmentary and, as a rule, are directly related to the questions under consideration. We did not try to give the complete list of origins. Concerning the history of the soliton theory and the relations with classical results, one can find much in the published books and survey works.

The continuation of this book [Ch11] is devoted to the invariant theory of soliton equations based on the loop groups and containes quite a few additional issues. The r-matrix technique and the general theory of τ-functions are among them. However many questions are not touched upon or hardly illustrated. Here we name some of them.

Quantum theory. Mathematical aspects of the quantum inverse problem method [F4], including the applications to the theory of Lie groups and the representation theory (see, for example, [Dr1]). Certain related questions are considered in [Ch11].

The *theory of τ functions* and the corresponding part of the representation theory of infinite dimensional groups. Recent applications of τ-functions and integrable equations to the conformal field theory, the 2D-gravity and other problems of modern mathematical physics.

Differentially-geometrical methods. We mean the results by Estabrook, Wahlquist, Lund, Regge and others (cf. comments to §1 Ch.II), their classic origins, and the continuation. The works by Griffiths on the general theory of differential forms and the results by Verdier on the chiral fields are related to this direction. We also mention an approach to the Riemann factorization problem based on the micro-differential operators (Mulase).

Besides this, there is the vast theory of *discrete equations* and *lattice models* started by Toda, Calogero, Moser, Olshanetsky, Perelomov, Kostant, Kuperschmidt and others. It is based on the group-theoretical and geometrical methods, the τ-function technique (cf., for example, [DJM] and [UT]), algebraic-geometric approach, and the hamiltonian r-matrix methods (see, [TF2]). We discuss some of them here and in [Ch11]. Of course, discrete variants of the inverse scattering problem (Ablowitz, Ladik, Flashka, MacLaughlin, Manakov and others) should not be forgotten.

In many publications multi-dimensional, integral and supersymmetric generalization of two-dimensional soliton equations are considered. The examples are the Kadomtsev-Petviashvili equation, the *duality equation* (cf. Ch.I §4.4) and some special cases of the *Einstein equation* (Belinskii, Zakharov, Neugebauer, Kramer, Nakamura, Takasaki, Wu and others). As to the integral soliton equations like the *Benny* or *Benjamin-Ono equations*, we mention the papers by Miura, Manin, Kuperschmidt, Zakharov, Case, Satsuma, Ablowitz, Kodama, Lebedev and Radul.

Concerning the algebraic theory of the soliton equations, this book and [Ch11] do not fully reflect today's situation. Say, one of the main applications in algebraic geometry will be not discussed at all, that is the proof of the so-called Novikov conjecture on the Schottky problem (characterization of the Jacobian varieties among Abelian varieties) by Mulase and Shiota. There are also deep relations with the topological field theories, and the $2D$-gravity.

Nevertheless, we hope that the present book (and its continuation [Ch11]), the

τ-function techniques from the papers by Date, Jimbo, Kashiwara, Miwa "Transformation groups of soliton equations" (see also the last chapter from [Kac]), the article [DrS1], and recent results on Virasoro and W-algebras in connection with KdV- like equations will give rather a well-balanced picture of algebraic methods of the (classical) soliton theory.

0.2. Chiral fields and Sin-Gordon equation.

One of the simplest ways to get an interesting differential equation is to impose constraints on a free particle. In the classical field theory, the analogous way is in restricting the values of free fields (vector functions on the space-time) to certain closed submanifols. Symmetries (automorphisms) of the latter induce the symmetries of corresponding equations which are called the *chiral symmetries* in physics papers. Quantities invariant relative to chiral symmetries are sometimes called simply the *invariants.* The equations of motion (the Euler-Lagrange equations) can be obtained by the method of Lagrange multipliers in both mechanics and the field theory.

By *principal chiral fields* (cf., for example, [STSF]) we mean free matrix fields with values in the manifold of invertible matrices. The chiral symmetries of such fields are generated by the action of the linear group GL_n on the left and on the right. In order to derive the equation of motion it is necessary to overcome the following difficulty: invertible matrices form an open (not closed) submanifold. For this purpose, there is a natural "algebraic-geometrical" recipe. One should consider free bi-matrix fields (g, f) (i.e., pairs of matrix valued functions g, f on the space-time), restrict the range to the closed subvariety $gf = I$, write down the corresponding equation of motion, and project it onto the first component g.

Let us denote the light cone coordinates in the two-dimensional space-time by x, t : $x = (x^0 + x^1)/2$, $t = (x^0 - x^1)/2$ (the letters ξ, η are in common use instead of x, t in physics). Choose the Lagrangian density (the density of the Lagrangian) of the free field (g, f) as

$$-\frac{1}{4}\operatorname{Sp}(g_x f_t + f_x g_t) = -\frac{1}{2}\operatorname{Sp}(g_{x^0} f_{x^0} + g_{x^1} f_{x^1}),$$

where Sp is the usual matrix trace, $g_x = \partial g(x,t)/\partial x$, $g_t = \partial g(x,t)/\partial t$ and so on. Then the Lagrangian density with the constraint $gf = \mathbf{I}$ is

$$L = -\frac{1}{4}\operatorname{Sp}(g_x f_t + f_x g_t + \Lambda(gf - \mathbf{I}))$$

for the matrix Lagrangian multiplier Λ. Calculating the variation $\delta \int L\, dx\, dt$ and setting the terms of δg, δf and $\delta\Lambda$ equal to zero, we obtain the following equations:

$$-2f_{xt} + \Lambda f = 0 = -2g_{xt} + g\lambda,$$
$$fg = \mathbf{I}.$$

Eliminating Λ and f, we come to the ***principal chiral field equation*** (PCF):

$$2g_{xt} = g_x g^{-1} g_t + g_t g^{-1} g_x. \tag{0.1}$$

This equation is consistent with the restriction of the values of g to an arbitrary Lie subgroup $G \subset GL_n$. To be more exact, if the initial data of the Cauchy problem of (0.1) take the values in G, then the same is true for the whole solution. In this case we call it a *G-field*. Let us suppose, for example, that $G = \{g \in GL_n,\ gg^* = \mathbf{I}\}$ where $*$ is an ***anti-involution*** of complex matrices, e.g., the Hermitian conjugation $\dagger$. Then the equation (0.1) is consistent with the restriction $g(x,t) \in G$. Indeed, given a field (solution) g, the function (field) $(g^*)^{-1}$ is a solution of (0.1) as well and we may apply the uniqueness theorem for differential equations. For an arbitrary group G, let us rewrite (0.1) in another form to make the consistency absolutely clear.

Set $U = g_x\, g^{-1}$, $V = g_t\, g^{-1}$. These functions are invariant with respect to the right multiplication of g by constant matrices. One has:

$$L = -\tfrac{1}{2}\,\mathrm{Sp}(g_x (g^{-1})_t) = \tfrac{1}{2}\,\mathrm{Sp}(UV).$$

It is natural to call U and V the ***left currents*** of g (with respect to the cone variables). Equation (0.1) results in the following ***PCF system*** for $U(x,t)$ and $V(x,t)$:

$$U_t + V_x = 0, \tag{0.2a}$$

$$U_t - V_x = [V, U], \tag{0.2b}$$

where the second equation is derived directly from the definition of U and V while the first is equivalent to (0.1). The field g can be uniquely recovered from the currents U, V in the connected domain of the plane $\{(x,t)\}$, if g is known at one (arbitrary) point (x', t').

If $g(x,t) \in G$, then U and V take their values in the Lie algebra $\mathfrak{g} = \mathrm{Lie}\, G$. The system (0.2) is compatible with the restriction to $\mathfrak{g}$, since it can be expressed in terms of the generators of $\mathfrak{g}$. Hence, $g(x', t') \in G \Rightarrow g(x,t) \in G$ for all x, t. In particular, $g^* g = \mathbf{I} \Rightarrow U^* + U = 0 = V^* + V$, i.e., the $*$-unitarity of g gives that the functions U and V are $*$-anti-hermitian.

The equations (0.2) can be interpreted as the compatibility condition of the system of equations depending on an auxiliary parameter λ:

$$\Phi_x - \frac{1}{1-\lambda} U\Phi = 0, \tag{0.3a}$$

$$\Phi_t - \frac{1}{1+\lambda} V\Phi = 0 \tag{0.3b}$$

for an invertible matrix valued function $\Phi(x,t;\lambda)$, or as the commutativity condition:

$$\left[\frac{\partial}{\partial x} - \frac{1}{1-\lambda}U, \frac{\partial}{\partial t} - \frac{1}{1+\lambda}V\right] = 0. \tag{0.4}$$

Equations (0.3) and (0.4) are called the *zero curvature representation for PCF.* In order to check the equivalence of (0.4) and (0.2) it is enough to expand the commutator and set the coefficients of $\lambda(1-\lambda)^{-1}(1+\lambda)^{-1}$ and $(1-\lambda)^{-1}(1+\lambda)^{-1}$ equal to zero. Note that the field g coincides with the value of Φ at $\lambda = 0$ up to a right constant factor.

Chiral fields on spheres (σ-model). Fields with values on the $n-1$ dimensional sphere, in short, S^{n-1}*-fields* are special cases of the PCF. They are defined as free fields $q(x,t) \in \mathbb{R}^n$ with values in $S^{n-1} = \{q \in \mathbb{R}^n, (q,q) = 1\}$, where (q,q) is the Euclidean scalar product. Choosing the Lagrangian density of the free $\mathbb{R}^n$ field proportional to (q_x, q_t), we obtain the equation of the S^{n-1} field:

$$q_{xt} + (q_x, q_t)q = 0, \qquad (q,q) = 1, \tag{0.5}$$

by means of the Lagrange multiplier method. It is also called the σ-model equation or the n-field equation (cf., for example, [Poh]). This equation is invariant under constant orthogonal transformations of q.

The equation (0.5) can be embedded in (0.1) in the following way (cf.[ZMi2]). Let us impose the condition $g^2 = \mathbf{I}$ on an O_n-field g. It is possible because g and g^{-1} satisfy (0.1) simultaneously. Then g can be put in the form $g = \mathbf{I} - 2P$, where $P^2 = P$. It follows from the orthogonality of g that the function $P(x,t)$ is an orthogonal projection on a subspace (which depends on x and t). Its dimension is a discrete invariant. When this dimension is one, we may set $Pz = (z,q)q$ for $z \in \mathbb{R}^n$ and suitable $q(x,t) \in S^{n-1}$. It is easy to see by a direct calculation that q satisfies (0.5). We will use U and V to make it absolutely clear.

Let us denote by $p \wedge q$ the skew symmetric matrix which acts on z as:

$$(p \wedge q)z = (z,p)q - (z,q)p,$$

where $p, q, z \in \mathbb{R}^n$. Then

$$U = g_x g^{-1} = -2P_x(1-2P) = -2q_x \wedge q,$$
$$V = g_t g^{-1} = 2q \wedge q_t$$

(we made use of the trivial relations $(q, q_x) = 0 = (q, q_t)$). Applying the formulas

$$(p \wedge q) = -q \wedge p,$$
$$(p \wedge q)_x = p_x \wedge q + p \wedge q_x,$$
$$(p \wedge q)_t = p_t \wedge q + p \wedge q_t,$$

we see that (0.2a) is equivalent to the equation

$$q_{xt} \wedge q = 0.$$

Thus the vector function q_{xt} is proportional to q. The coefficient of proportionality can be found easily and has the form exactly as in (0.5) (multiply the equation $q_{xt} = c(x,t)q$ by $(\cdot, q)$ and transform the relation $c = (q_{xt}, q)$ using the constraint $(q,q) = 1$).

From S^2-fields to Sin-Gordon equation. Solutions of (0.5) (or (0.1)) go again to solutions under the change of variables $(dx, dt) \mapsto (h\,dx, k\,dt)$, where functions $h, k > 0$ depend only on x and on t respectively. Hence, taking the conservation laws $(q_x, q_x)_t = 0 = (q_t, q_t)_x$ into account, we can normalize $q(x,t)$ by the condition $(q_x, q_x) = 1 = (q_t, q_t)$ without losing much generality. These fields q are called *S^{n-1}-fields in normalized coordinates.*

Now set $n = 3$. Then the relations

$$(q,q) = 1 = (q_x, q_x) = (q_t, q_t) \tag{0.6}$$

are sufficient to give

$$q_{xt} + (q_x, q_t)q = 0,$$

if we discard the trivial case when q_x and q_t are proportional. Really, q, q_x and q_t form a basis in $\mathbb{R}^3$. Hence the vector function q_{xt} is to be proportional to q, because it is orthogonal to q_x and q_t (it follows from (0.6) that $(q, q_x) = 0 = (q, q_t)$, $(q_{xt}, q_x) = 0 = (q_{xt}, q_t)$).

If we regard q_x and q_t as unit vector fields (at the point $q(x,t)$) on S^2, then $[q_x, q_t] = 0$. Conversely, the commutativity of unit vector fields on S^2 means that they have the form q_x, q_t for a proper q. The coordinates (x,t) are transformed by the map q into a coordinate net on S^2 with the following defining property. The opposite sides of any coordinate rectangle have the same length. Such nets are due to Tchebychef, who established their connection with the Sin-Gordon equation. Let us explain his statement.

Set $(q_x, q_t) = \cos\alpha$ (i.e., denote the *net angle* corresponding to $q(x,t)$ by α). The most direct way of getting the equation for α is to express all the derivatives of q in terms of the basis $\{q, q_x, q_t\}$. We have:

$$q_{xx} = -q - \frac{\alpha_x}{\sin\alpha} q_t + \alpha_x \cot\alpha\, q_x, \tag{0.7a}$$

$$q_{tt} = -q - \frac{\alpha_t}{\sin\alpha} q_x + \alpha_t \cot\alpha\, q_t. \tag{0.7b}$$

Thus, $(q_{xx}, q_{tt}) = 1 - \alpha_x \alpha_t \cos\alpha$. On the other hand, we obtain, using (0.5):

$$\begin{aligned}(q_{xx}, q_{tt}) &= (q_x, q_t)_{xt} - (q_{xxt}, q_t) = \\ &= 1 - \sin^2\alpha - \alpha_x\alpha_t \cos\alpha - \alpha_{xt}\sin\alpha.\end{aligned}$$

Comparing these equations, we arrive at the *Sin-Gordon equation*

$$\alpha_{xt} + \sin\alpha = 0. \tag{0.8}$$

0.3. Generalized Heisenberg magnet and VNS equation.

In the equation of the principal chiral fields (0.1) the variables x and t apper on equal footing. It is also related to system (0.2). The function V_t cannot be expressed without integration in terms of U, V and their derivatives by x. But in mechanics and the field theory the so-called evolution equations play an important role. They are solved with respect to the derivatives by the time variable t, in constrast to (0.1), (0.2). Among evolution equations,thè Hamiltonian equations are especially interesting which are constructed by a Poisson bracket (symplectic structure) on the phase space and by a Hamiltonian, that is a function or a functional on this space.

As the phase space, let us take the space of matrix-valued functions $U(x) = (u_p^q(x))$ of the variable $x \in \mathbb{R}$ ($1 \leqq p, q \leqq n$ are the indices of rows and columns). Set

$$\{u_p^q(x), u_r^s(y)\} = (u_p^q(x)\delta_r^q - u_r^q(y)\delta_p^s)\delta(x-y),$$

where δ_p^q is the Kronecker symbol and δ is Dirac's delta function. Informally speaking, $u_p^q(x)$ corresponds to the matrix $I_p^q = (\delta_i^p \delta_q^j)$ (i, j are the indices of rows and columns), "located" at the point x. The Poisson brackets among $u_p^q(x)$'s are generated by the corresponding commutators of the matrices I_p^q at "coinciding" points x, y and are trivial for $x \neq y$.

The simplest interesting Hamiltonian is

$$H = \frac{1}{2}\int \mathrm{Sp}(U'(x)\,{}^tU'(x))\,dx,$$

where ^{t} is the transposition and, temporarily, the notation $U' \stackrel{\mathrm{def}}{=} U_x = dU/dx$ is used. We will calculate the corresponding equation of motion

$$U_t(x) = \{H, U(x)\},$$

which defines the evolution of U with respect to t. Since

$$\{(u_p^q)'(x), u_r^s(y)\} = \delta'(x-y)(u_p^s(y)\delta_r^q - u_r^p(y)\delta_p^s),$$

one has:

$$\{H, u_r^s(y)\} = \frac{1}{2}\sum_{p,q=1}^{n}\int dx\{(u_q^p)'(x)(u_q^p)'(x), u_r^s(y)\} =$$

$$= \sum_{p,q=1}^{n}\int dx\delta'(x-y)(u_p^s(y)\delta_r^q - u_r^q(y)\delta_p^s)(u_q^p)'(x) =$$

$$= -\sum_{p,q=1}^{n}\int dx\delta(x-y)(u_p^s(y)\delta_r^q - u_r^q(y)\delta_p^s)(u_q^p)''(x) =$$

$$= \sum_{q=1}^{n} u_r^q(y)(u_q^s)''(y) - (u_r^q)''(y)u_q^s(y).$$

Thus,

$$U_t = [U, U_{xx}]. \tag{0.9}$$

The Poisson bracket constructed above is degenerate. Its kernel (center, commutant) contains the elements of the center of the universal enveloping algebra of $\mathfrak{gl}_n$ (the Casimir operators), where $u_p^q(x)$ are substituted for I_p^q. In particular, functionals $\mathrm{Sp}(U^m)$ lie in the center of the bracket and therefore do not depend on t (i.e., are integrals of (0.9)), where $m = 1, 2, \ldots$. This allows us to fix the conjugacy class of the matrix U. Let

$$U = F_0 U_0 F_0^{-1}, \qquad U_0 = c_1 \sum_{i=1}^{p} I_i^i + c_2 \sum_{i=p+1}^{n} I_i^i, \tag{0.10}$$

where $1 \leqq p < n$, F_0 is a certain matrix (depending on x and t) and $c_1, c_2 \in \mathbb{C}$, $c \stackrel{\text{def}}{=} c_1 - c_2 \neq 0$. The equation (0.9) with the constraint (0.10) is called the *generalized Heisenberg magnet equation* (GHM).

It is derived from (0.10), that $[U,[U,[U,X]]] = c^2[U,X]$ for an arbitrary matrix X. Since $U_x = [(F_0)_x F_0^{-1}, U]$, then $[U,[U,U_x]] = c^2 U_x$. Using the last relation we can establish the equivalence of (0.9) (under constraint (0.10)) and the equation (the *zero curvature representation*)

$$\left[\frac{\partial}{\partial x} - kU, \frac{\partial}{\partial t} - c^2k^2U - k[U, U_x]\right] = 0, \tag{0.11}$$

which holds identically for any k. This is a generalization of the analogous representation in the case $n = 2$ from [ZT, TF2].

Reduction to vector nonlinear Schrödinger equation. Supposing the constraint (0.10) is satisfied, set

$$\mathfrak{gl}_n^{\pm} = \{X \in \mathfrak{gl}_n \mid [U_0, X] = \pm cX\},$$
$$\mathfrak{gl}_n^{0} = \{X \in \mathfrak{gl}_n \mid [U_0, X] = 0\},$$
$$\mathfrak{gl}_n' = \mathfrak{gl}_n^{+} \oplus \mathfrak{gl}_n^{-} = [U_0, \mathfrak{gl}_n].$$

Then $[\mathfrak{gl}_n^{\pm}, \mathfrak{gl}_n^{\pm}] = 0$, $[\mathfrak{gl}_n^{\pm}, \mathfrak{gl}_n^{\mp}] \subset \mathfrak{gl}_n^{0}$. In Chapter I (Corollary 1.1) we will show that an arbitrary solution of (0.9) associated with U_0 is written in the form $U = F_0 U_0 F_0^{-1}$ for a solution $F_0(x,t)$(unique up to the right multiplication by a constant matrix) of the equation

$$F_0^{-1}(F_0)_t = [U_0, (F_0^{-1}(F_0)_x)_x] - \frac{1}{2}[F_0^{-1}(F_0)_x, [U_0, F_0^{-1}(F_0)_x]] \tag{0.12}$$

subject to the constraint $F_0^{-1}(F_0)_x(x,t) \in \mathfrak{gl}_n'$. Here we only check that $U = F_0 U_0 F_0^{-1}$ satisfies (0.9) if (0.12) and the last constraint hold true. Indeed,

$$U_{xx} = F_0\big([F_0^{-1}(F_0)_x, [F_0^{-1}(F_0)_x, U_0]] + [(F_0^{-1}(F_0)_x)_x, U_0]\big)F_0^{-1},$$
$$U_t = F_0[F_0^{-1}(F_0)_t, U_0]F_0^{-1}, \text{ and } [[F_0^{-1}(F_0)_x, [U_0, F_0^{-1}(F_0)_x]], U_0] = 0,$$

which implies (0.9).

Set $F_0^{-1}(F_0)_x = R = R^{(+)} + R^{(-)}$, where $R^{(+)}(x,t) \in \mathfrak{gl}_n^{+}$, $R^{(-)}(x,t) \in \mathfrak{gl}_n^{-}$ and F_0 satisfies (0.12). Then, using (0.12), we get

$$\begin{aligned} R_t =&(F_0^{-1}(F_0)_t)_x + [R, F_0^{-1}(F_0)_t] = \\ =&[U_0, R_{xx}] - \frac{1}{2}[R_x, [U_0, R]] + \frac{1}{2}[[R, [U_0, R]], R] - \frac{1}{2}[R, [U_0, R_x]] - [[U_0, R], R]. \end{aligned}$$

Taking the commutator with U_0, we find that

$$\begin{aligned} [U_0, R_t] =&cR_t^{(+)} - cR_t^{(-)} = \\ =&c^2 R_{xx}^{(+)} + c^2 R_{xx}^{(-)} + c^2([[R^{(+)}, R^{(-)}], R^{(-)}] + [[R^{(-)}, R^{(+)}], R^{(+)}]). \end{aligned}$$

Thus, we come to the system

$$\frac{1}{c}R_t^{(+)} = R_{xx}^{(+)} + [[R^{(-)}, R^{(+)}], R^{(+)}], \tag{0.13a}$$
$$-\frac{1}{c}R_t^{(-)} = R_{xx}^{(-)} + [[R^{(+)}, R^{(-)}], R^{(-)}], \tag{0.13b}$$

where the second equation is obtained from the first by the formal conjugation $R^{(\pm)} \to \varepsilon^{\pm} R^{(\mp)}$, $c \to -c$ for $\varepsilon^{\pm} \in \mathbb{C}$ such that $\varepsilon^{+}\varepsilon^{-} = 1$.

Let us take now $p = 1$ and set

$$R^{(+)} \stackrel{\text{def}}{=} \sum_{i=1}^{n-1} r_i^{(+)} I_1^{i+1}, \qquad R^{(-)} \stackrel{\text{def}}{=} \sum_{i=1}^{n-1} r_i^{(-)} I_{i+1}^{1}.$$

Then

$$[[R^{(+)}, R^{(-)}], R^{(-)}] = -2\Big(\sum_{i-1}^{n-1} r_i^{(+)} r_i^{(-)}\Big) R^{(-)},$$

$$[[R^{(-)}, R^{(+)}], R^{(+)}] = -2\Big(\sum_{i-1}^{n-1} r_i^{(+)} r_i^{(-)}\Big) R^{(+)}.$$

Let us use the vector notations :

$$\boldsymbol{r}^{(+)} = {}^t(r_i^{(+)}), \qquad \boldsymbol{r}^{(-)} = {}^t(r_i(-)), \qquad \sum_{i-1}^{n-1} r_i^{(+)} r_i^{(-)} = (\boldsymbol{r}^{(+)}, \boldsymbol{r}^{(-)}),$$

where $(r_i^{(\pm)}) \stackrel{\text{def}}{=} (r_1^{(\pm)}, \ldots, r_{n-1}^{(\pm)})$ and ^{t} is transposition. In the vector form, (0.13) is rewritten as

$$\frac{1}{c}\boldsymbol{r}_t^{(+)} = \boldsymbol{r}_{xx}^{(+)} - 2(\boldsymbol{r}^{(+)}, \boldsymbol{r}^{(-)})\boldsymbol{r}^{(+)}, \tag{0.14a}$$

$$-\frac{1}{c}\boldsymbol{r}_t^{(-)} = \boldsymbol{r}_{xx}^{(-)} - 2(\boldsymbol{r}^{(+)}, \boldsymbol{r}^{(-)})\boldsymbol{r}^{(-)}. \tag{0.14b}$$

To continue we fix $\omega_j = \pm 1, \omega_1 = +1$ and set $X^* = \Omega\, {}^t\overline{X}\Omega$, where $\overline{X}$ means the complex conjugate,

$$\Omega = \operatorname{diag}(\omega_j) = \sum_{j=1}^{n} \omega_j I_j^j, \; \widetilde{\Omega} = \sum_{j=1}^{n-1} \omega_{j+1} I_j^j.$$

Let $c_1, c_2 \in i\mathbb{R}$, $c = c_1 - c_2 = \pm i$. We will assume that the solution U is $*$-anti-hermitian. Respectively, we can take $*$-unitary F_0 and $*$-anti-hermitian R. Then $\boldsymbol{r}^{(-)} = -\widetilde{\Omega}\overline{\boldsymbol{r}}^{(+)}$ and (0.14a) and (0.14b) are equivalent. Setting $\boldsymbol{r} \stackrel{\text{def}}{=} \boldsymbol{r}^{(\pm)}, \sigma = \pm c^{-1}$, we arrive at

$$\sigma \boldsymbol{r}_t = \boldsymbol{r}_{xx} + 2(\boldsymbol{r}, \widetilde{\Omega}\overline{\boldsymbol{r}})\boldsymbol{r}, \tag{0.15}$$

which is called the *vector (multi-component) nonlinear Schrödinger equation* (briefly, VNS). In particular, let $n = 2$, $\sigma = i$, $\widetilde{\Omega} = \omega = \pm 1$. Then we come to the *(scalar) nonlinear Schrödinger equation* $\sigma r_t = r_{xx} + 2\omega(r, \bar{r})r$. In this case, the calculation above establishes the equivalence of NS for $\omega = 1$ and the equation of *Heisenberg magnet*:

$$\mathbf{s}_t = \mathbf{s} \times \mathbf{s}_{xx}, \qquad (\mathbf{s}, \mathbf{s}) = 1.$$

Here $c_1 = -c_2 = i/2$, $\mathbf{s} = {}^t(s_1, s_2, s_3)$, functions $s_j(x,t) \in \mathbb{R}$ are defined by means of the expansion $-2iU = \sum_{j=0}^{3} s_j \sigma_j$ with respect to the Pauli matrices $\{\sigma_j\}$, and $\times$ is the vector product.

0.4. Four key constructions.

Restricting ourselves to the principal chiral fields, let us briefly outline the main methods of the study of the two-dimensional soliton equations in this book.

A) Let the pair U and V be a solution of the system (0.2). Find invertible matrix solutions $\Phi_1(x,t)$ and $\Phi_2(x,t)$ of the system (0.3) for the two fixed values of the parameter $\lambda_1, \lambda_2 \neq \pm 1, \infty$. Take two arbitrary spaces $K_1^0, K_2^0 \subset \mathbb{C}^n$ of complementary dimensions ($\dim K_1^0 + \dim K_2^0 = n$) and construct the following spaces depending on x, t: $K_1 = \Phi_1 K_1^0$, $K_2 = \Phi_2 K_2^0$. Let $P(x,t)$ be the projector onto the space K_1 along K_2. If K_1 and K_2 are in general position, then the functions

$$\widetilde{U} = U + (\lambda_1 - \lambda_2)P_x,$$
$$\widetilde{V} = V + (\lambda_2 - \lambda_1)P_t$$

satisfy (0.2). This is the simplest example of the *Bäcklund-Darboux transformation* for the PCF. The pair $\{\widetilde{U}, \widetilde{V}\}$ constructing by constant diagonal $U = U_0, V = V_0$ (taking as an initial solution) is called the *one-soliton* pair. The corresponding g is a one-soliton PCF.

B) Let us assume that U is equivalent to a constant diagonal matrix $U_0 \stackrel{\text{def}}{=} \operatorname{diag}(\mu_1, \ldots, \mu_n)$, where $\{\mu_j\} \subset \mathbb{C}$ (i.e., U is conjugated to U_0). Then we can construct a formal solution of (0.3) in the following form:

$$\Phi = \Big(\sum_{s=0}^{\infty} \Phi_s (1-\lambda)^s\Big) \exp\big((1-\lambda)^{-1} U_0 x\big), \qquad \Phi_0(x,t) \in GL_n.$$

As to the computation of $\{\Phi_s\}$, we have to integrate. The coefficients $\{\Phi_s\}$ are not uniquely determined. However, if all $\{\mu_j\}$ are pairwise distinct, an arbitrary solution $\widetilde{\Phi}$ has the form ΦC where C is a constant purely diagonal series of $(1-\lambda)$.

Therefore in this case the coefficients in the expansions with respect to $(1-\lambda)$ of $(\log \operatorname{diag} \Phi)_x$ and $(\log \operatorname{diag} \Phi)_t$ are polynomials of matrix elements (entries) of U, V, and their derivatives with respect to x of suitable orders. Hence the (trivial) identity

$$\frac{\partial}{\partial t}(\log \operatorname{diag} \Phi)_x = \frac{\partial}{\partial x}(\log \operatorname{diag} \Phi)_t,$$

gives an infinite series of ***local conservation laws*** for the system (0.2).

Under proper analytical assumptions, integrals of $(\log \operatorname{diag} \Phi_s)_x$ with respect to x from $-\infty$ to $+\infty$ exist and are ***integrals*** of the system (0.2) (i.e., are independent of t).

C) *Inverse scattering technique* for (0.2) is outlined as follows. Let

$$U_0 = \operatorname{diag}(\mu_1, \ldots, \mu_n), \ V_0 = \operatorname{diag}(\nu_1, \ldots, \nu_n) \ \text{for constant} \ \{\mu_j, \nu_j\},$$

$S(\lambda)$ be a matrix depending only on $\lambda \in \mathbb{R} \cup \infty$, and $[S(+1), U_0] = 0 = [S(-1), V_0]$. For the sake of simplicity, we suppose that $\{\mu_j, \nu_j\}$ are purely imaginary ($\in i\mathbb{R}$). Set

$$\Xi = \exp\Big(\frac{1}{1-\lambda}U_0 x + \frac{1}{1+\lambda}V_0 t\Big),$$
$$\widetilde{S} = \Xi S \Xi^{-1}.$$

Let us assume that there exist matrix functions $\widetilde{\Phi}$, $\widetilde{\Psi}$ (depending on x, t, λ) with the following properties:

a) $\widetilde{\Phi}\widetilde{\Psi} = \widetilde{S}$;

b) $\widetilde{\Phi}$(*resp.*, $\widetilde{\Psi}$) is analytically continued to the upper (resp., lower) half plane;

c) the continuations are invertible;

d) $\widetilde{\Phi}(\lambda = \infty) = \mathbf{I}$.

If such $\widetilde{\Phi}$ and $\widetilde{\Psi}$ exist, they are uniquely determined by $\widetilde{S}$. In this case the pair $\widetilde{\Phi}, \widetilde{\Psi}$ is called a solution of the ***regular Riemann-Hilbert problem*** for $\widetilde{S}$.

Set $\Phi = \widetilde{\Phi}\Xi$, $\Psi = \widetilde{\Psi}\Xi$. Then the function $\Phi_x\Phi^{-1} = -\Psi^{-1}\Psi_x$ is meromorphically continued to the whole complex plane with the unique "pole" at $\lambda = 1$ and "zero" at $\lambda = \infty$. Consequently,

$$\Phi_x\Phi^{-1} = \frac{1}{1-\lambda}U(x,t), \ \Psi_t\Psi^{-1} = \frac{1}{1+\lambda}V(x,t).$$

for matrix functions $U, V(x,t)$ independent of λ. These functions U and V satisfy (0.2) (cf.(0.3)) and are equivalent to U_0 and V_0.

If the initial function S is subject to proper analytical requirements, then the functions U, V will approach to U_0, V_0 for $x \to \pm\infty$ (and for arbitrary t).

D) Let Γ be a Riemann surface of genus g on which there exists a meromorphic function λ of order n. Let us assume that the poles $R_1, \ldots, R_n$ of the function λ and the zeroes $\{R_j^\pm\}$ of functions $1 \mp \lambda$ are all simple. Choose a divisor $D \geq 0$ of degree $g+n-1$ (i.e., a set of $g+n-1$ points) which does not containe the points from the set $\{R_j, R_j^\pm, 1 \leqq j \leqq n\}$. For x, t in a open subset (everywhere dense in $\mathbb{R}^2$), there exist unique *Baker functions* $\phi_1, \ldots, \phi_n$ which are meromorphic apart from the points $\{R_j^\pm\}$ with the pole divisor D and have the following properties:
a) the functions $\exp(-\mu_i(1-\lambda)^{-1}x)\phi_j$ are holomorphic at the point R_i^+ for $1 \leq i \leq n$,
b) $\exp(-\nu_i(1+\lambda)^{-1}t)\phi_j$ are holomorphic at $\{R_i^-\}$,
c) $\phi_i(R_j) = \delta_i^j$ $(1 \leqq i, j \leqq n)$.

Let us denote the vector function ${}^t(\phi_1, \ldots, \phi_n)$ by $\boldsymbol{\phi}$. Then there exist unique matrix functions $U, V(x,t)$ such that

$$\boldsymbol{\phi}_x = \frac{1}{1-\lambda} U \boldsymbol{\phi}, \quad \boldsymbol{\phi}_t = \frac{1}{1+\lambda} V \boldsymbol{\phi}.$$

This pair $\{U, V\}$ satisfies (0.2) and is equivalent to the pair $\{U_0, V_0\}$ (see above). The matrices U, V (*algebraic-geometric currents*) can be expressed in terms of the Riemann theta function of the curve Γ.

0.5. Basic notations.

$\mathbb{C}, \mathbb{R}(\mathbb{C}^*, \mathbb{R}^*)$	—	the (nonzero) complex, real numbers;
$\mathbb{Z}, (\mathbb{Z}_+), \mathbb{Z}_n$	—	the (non-negative) integers, integers modulo n;
$\mathrm{Re}\,\alpha, \mathrm{Im}\,\alpha (\alpha \in \mathbb{C})$	—	the real, imaginary part of α;
$GL_n, \mathfrak{gl}_n$	—	the complex general linear group, its Lie algebra;
$GL_n(\mathbb{R}), \mathfrak{gl}_n(\mathbb{R})$	—	the real general linear group, its Lie algebra;
$U_n, \mathfrak{u}_n$	—	the unitary group, its Lie algebra;
$O_n, \mathfrak{o}_n$	—	the orthogonal group, its Lie algebra;
$SL_n, \mathfrak{sl}_n$	—	the special linear group, its Lie algebra;
$\otimes, \bigwedge^p$	—	tensor product, wedge product;
$\mathrm{Sp}, \det$	—	matrix trace, determinant;
$(p,q), \quad p,q \in \mathbb{C}^n, \mathbb{R}^n$	—	the euclidean or hermitian scalar product;
${}^t(\cdot), \overline{(\cdot)}, (\cdot)^\dagger$	—	transposition, complex, hermitian conjugation;
$(p \wedge q)z = (z,p)q - (z,q)p$	—	the $\mathfrak{u}_n$-valued "vector" product;
$\lvert z\rvert^2 = z^\dagger z = (z,z)$	—	the hermitian norm of $z \in \mathbb{C}^n$;

$\delta_i^j, \delta(x)$	— Kronecker's symbol, Dirac's delta ;
$A = (a_i^j) = \sum_{i,j} a_i^j I_i^j$	— a_i^j is the (i,j) element of a matrix A;
$\boldsymbol{a} = {}^t(a_i) = \sum_i a_i e^i$	— a_i is the i-th component of a vector $\boldsymbol{a}$;
$\operatorname{diag} A = \sum_i a_i^i I_i^i$	— the diagonal part of A;
$e^l = {}^t(\delta_i^l)$	— the standard bases of $\mathbb{C}^n$;
$I_k^l = (\delta_i^k \delta_l^j)$, $\mathbf{I}$	— the standard bases of $\mathfrak{gl}_n$, matrix unit;
$[A]_p = \sum_{i,j=1}^p a_i^j I_i^j$	— the cross of the first p rows and columns;
$l_p(A) = \mathrm{Sp}[A]_p$	— the trace of the matrix $[A]_p$;
$m_p(A) = \det[A]_p$	— the p-th principal minor of A;
$\boldsymbol{a}^p = {}^t(a_i^p)$	— the p-th column of A;
$g_x = \dfrac{\partial g}{\partial x}, g_t = \dfrac{\partial g}{\partial t}$	— the derivatives of g;
$\mathbb{C}[\lambda], \mathbb{C}[[\lambda]]$	— the ring of polynomials of λ, formal power series;
$\mathbb{C}(\lambda), \mathbb{C}((\lambda))$	— their quotient fields;
Φ, Ψ, ϕ, ψ	— the λ-analitical solutions of spectral problem ;
E_+, E_-	— the solutions normalized at $x \to \pm\infty$;
$T, t_p = m_p(T)$	— the monodromy matrix and its minors;
(1.2), §1.2	— the second formula, item of §1;
Theorem 1.2	— the second theorem of §1.

CHAPTER I

CONSERVATION LAWS & ALGEBRAIC-GEOMETRIC SOLUTIONS

We call relations $\zeta_t = \eta_x$ *conservation laws* either for system (0.2) describing the currents of the principal chiral fields or for the generalized Heisenberg magnet (0.9), where ζ and η are determined in terms of U and V from the corresponding equations. We will exclusively consider *local* conservation laws with the *densities* ζ that are polynomials of the entries of U and their derivatives with respect to x. As to η, it is of the same type for GHM and also depends linearly on the entries of V in the case of PCF. Supposing that the equation $\eta(x_+, t) = \eta(x_-, t)$ is satisfied for constant $x_+ > x_-$, the functional

$$I = \int_{x_-}^{x_+} \zeta(x)\, dx$$

of the entries of U is an *integral of motion* of the corresponding equation because

$$I_t = \frac{dI}{dt} = \int_{x_-}^{x_+} \eta_x(x)\, dx = \eta(x_+) - \eta(x_-) = 0.$$

The results of §1 of this chapter on local conservation laws are adapted to the case $x_\pm = \infty_\pm$ which makes them compatible with Ch.II. The corresponding algebraic machinery should be changed a little if $x_\pm$ are finite (cf., e.g., [Bo, DrS, RSTS]). We also introduce a formal analogue of the *resolvent* of equation (0.3a). The resolvent is closely related to local conservatin laws and the "higher" equations of the GHM type. We will establish a direct analytic relation between the resolvent and the densities of local conservation laws in Ch.II, using variational derivatives.

The next §2 is somewhat isolated in this book. We discuss two constructions of the generalized Lax equations, based on the zero curvature representations and on the Lax pairs respectively. These results are used in this book mostly for the Sin-Gordon equation (0.8) and the nonlinear vector Schrödinger equation (0.15). Quite a few statements are given as excersises. However a certain acquaintance with the Lax equations is important to understand the main ideas of the soliton theory and will be useful when reading other paragraphs.

In §3, the zero curvature representation results in the construction of ***algebraic-geometric*** ("finite-zone") solutions of the basic equations (0.2) and (0.9). In particular, we prove that the manifold of anti-hermitian solutions is connected. Algebraic-geometric solutions of the main reduction equations (the equations of the S^{n-1}-fields (0.5), VNS (0.15), and Sin-Gordon) are discussed in §4 . The reality conditions are considered systematically. We would like to demonstrate the variety of the methods of "finite-zone integration". In order to show universality of the developed technique, we construct algebraic-geometric solutions of the four-dimensional Euclidean duality equation.

The finite-zone solutions from §3,4 are based on the Baker functions of smooth algebraic curves (functions with exponential singularities). We use theta-functions mainly for illustrating purposes. Respectively, the divisors and distributions play the main role (instead of explicit formulae in terms of theta-functions). Baker functions are more convenient to establish the main properties of the algebraic-geometric solutions and are much better to deal with for the curves with arbitrary singularities than theta-functions are (see [Ch11]). It is worth mentioning that the algebraic-geometric integration replaces the Jost and τ-functions by the Baker and theta-functions respectively.

§1. Local conservation laws

In §1.1 we will construct formal Jost functions for the spectral problem associated with either PCF (0.2) or GHM (0.9). In §1.2 we formulate the main results on the locality of the conservation laws and the resolvents for these equations, which are proved in §1.3. In the next §1.4 we discuss direct methods of calculation of the resolvents and densities of conservation laws and applications to GHM. The last §1.5 is devoted to relations between various series of conservation laws modulo exact derivatives (including that for unitary or orthogonal PCF and S^{n-1}-fields). We use only the simplest properties of linear differential equations (cf., e.g., [A2]).

1.1. Formal Jost functions.

Here and below, depending on the context, either U, V satisfy system (0.2) or U is a solution of (0.9) under the constraint from (0.10). Let us assume that the eigenvalues $\mu_1, \dots, \mu_n$ of the matrix U do not depend on x (though this assumption is not strictly necessary — see [Ch6]). We assume that U is diagonizable, i.e.(cf. the Introduction),

$$UF_0 = F_0U_0,$$

$$\text{where } U_0 = \operatorname{diag}(\mu_1, \dots, \mu_n) = \sum_{i=1}^{n} \mu_i I_i^i, \qquad F_0(x,t) \in GL_n.$$

As to the GHM equation, (0.10) gives that

$$\mu_1 = \dots = \mu_p = c_1, \qquad \mu_{p+1} = \dots = \mu_n = c_2.$$

Set

$$\mathfrak{gl}_n^0 = \{ X \in \mathfrak{gl}_n, [U_0, X] = 0 \}, \qquad \mathfrak{gl}_n' = [U_0, \mathfrak{gl}_n].$$

Then $\mathfrak{gl}_n = \mathfrak{gl}_n^0 \oplus \mathfrak{gl}_n'$ and an arbitrary element $X \in \mathfrak{gl}_n$ can be expressed uniquely in the form $X = X^0 + X'$ where $X^0 \in \mathfrak{gl}_n^0$ and $X' \in \mathfrak{gl}_n'$. We will often regard $\mathfrak{gl}_n$ as an algebra, not only as a Lie algebra.

It will be convenient to introduce

$$k = \frac{1}{1-\lambda}, \qquad k' = \frac{1}{1+\lambda} = \frac{k}{2k-1}.$$

Then equation (0.3a) is rewritten in the form

$$\Phi_x = kU\Phi. \tag{1.1}$$

System (0.2) is the compatibility condition of (1.1) and the equation

$$\Phi_t = k'V\Phi. \tag{1.2A}$$

As to equation (0.9), it is the compatibility condition of (1.1) and

$$\begin{gathered}\Phi_t - (c^2k^2U + k[U, U_x])\Phi,\\ c = c_1 - c_2 = \mu_1 - \mu_n,\end{gathered} \tag{1.2B}$$

by virtue of (0.11). Later on, $Q = F_0^{-1}(F_0)_x$.

Proposition 1.1. a) *There exists an invertible formal solution of equation (1.1) in the form*

$$\Phi = (\Phi_0 + \Phi_1 k^{-1} + \cdots + \Phi_s k^{-s} + \cdots)\exp(kU_0x),\ \ \Phi_s(x,t) \in \mathfrak{gl}_n. \tag{1.3}$$

Here $\exp(kU_0x)$ *is understood formally (all natural properties with respect to multiplication and differentiation hold), the invertibility of* Φ *means the invertibility of the formal series* $\sum_{s=0}^{\infty}\Phi_s k^{-s}$ *that is equivalent to the invertibility of* Φ_0.

b)*An arbitrary (not necessarily invertible) solution of (1.1) of type (1.3) is the product* ΦC *for a proper series independent of* x:

$$C = C_0 + C_1k^{-1} + \cdots + C_sk^{-s} + \cdots, \qquad [C_s, U_0] = 0. \tag{1.4}$$

Proof. We set $\Phi = F_0\Psi$, $\Phi_s = F_0\Psi_s$, where $U = F_0U_0F_0^{-1}$ (see above). Then equation (1.1) transforms into the equation

$$\Psi_x + Q\Psi = kU_0\Psi \tag{1.5}$$

for $\Psi = \sum_{s=0}^{\infty}\Psi_s k^{-s}\exp(kU_0x)$, which, in its turn, leads to the system

$$\begin{aligned}&[U_0, \Psi_0] = 0,\\ &[U_0, \Psi_s] = Q\Psi_s + (\Psi_s)_x, \qquad s = 0, 1, 2, \ldots.\end{aligned}$$

The invertibility of Φ is equivalent to the invertibility of Ψ_0. Replacing Ψ_s by $\Psi_s^0 + \Psi_s'$ (see the beginning of this section), we obtain that $\Psi_0' = 0$ and

$$(\Psi_s^0)_x = -(Q\Psi_s)^0, \tag{1.6^0}$$

$$(\Psi_s')_x + (Q\Psi_s)' = [U_0, \Psi_{s+1}']. \tag{1.6$'$}$$

Supposing that Ψ_j for $j < s$ and Ψ'_s have been already determined, we will find Ψ^0_s and Ψ'_{s+1} as follows. First (1.6^0) can be solved because it is a linear inhomogeneous differential equation for Ψ^0_s (with coefficients depending on Q, Ψ'_s). Substituting Ψ^0_s into $(1.6')$, we get Ψ_{s+1}. Thus assertion a) is proved.

In the above procedure, Ψ'_{s+1} was determined uniquely, but Ψ^0_s was not. One can add to the latter a solution of the homogeneous equation

$$(\Psi^0_{s-1})_x + (Q\Psi^0_{s-1})^0 = 0.$$

When $s = 1$, we see that $\Psi^0_0 = \Psi_0$ is unique up to a matrix independent of x and commutative with U_0 as the right factor.

Let us fix an invertible solution Ψ of (1.5). Then arbitrary solution $\dot{\Psi}^0_s$ of equation (1.6^0) (involving $\Psi_0, \dots, \Psi_{s-1}, \Psi'_s$) is represented as

$$\dot{\Psi}^0_s = \Psi^0_s + \Psi_0 \dot{C}_s,$$

where $\dot{C}_s(t) \in \mathfrak{gl}^0_n$. Therefore,

$$\sum_{j=0}^{s-1} \Psi_j k^{-j} + (\Psi'_s + \dot{\Psi}^0_s)k^{-s} = \Big(\sum_{j=0}^{s} \Psi_j k^{-j}\Big)(1 + \dot{C}_s k^{-s})$$

modulo $O(k^{-s-1}) \stackrel{\text{def}}{=} (\cdot)k^{-s-1}$. Hence the induction on s proves claim b) for Ψ and, consequently, for $\Phi = F_0\Psi$. □

Proposition 1.2. *Let us consider either system (1.1,2A) or (1.1,2B) corresponding to PCF and GHM respectively. There exists an invertible solution* $\check{\Phi}(x,t)$ *of type (1.3A) for PCF and of type (1.3B) for GHM:*

$$\check{\Phi}(x,t) = (\Phi_0 + \Phi_1 k^{-1} + \cdots + \Phi_s k^{-s} + \cdots)\exp(kU_0 x), \tag{1.3A}$$

$$\check{\Phi}(x,t) = \Big(\sum_{s=0}^{\infty} \Phi_s k^{-s}\Big)\exp(kU_0 x + c^2k^2U_0 t). \tag{1.3B}$$

This solution is unique up to a constant (independent of x, t*) invertible right factor* C *in the form (1.4).*

Proof. Let us use the zero curvature representations (0.4) and (0.11). Given an arbitrary solution $\check{\Phi}$ of equation (1.1) of type either (1.3A) or (1.3B), the function $\dot{\Phi}$

$$\text{A)}\ \dot{\Phi} = \check{\Phi}_t - k'V\Phi,$$
$$\text{B)}\ \dot{\Phi} = \check{\Phi}_t - c^2k^2U\check{\Phi} - k[U, U_x]\check{\Phi},$$

is a solution of (1.1) as well. Moreover $\dot{\Phi}$ is a formal series of the same kind as $\check{\Phi}$.

Here k' in A) is to be replaced by the corresponding series in k^{-1}. As to case B), we have to prove that there are no terms with $k^s, s \geq 0$, in the expansion of $\dot{\Phi}$. This will be checked below in §1.4. For the moment we shall assume this to be true. Let $\check{\Phi}$ be invertible.

Using Proposition 1.1b), we obtain that $\dot{\Phi} = \check{\Phi}C$, where $C = C(t)$ is from (1.4). We can find an invertible formal solution of the equation $S_t + CS = 0$ in the series $S(t)$ of type (1.4), successively integrating the equations for the coefficients of S. Then the formal series $\check{\Phi}S$ is a desired solution of the corresponding systems (S commutes with the exp-factor). The uniqueness (up to a series from (1.4)) of $\check{\Phi}$ follows from that for S. □

The functions constructed in Propositions 1.1 and 1.2 are called *formal Jost functions.*

Corollary 1.1. *One can choose the matrix F_0 satisfying $UF_0 = F_0U_0$ (see above) such that $F_0^{-1}(F_0)_x(x,t) \in \mathfrak{gl}'_n$, and*

A) $2(F_0)_tF_0^{-1} = V$ *in the case of PCF,*

B) $F_0^{-1}(F_0)_t = [U_0, ((F_0)^{-1}(F_0)_x)_x] - \frac{1}{2}[F_0^{-1}(F_0)_x, [U_0, F_0^{-1}(F_0)_x]]$ *for GHM.*

In particular, system (0.2) (in the case of A) becomes equivalent to the following single equation for F_0:

$$(F_0^{-1}(F_0)_x)_t = \tfrac{1}{2}[U_0, F_0^{-1}(F_0)_t].$$

Proof. We take $F_0 = \check{\Phi}_0$ for $\check{\Phi}$ from Proposition 1.2. Then

$$(\check{\Phi})_t = k'V\check{\Phi} = k(2k-1)^{-1}V\check{\Phi}$$

and $(\check{\Phi}_0)_t = \frac{1}{2}V\check{\Phi}_0$ for PCF. We postpone the corresponding calculation in the case of GHM till §1.4. □

We note that there are two ways of handling the exp-factors more accurately. One can eliminate them and rewrite everything in terms of $\widehat{\Phi} \stackrel{\text{def}}{=} \sum_{s=0}^{\infty} \Phi_s k^{-s}$ instead of Φ. The second way is to "legitimate" the exp-factors describing formally all their properties. Actually it means that we treat them as functions defined in a "small" punctured neighbourhood of $k^{-1} = 0$.

1.2. Basic constructions.

First we introduce $\mathbb{C}[\widetilde{U}]$, the ring of *differential polynomials* of the entries (matrix elements) of U from §1.1 with respect to x. The arguments of these polynomials (with coefficients in $\mathbb{C}$) are derivatives with respect to x of order ≥ 0 of the entries of U. We define $\mathbb{C}[\widetilde{Q}]$ for $Q = F_0^{-1}(F_0)_x$ (cf. §1.1) in the same way. Differential polynomials in $\mathbb{C}[\widetilde{Q}]$ without constant terms form an ideal denoted by $\mathbb{C}_0[\widetilde{Q}] \subset \mathbb{C}[\widetilde{Q}]$. Respectively, let $\mathbb{C}_0[\widetilde{U}]$ be the kernel of the homomorphism $\mathbb{C}[\widetilde{U}] \to \mathbb{C}$ which maps U to U_0. When coefficients of polynomials are in a certain given $\mathbb{C}$-linear space of functions $\mathcal{A}$, we write $\mathcal{A}[\widetilde{U}]$, $\mathcal{A}[\widetilde{Q}]$. We use the same notations for other matrix functions (or their collections) instead of U and Q.

Theorem 1.1. a) *A differential polynomial of the entries of $\Psi_0, \Psi_1, \dots$ with coefficients in a certain $\mathcal{A}$ (i.e., an element of the space $\mathcal{A}[\widetilde{\Psi}_0, \widetilde{\Psi}_1, \dots, \widetilde{\Psi}_s, \dots]$) lies in $\mathcal{A}[\widetilde{Q}]$, if it does not depend on the particular choice of the solution Ψ of equation (1.5) of type (1.3).*

b) *An element of the ring $\mathbb{C}[\widetilde{\Phi}_0, \widetilde{\Phi}_1, \dots]$ which is invariant with respect to all the automorphisms corresponding to the substitutions $\Phi \to \Phi C$ belongs to $\mathbb{C}[\widetilde{U}]$, where Φ is a solution of (1.1) of type (1.3), C is from (1.4).*

The proof of this theorem will be given in the next section. Now we will discuss its applications.

Let us assume that the numbers μ_s (the eigenvalues of U and U_0) are ordered in the following way: $\mu_1 = \dots = \mu_p$, $\mu_s \neq \mu_p$ for $s > p$ (p is the multiplicity of μ_1). Given an arbitrary matrix $A = (a_i^j)$, let

$$[A]_p \stackrel{\text{def}}{=} (a_i^j;\ 1 \leqq i, j \leqq p), \qquad m_p(A) \stackrel{\text{def}}{=} \det([A]_p).$$

We will use the same notation $[A]_p$ for the matrix in $\mathfrak{gl}_n$ that is the extension of $[A]_p$ by zeros (for $i > p$, $j > p$). If $m_p(A) \neq 0$, let us define

$$l_p(B; A) \stackrel{\text{def}}{=} \operatorname{Sp}([A]_p^{-1}[BA]_p)$$

for an arbitrary matrix $B \in \mathfrak{gl}_n$ (Sp is the trace of a matrix).

Let Ψ be an invertible solution of equation (1.5) of type (1.3). Given an arbitrary matrix B, we will consider $l_p(B; \Psi) = l_p(B; \widehat{\Psi})$, where

$$\widehat{\Psi} \stackrel{\text{def}}{=} \Psi \exp(-kU_0 x) = \sum_{s=0}^{\infty} \Psi_s k^{-s}.$$

The exp-factor does not matter because it commutes with $\mathfrak{gl}_n^0$. We will also consider $m_p(\Psi) = m_p(\widehat{\Psi}) \exp(pk\mu_1 x)$. Here $m_p(\widehat{\Psi})$ is a power series in k^{-1} with the constant term $m_p(\Psi_0) \neq 0$. Set

$$\log(m_p(\Psi)) = pk\mu_1 x + \log(m_p(\Psi_0)) + \log(m_p(\Psi_0)^{-1} m_p(\Psi)),$$

where the last logarithm is regarded as the formal series $\log(1+z) = z - z^2/2 + \cdots$, $\log(m_p(\Psi_0))$ is defined for a certain fixed branch of the usual logarithmic function.

We introduce the formal power series ζ in k^{-1}:

$$\zeta + p\mu_1 k = (\log(m_p(\Psi)))_x = \frac{(m_p(\Psi))_x}{m_p(\Psi)} = l_p(kU_0 - Q; \Psi). \tag{1.7}$$

We also define the following matrices:

$$A = k' F_0^{-1} V F_0 - F_0^{-1}(F_0)_t \qquad \text{for PCF,}$$
$$B = c^2 k^2 U_0 + k[U_0, [Q, U_0]] - F_0^{-1}(F_0)_t \qquad \text{for GHM,}$$

and set

$$l_p(A; \Psi) = \eta, \qquad l_p(B; \Psi) = \eta + pc^2 k^2 \mu_1. \tag{1.7A,B}$$

Theorem 1.2. *The above ζ and η do not depend on the choice of the concrete solution Ψ of equation (1.5) and are formal power series in k^{-1}. The coefficients of the expansion of ζ in k^{-1} belong to $\mathbb{C}_0[\widetilde{Q}]$. The coefficients of η are contained in $\mathcal{A}[\widetilde{Q}]$ and $\mathcal{B}[\widetilde{Q}]$ respectively (for PCF and GHM), where $\mathcal{A}$ is linearly generated by the entries of $F_0^{-1} V F_0$ and $F_0^{-1}(F_0)_t$, $\mathcal{B}$ is generated by the entries of $F_0^{-1}(F_0)_t$. These series are connected by the following conservation law*

$$\zeta_t = \eta_x. \tag{1.8}$$

Proof. The independence of ζ and η of the choice of Ψ follows from Proposition1.1b) after the substitution $\Phi = F_0 \Psi$, because the replacement of Ψ by ΨC for C of type (1.4) does not change the series $l_p((\cdot); \Psi)$. Hence we can take $\check{\Psi} = F_0^{-1}\check{\Phi}$ as Ψ, where $\check{\Phi}$ from Proposition1.2 is either of type (1.3A) or of type (1.3B). Then (see (1.7)) $(\log m_p(\Psi))_t = (m_p(\check{\Psi}))_t / m_p(\check{\Psi})$ equals either $l_p(A; \check{\Psi}) = \eta$ or $l_p(B; \check{\Psi}) = \eta + pc^2 k^2 \mu_1$. Use the relations $\check{\Psi}_t = A\check{\Psi}$ for PCF and $\check{\Psi}_t = B\check{\Psi}$ for GHM which are derived from (1.2A) and (1.2B) respectively. In paritcular, η for GHM turns out to be a power series in k^{-1} (it is obvious for PCF), because

$$(\log m_p(\check{\Psi}))_t = (\log m_p(\widehat{\Psi}))_t + pc^2 k^2 \mu_1,$$

where $\check{\Psi} = \widehat{\Psi} \exp(kU_0 x + c^2 k^2 U_0 t)$ (cf. (1.3B)).

Comparing (1.7) and the expression obtained for η, we get the conservation law (1.8). Similarly, the independence of ζ, η of Ψ and the above formulae for A and B give that $\zeta_s \in \mathbb{C}[\widetilde{Q}]$, $\eta_s \in \mathcal{A}[\widetilde{Q}], \mathcal{B}[\widetilde{Q}]$ for the coefficients of the series ζ and η (use Theorem1.1a)). If $Q \equiv 0$, then $\zeta = 0$. Hence the coefficients of ζ belong to $\mathbb{C}_0[\widetilde{Q}]$. $\square$

Exercise 1.1. *If F_0 is chosen as in Corollary1.1, then the $\mathbb{C}$-space $\mathcal{A}$ of Theorem1.2 is generated by the entries of $F_0^{-1}VF_0$. Moreover we can take $\mathbb{C}$ as $\mathcal{B}$.* □

Let us fix $D = \mathrm{diag}(\delta_i)$ from the center of $\mathfrak{gl}_n^0$ (i.e., satisfying the condition that $\delta_i = \delta_j$ if $\mu_i = \mu_j$) and an invertible solution Φ of equation (1.1) of type (1.3). The formal power series $F^{\mathcal{D}} = \Phi\mathcal{D}\Phi^{-1}$ in k^{-1} is called the *formal resolvent*. There are no exp-factors in $F^{\mathcal{D}}$ because they commute with $\mathcal{D}$.

Let us set $l_p(A) = \mathrm{Sp}([A]_p) = \mathrm{Sp}(A[I]_p)$ for an arbitrary matrix A, where $[I]_p = \sum_{s=1}^{p} I_s^s$ (in the notations of the Introduction). Using the formal differentiation $\partial/\partial k^{-1}$, we define the series α by the following relations:

$$\alpha - k^2 p\mu_1 = \left(l_p\left(\Phi^{-1}\frac{\partial\Phi}{\partial k^{-1}}\right)\right)_x = -k^2 l_p(\Phi^{-1}U\Phi) = -k^2\,\mathrm{Sp}(UF^p), \tag{1.9}$$

where $F^p \stackrel{\mathrm{def}}{=} F^{\mathcal{D}}$ for $\mathcal{D} = [I]_p$.

The second equality can be checked by means of the formula

$$\left(\frac{\partial\Phi}{\partial k^{-1}}\right)_x = \frac{\partial(\Phi_x)}{\partial k^{-1}} = kU\left(\frac{\partial\Phi}{\partial k^{-1}}\right) - k^2U\Phi.$$

The exponential factors in (1.9) are canceled. Moreover, α is a formal power series in k^{-1}, since $\partial\exp(kU_0x)/\partial k^{-1} = -k^2\exp(kU_0x)U_0x$, and $\partial k/\partial k^{-1} = -k^2$.

We also introduce the series

$$\beta = -(k')^2 l_p(\Phi^{-1}V\Phi) = -(k')^2\,\mathrm{Sp}(VF^p), \tag{1.9A}$$

$$\beta = -\,\mathrm{Sp}((2c^2k^3U + k^2[U,U_x])F^p) + 2pc^2k^3\mu_1, \tag{1.9B}$$

for PCF (A) and GHM (B) respectively.

Theorem 1.3. *The series $F^{\mathcal{D}}$ does not depend on the concrete choice of Φ. The entries of its coefficients in the expansion with respect to k^{-1} belong to $\mathbb{C}[\widetilde{U}]$. The coefficients of the series α and β, connected by the conseravation law*

$$\alpha_t = \beta_x, \tag{1.10}$$

belong respectively to $\mathbb{C}_0[\widetilde{U}]$ and either to $\mathcal{A}[\widetilde{U}]$ (PCF) or to $\mathcal{B}[\widetilde{U}]$ (GHM), where $\mathcal{A}$ is generated by the entries of V (PCF) and $\mathcal{B} = \mathbb{C}$ for GHM.

Proof. The statements on the coefficients of $F^{\mathcal{D}}$, α, β follow from Theorem1.1b) and formulae (1.9A,B). As in the proof of Theorem1.2, we may take the series $\check{\Phi}$ from Proposition1.2 as Φ. Then

$$\beta = \left(l_p\left(\check{\Phi}^{-1}\frac{\partial\check{\Phi}}{\partial k^{-1}}\right)\right)_t \qquad \text{for PCF,}$$

$$\beta - 2k^3t\mu_1 = \left(l_p\left(\check{\Phi}^{-1}\frac{\partial\check{\Phi}}{\partial k^{-1}}\right)\right)_t \qquad \text{for GHM.}$$

It gives (1.10). □

Exercise 1.2. *Under the hypotheses of Theorem 1.3, the coefficients of* $F^{\mathcal{D}}$ *in* k^{-1} *are differential polynomials of* U, *i.e., are polynomials in* $U, U_x, U_{xx}, \ldots$. (Rewrite (1.6) in terms of Φ, using U only). □

Exercise 1.3*. *Suppose that the derivatives of* Q *with respect to* x *of arbitrary order* $\geqq 0$ *are absolutely integrable from* $x = -\infty$ *to* $x = \infty$ *and* $U \to U_0$ *when* $x \to \infty$ *and* $x \to -\infty$ *(for PCF, we need to assume additionally that the limits* $V(+\infty)$, $V(-\infty)$ *exist and coincide). Then the coefficients of the power series* $\int_{-\infty}^{+\infty} \zeta(x)\,dx$, $\int_{-\infty}^{+\infty} \alpha(x)\,dx$ *in* k^{-1} *give integrals of motion of equations (0.2) and (0.9).* (For a stronger statement, see Ch.II.) □

1.3. Formal jets.

Turning to to the proof of Theorem1.1, first we deduce statement b) from statement a).

Lemma 1.1. *We can choose matrix* $F_0 = F_0(U)$ *satisfying the relation* $UF_0 = F_0U_0$ *such that its entries (matrix elements) are polynomials of the entries of* $U = (u_p^q)$ *(i.e., belong to the ring of polynomials* $\mathbb{C}[U] = \mathbb{C}[u_1^1, \ldots, u_p^q, \ldots, u_n^n]$ *of* $\{u_p^q\}$*). Given an arbitrary matrix* $\dot{U} \in \mathfrak{gl}_n$, *there exists an appropriate construction of the matrix* F_0 *of the above type such that* $F_0^{-1} = f^{-1}G$ *where* f *and the entries of* G *belong to* $\mathbb{C}[U]$ *and* $f(\dot{U}) = f(\ldots, \dot{u}_p^q, \ldots) \neq 0$.

Proof. The colums of F_0 can be taken as follows:

$$P(U)(U-\mu_1)^{-1}z_1, P(U)(U-\mu_2)^{-1}z_2, \ldots P(U)(U-\mu_n)^{-1}z_n,$$

where $z_1, z_2, \ldots, z_n \in \mathbb{C}^n$, and $P(U)$ is the minimal polynomial of U (the product of the pairwise distinct $U - \mu_s \overset{\text{def}}{=} U - \mu_s I$). Given $\dot{U}$, we can choose the vectors z_s ensuring the linear independence of the columns of F_0 at $\dot{U}$. □

According to the lemma, there exists a matrix F_0 such that the entries of $Q = F_0^{-1}(F_0)_x$ belong to $f^{-1}\mathbb{C}[\widetilde{U}]$ for a suitable polynomial $f \in \mathbb{C}[U]$. Then an arbitrary element of the ring $\mathbb{C}[\widetilde{Q}]$ belongs to $f^{-l}\mathbb{C}[\widetilde{U}]$ for an appropriate $l \in \mathbb{Z}_+$. On the other hand, $\mathbb{C}[\widetilde{\Phi}_0, \widetilde{\Phi}_1, \ldots] \subset \mathbb{C}[\widetilde{U}, \widetilde{\Psi}_0, \widetilde{\Psi}_1, \ldots]$ because $\Phi = F_0\Psi$ and the entries of F_0 belong to $\mathbb{C}[U]$. Therefore an arbitrary $a \in \mathbb{C}[\widetilde{\Phi}_0, \widetilde{\Phi}_1, \ldots]$ independent of the choice of Φ (and $\Psi = F_0^{-1}\Phi$) must belong to $\mathbb{C}[\widetilde{U}, \widetilde{Q}]$ (assertion a) of the theorem).

Here F_0 and $Q = F_0^{-1}(F_0)_x$ can be arbitrary (if, of course, F_0 satisfies the condition of the lemma) and play an auxiliary role. Hence, $a \in \cap_Q \mathbb{C}[\widetilde{U}, \widetilde{Q}]$ and $a \in \cap_f (f^{-l}\mathbb{C}[\widetilde{U}])$, where f runs over the set of f for all possible constructions of F_0 ($l \in \mathbb{Z}_+$ is determined by f and a). We obtain that $a = a_1/f_1 = \cdots = a_r/f_r$, where $f_s \in \mathbb{C}[U]$, $a_s \in \mathbb{C}[\widetilde{U}]$, and there exist polynomials $h_s \in \mathbb{C}[U]$ such that $\sum_{s=1}^{r} f_s h_s = 1$. Therefore

$$a = \frac{a_1 h_1 + \cdots + a_r h_r}{f_1 h_1 + \cdots + f_r h_r} \in \mathbb{C}[\widetilde{U}].$$

Thus b) follows from a).

Proof of a). Analyzing the proof of Proposition1.1, we see that the ring of functions of x and t can be replaced by an arbitrary ring over $\mathbb{C}$ (containing the entries of Q) with a differentiation "$\partial/\partial x$" such that any linear inhomogeneous differential equation with coefficients in this ring has a solution.

We denote the ring of $\mathbb{C}$-valued functions of x by $\mathcal{X}$ (the dependence on t is not important). Set

$$\widetilde{\mathcal{X}} = \{\widetilde{f} = \sum_{i=0}^{\infty} f^{(i)} \frac{y^i}{i!}, f^{(i)} \in \mathcal{X}\}.$$

The ring $\mathcal{X}$ is naturally included into $\widetilde{\mathcal{X}}$:

$$f \mapsto \widetilde{j}(f) \stackrel{\text{def}}{=} \sum_{i=0}^{\infty} \frac{\partial^i f}{\partial x^i} \frac{y^i}{i!}.$$

This map is, in fact, the Taylor series of f. Under this embedding, the differentiation with respect to y is a continuation of the differentiation of functions from $\mathcal{X}$ by x. The product fg of functions in $\mathcal{X}$ goes into a series $\widetilde{j}(fg)$ whose coefficients are expressed in terms of the coefficients of $\widetilde{j}(f)$ and $\widetilde{j}(g)$ by the Leibniz rule. Extended to arbitrary elements of $\widetilde{\mathcal{X}}$ these formulas make this vector space a ring.

Formally, we can introduce $\widetilde{\mathcal{X}}$ as the space of infinite series $(f^{(0)}, f^{(1)}, \dots, f^{(i)}, \dots)$ for arbitrary $f^{(i)} \in \mathcal{X}$. The differentiation shifts the series by one to the left (replaces $f^{(i)}$ by $f^{(i+1)}$). For a function $f \in \mathcal{X}$, $\widetilde{j}(f)$ is considered as a formal set of the derivatives $f^{(i)} = \partial^i f/\partial x^i$, $i \geqq 0$. The addition and multiplication in $\widetilde{\mathcal{X}}$ are defined uniquely from the condition that the map $f \mapsto \widetilde{j}(f)$ is a homomorphism. This ring $\widetilde{\mathcal{X}}$ is called the *ring of formal jets*.

The algebra $\mathfrak{gl}_n(\widetilde{\mathcal{X}})$ of matrices with entries in $\widetilde{\mathcal{X}}$ contains the algebra $\mathfrak{gl}_n(\mathcal{X})$ of the matrix-valued functions of x in a natural way. Algebraically, $\mathfrak{gl}_n(\mathcal{X}) = \mathfrak{gl}_n \otimes_{\mathbb{C}} \mathcal{X}$, $\mathfrak{gl}_n(\widetilde{\mathcal{X}}) = \mathfrak{gl}_n \otimes_{\mathbb{C}} \widetilde{\mathcal{X}}$. We also introduce

$$\mathfrak{gl}'_n(\widetilde{\mathcal{X}}) = \mathfrak{gl}'_n \otimes_{\mathbb{C}} \widetilde{\mathcal{X}} \supset \mathfrak{gl}'_n(\mathcal{X}),$$
$$\mathfrak{gl}^0_n(\widetilde{\mathcal{X}}) = \mathfrak{gl}^0_n \otimes_{\mathbb{C}} \widetilde{\mathcal{X}} \subset \mathfrak{gl}^0_n(\mathcal{X}).$$

It is obvious that the inclusion of $\mathfrak{gl}_n(\mathcal{X})$ into $\mathfrak{gl}_n(\widetilde{\mathcal{X}})$ is a homomorphism of algebras. Here we consider $\mathfrak{gl}_n$ and $\mathfrak{gl}_n(\widetilde{\mathcal{X}})$ as associative algebras (with the usual matrix multiplication).

The formulation and the proof of Proposition 1.1 remain the same for $\widetilde{j}(U)$ instead of U and for $\widetilde{j}(F_0)$ and $\widetilde{j}(Q)$ instead of F_0 and Q. Now it is necessary to take $\widetilde{\Phi}, \widetilde{\Psi}$ with the coefficients $\widetilde{\Phi}_s, \widetilde{\Psi}_s \in \mathfrak{gl}_n(\widetilde{\mathcal{X}})$. The independence of C_x of x is to be replaced by the condition that the coefficients $C_s^{(i)}$ of the expansion of the series $\widetilde{C}_s \in \mathfrak{gl}_n^0(\widetilde{\mathcal{X}})$ with respect to y vanish for $i > 0$, $s \geqq 0$. "Constants" in formal jets mean series whose components are zero except the first one (on the left).

The invertibility of $\widetilde{\Phi}$ and $\widetilde{\Psi}$ is equivalent to the invertibility of $\Phi_0^{(0)}$ and $\Psi_0^{(0)}$. The ring $\mathbb{C}[\widetilde{\Psi}_0, \widetilde{\Psi}_1, \ldots]$ is identified naturally with the ring of polynomials of the entries of $\Psi_s^{(i)}$, where $\widetilde{\Psi}_s = \sum_{i=0}^{\infty} \Psi_s^{(i)}(y^i/i!)$. In the formal jets, one can always replace a differential equation by the corresponding set of polynomial recurrent relations. Let us demonstrate it, calculating the solutions of equation (1.5) over $\widetilde{\mathcal{X}}$.

Set $\Psi_s^{(i)} = (\Psi_s^{(i)})' + (\Psi_s^{(i)})^0$, where $(\Psi_s^{(i)})' \in \mathfrak{gl}_n'(\mathcal{X})$ and $(\Psi_s^{(i)})^0 \in \mathfrak{gl}_n^0(\mathcal{X})$. Given s, let us suppose that the matrices $(\Psi_s^{(i)})'$ have been already found for $0 \leqq i \leqq l$ (see (1.6^0) and (1.6′)). Then

$$(1.11^0) \qquad (\Psi_s^{(i+1)})^0 = -\sum_{j=0}^{l} \binom{l}{j} \left(\frac{\partial^j Q}{\partial x^j} \Psi_s^{(l-j)}\right)^0, 0 \leq i \leq l,$$

and $(\Psi_{s+1}^{(i)})'$ for $0 \leqq i \leqq l-1$ are determined from the equations

$$(1.11') \qquad [U_0, (\Psi_{s+1}^{(i)})'] = \Psi_s^{(i+1)} + \sum_{j=0}^{l} \binom{l}{j} \frac{\partial^j Q}{\partial x^j} \Psi_s^{(l-j)}.$$

In these formulae, $(\Psi_0^{(0)})' = 0$ and $(\Psi_s^{(0)})^0$ are arbitrary matrices in $\mathfrak{gl}_n^0(\mathcal{X})$, $s \geqq 0$, $l \geqq 0$.

Lemma1.2. *If $a \in \mathbb{C}[\widetilde{\Psi}_0, \widetilde{\Psi}_1, \ldots]$ does not depend on the choice of the solution Ψ of equation (1.5), then a can be considered as a functional of the coefficients of the solution $\widetilde{\Psi}$ of equation (1.5) in formal jets for $\widetilde{j}(Q)$ instead of Q. It does not depend on the particular choice of $\widetilde{\Psi}$ as well.*

Proof. A polynomial a is a linear combination of the entries of Ψ_s^0 with coefficients expressed in terms of the entries of $\partial^j Q/\partial x^j$. Replacing Ψ_s^0 by $(\Psi_s^{(0)})^0$, we obtain the required extension. This procedure is well-defined because the initial value $\Psi_s^0(x_0)$ (for a certain given $x_0 \in \mathbb{R}$) can be fixed arbitrary. □

Set $(\Psi_0^{(0)})^0 = 1$, $(\Psi_s^{(0)})^0 = 0$ for $s > 0$. Then coefficients $\Psi_s^{(i)}$ of the corresponding solution $\widetilde{\Psi}^{\cdot}$ of system (1.11) belong to $\mathfrak{gl}_n(\mathbb{C}[\widetilde{Q}])$. Due to Lemma1.2, we can use this $\widetilde{\Psi}^{\cdot}$ in calculating a. Hence $a \in \mathbb{C}[\widetilde{Q}]$. The proof is completely the same for an arbitrary $\mathbb{C}$-space of coefficients $\mathcal{A}$. This completes the proof of Theorem1.1. □

Note that claim b) can be derived from a) (cf. the beginning of this paragraph) in a different way. We can use the relation $(F^{\mathcal{D}})_x = [kU, F^{\mathcal{D}}]$. We also remark that a) itself is a consequence of a certain analogue of the following fundamental theorem of the Galois theory for jets.

Exercise1.4*. *Let $\mathcal{D}$ be a ring over $\mathbb{C}$ with a differentiation ∂, $\mathcal{D}'$ an extension of $\mathcal{D}$ with a differentiation ∂' which extends ∂. Suppose that $\mathcal{D}'$ is generated by $a_1, \ldots, a_m \in \mathcal{D}'$ over $\mathcal{D}$ and that $a_1, \ldots, a_m$ satisfy a system of differential equations of first order with coefficients in $\mathcal{D}$ which is solved with respect to $\partial' a_i$. Then $\mathcal{D}'$ can be included into the ring $\widetilde{\mathcal{D}}$ of formal jets over $\mathcal{D}$. Given an arbitrary $a \in \mathcal{D}'$ such that $a \notin \mathcal{D}$, there exists an automorphism of $\widetilde{\mathcal{D}}$ preserving the differentiation that fixes all elements of $\widetilde{j}(\mathcal{D}) \subset \widetilde{\mathcal{D}}$ and does not fix the image of a in $\widetilde{\mathcal{D}}$.* □

Exercise1.5. *Choosing F_0 as in Lemma1.1, the coefficients of the expansion of ζ (Theorem1.2) with respect to k^{-1} belong to the kernel of the homomorphism taking U to U_0 from the ring $\mathbb{C}[f^{-1}, \widetilde{U}]$ of elements of $\mathbb{C}[\widetilde{U}]$ divided by $f, f^2, \ldots$ onto $\mathbb{C}$, where $f = \det F_0 \in \mathbb{C}[U]$ (i.e., $\partial^j U/\partial x^j \mapsto 0$, $j > 0$). The coefficients of η belong to $\mathbb{C}[f^{-1}, \widetilde{U}]$ for GHM and, in addition, linearly depend on the matrix elements of V in the case of PCF.* □

1.4. Direct calculations (Riccati equation).

We begin with the calculating the first two coefficients of the resolvent $F^{\mathcal{D}}$ (see Theorem1.2 and Exercise1.2). Let $[\mathcal{D}, \mathfrak{gl}_n^0] = 0$. We can express $\mathcal{D}$ as $d(U_0)$ for an appropriate polynomial $d(z)$. We use the notation $\mathrm{ad}_Y X = [Y, X]$ for $X, Y \in \mathfrak{gl}_n$. There exists a polynomial $q(z)$ without a constant term such that $q(\mathrm{ad}_{U_0})X = X'$, where $X = X' + X_0$ is the decomposition of §1.1. For instance, if U_0 satisfies condition (0.10) (the case of GHM), then we can take $q = z^2/c^2$, $d = (\delta_1 - \delta_n)z/c+$ const, where $\mathcal{D} = \mathrm{diag}(\delta_i)$, $\delta_1 = \cdots = \delta_p$, $\delta_{p+1} = \cdots = \delta_n$.

It results from equation (1.1) that

$$\Phi_0 U_0 \Phi_0^{-1} = U, \qquad U_x = [(\Phi_0)_x \Phi_0^{-1}, U], \tag{1.12a}$$

$$(\Phi_0)_x \Phi_0^{-1} = [U, \Phi_1 \Phi_0^{-1}], \qquad (\Phi_1)_x \Phi_0^{-1} = [U, \Phi_2 \Phi_0^{-1}]. \tag{1.12b}$$

Set $F^{\mathcal{D}} = \Phi \mathcal{D} \Phi^{-1} = \sum_{i=0}^{\infty} (F^{\mathcal{D}})_i k^{-i}$. Then

$$(F^{\mathcal{D}})_0 = \Phi_0 \mathcal{D} \Phi_0^{-1}, \qquad (F^{\mathcal{D}})_1 = [\Phi_1 \Phi_0^{-1}, d(U)].$$

Using (1.12), we arrive at

$$\begin{aligned}(F^{\mathcal{D}})_0 &= d(U),\\ (F^{\mathcal{D}})_1 &= [q(\mathrm{ad}_U)(\Phi_1\Phi_0^{-1}), d(U)] =\\ &= [\widehat{q}(\mathrm{ad}_U)((\Phi_0)_x\Phi_0^{-1}), d(U)] =\\ &= -[\widehat{q}^2(\mathrm{ad}_U)(U_x), d(U)],\end{aligned}$$

where $\widehat{q}(z) = q(z)/z$, $\widehat{q}^2(z) = (\widehat{q}(z))^2$. If U_0 satisfies (1.10), $\mathcal{D} = c^2U_0$, $\widehat{q} = z/c^2$, and $d = c^2z$, then we obtain the relations

$$\begin{aligned}(F^{\mathcal{D}})_0 &= c^2U,\\ (F^{\mathcal{D}})_1 &= -\frac{1}{c^2}[[U,[U,U_x]],U] = [U,U_x].\end{aligned}$$

Let us discuss applications of the last two formulae to the GHM equation. Given an arbitrary solution

$$\check{\Phi} = (\sum_{i=0}^{\infty}\Phi_ik^{-i})\exp(kU_0x + c^2k^2U_0t)$$

of equation (1.1), one has :

$$\begin{aligned}\check{\Phi}_t\check{\Phi}^{-1} &= k^2\check{\Phi}c^2U_0\check{\Phi}^{-1} + O(1) =\\ &= c^2k^2U + k[U,U_x] + O(1),\end{aligned}$$

where $O(1)$ is a power series of k^{-1}. We used this relation for the proof of Proposition1.2.

If $\check{\Phi}$ satisfies equation (1.3B)(Proposition1.2) in addition to (1.1), then

$$(\Phi_0)_t + c^2\Phi_2U_0 = c^2U\Phi_2 + [U,U_x]\Phi_1.$$

Consequently,

$$\begin{aligned}\Phi_0^{-1}(\Phi_0)_t &= c^2[U_0,\Phi_0^{-1}\Phi_2] + c^2[\Phi_0^{-1}\Phi_2,U_0]\Phi_0^{-1}\Phi_1\\ &= c^2(\Phi_0^{-1}\Phi_1)_x.\end{aligned}$$

Here and below we apply (1.12b) conjugated by Φ_0. We arrive at the equations

$$(\Phi_0^{-1}(\Phi_0)_t)^0 = -c^2((\Phi_0^{-1}(\Phi_0)_x)(\Phi_0^{-1}\Phi_1)')^0, \tag{1.13^0}$$

$$[U_0,\Phi_0^{-1}(\Phi_0)_t] = c^2(\Phi_0^{-1}(\Phi_0)_x)_x. \tag{1.13$'$}$$

We have used the inclusion $\mathfrak{gl}'_n \cdot \mathfrak{gl}^0_n \subset \mathfrak{gl}'_n$. Since $(\Phi_0^{-1}\Phi_1)'$ is expressed in terms of $\Phi_0^{-1}(\Phi_0)_x$ (cf. (1.12b)), then (1.13) allows us to rewrite $\Phi_0^{-1}(\Phi_0)_t$ in terms of $\Phi_0^{-1}(\Phi_0)_x$. With $\Phi_0 = F_0$, relation (1.13) turns into the formula from Collorary1.1 which was necessary for the reduction from GHM to VNS (see the Introduction).

We will now calculate the constant term α_0 of the series α (see (1.9)), using the above calculation (U_0 satisfies (0.10) as above). First note that

$$\alpha_0 = -\operatorname{Sp}(U(F^p)_2) = -c\operatorname{Sp}((F^p)_0(F^p)_2),$$

where we replaced U_0 by $c[1]_p + \mu_2 1$ and used the equation $\operatorname{Sp}(F^p) = p$. Since $\operatorname{Sp}(F^pF^p) = \operatorname{Sp}([1]_p[1]_p) = p$, one has that $\operatorname{Sp}((F^p)_0(F^p)_2) = -\frac{1}{2}\operatorname{Sp}((F^p)_1(F^p)_1)$. However we know $(F^p)_1$ from the above consideration: $(F^p)_1 = [U, U_x]/c^3$. Applying (1.12a), we obtain

$$\alpha_0 = -\frac{1}{2c^3}\operatorname{Sp}(U_xU_x) = \frac{1}{2c}\operatorname{Sp}(Q'Q'). \tag{1.14}$$

Here $Q = F_0^{-1}(F_0)_x$, $Q = Q' + Q^0$.

Riccati equation. There are more systematic ways to calculate $F^{\mathcal{D}}$, ζ, η in terms of Q. Let $\Psi = F_0^{-1}\Phi$ be a solution of equation (1.5) of the form (1.3). We remind that $\Psi = \Psi' + \Psi^0$, $[U_0, \Psi^0] = 0$ (cf. the beginning of this section). Set

$$\Psi(\Psi^0)^{-1} = \Pi = \sum_{i=0}^{\infty} \Pi_i k^{-i},$$

where $\Psi^0 = (\sum_{i=0}^{\infty} \Psi_i^0 k^{-i})\exp(kU_0x)$ (note that the exponential factor is canceled in Π). The series Π does not depend on the choice of Ψ and $\Pi^0 = 1$. According to Theorem1.1, the entries of Π_i belong to $\mathbb{C}[\widetilde{Q}]$. We can calculate them explicitly.

Proposition1.3. *The formal power series* $\Pi = \sum_{i=0}^{\infty} \Pi_i k^{-i}$ *with the condition* $\Pi^0 = 1$ *is uniquely determined by the equation*

$$\Pi_x = k[U_0, \Pi] + \Pi(Q\Pi)^0 - Q\Pi. \tag{1.15}$$

Proof. Equation (1.15) results directly from (1.5):

$$\begin{aligned}\Pi_x &= (\Psi(\Psi^0)^{-1})_x = \\ &= (kU_0 - Q)\Pi - \Pi\Psi_x^0(\Psi^0)^{-1} = \\ &= k[U_0, \Pi] + \Pi(Q\Pi)^0 - Q\Pi,\end{aligned}$$

because $\Psi_x^0 = kU_0\Psi^0 - (U\Psi)^0$. Expanding (1.15) in the power series with respect to k^{-1}, we see that $\Pi_0 = 1$, $\Pi_s^0 = 0$ for $s \geqq 1$, and

$$[U_0, \Pi_s] = (\Pi_{s-1})_x - \sum_{r=0}^{s-1} \Pi_r (Q\Pi_{s-r-1})^0 + Q\Pi_{s-1}. \tag{1.16}$$

The Π_s are calculated recurrently from these relations. □

The resolvent $F^{\mathcal{D}}$ is connected with Π by the formula

$$F^{\mathcal{D}} = F_0 \Pi \mathcal{D} \Pi^{-1} F_0^{-1}.$$

We can rewrite (1.7), (1.7A,B) in terms of Π as well:

$$\zeta = l_p(kU_0 - Q; \Pi) - p\mu_1 k, \tag{1.17}$$

$$\eta = l_p(A; \Pi), \qquad \eta = l_p(B; \Pi) - pc^2k^2\mu_1. \tag{1.17A,B}$$

We have used the invariance of $l_p(X;Y)$ when Y is multiplied by matrices from $\mathfrak{gl}_n^0$ on the right.

As an example, we will calculate the first two coefficients ζ_1, ζ_2 of the series ζ in the case $p = 1$. We denote the first column of Π by $\pi^1 = \sum_{i=0}^{\infty}(\pi_i^1)k^{-i}$. Then the following results from (1.16):

$$(U_0 - \mu_1)\pi_s^1 = (\pi_{s-1}^1)_x - \sum_{r=0}^{s-1} (Q\pi_r^1)_{(1)} \pi_{s-r-1}^1 + Q\pi_{s-1}^1.$$

Here $z_{(1)}$ means the first component of the vector z. One has: $\pi_0^1 = {}^t(1, 0, \dots, 0)$, $\pi_1^1 = {}^t(0, b_2, \dots, b_n)$ for $b_s = q_s^1/(\mu_s - \mu_1)$, $s \geqq 2$, where $Q = (q_s^r)$. Finally, $\zeta = -(Q\pi^1)_{(1)}$ and

$$\zeta_0 = -q_1^1, \qquad \zeta_1 = \sum_{s=2}^{n} \frac{q_1^s q_s^1}{\mu_1 - \mu_s}. \tag{1.18}$$

Comparing (1.18) with formula (1.14), we see that $\alpha_0 = \zeta_1$ for $p = 1$. The connection of α and ζ generalizing this equality will be established in the next section. Turning to the reduction of GHM for $p = 1$ to VNS (0.14) (see the Introduction), we establish the formulae

$$\zeta_0 = 0, \qquad \zeta_1 = \frac{1}{c}\sum_{i=1}^{n-1} r_i^{(+)} r_i^{(-)} = \frac{1}{c}(\boldsymbol{r}^{(+)}, \boldsymbol{r}^{(-)}),$$

substituting

$$Q = F_0^{-1}(F_0)_x = \sum_{i=1}^{n-1} r_i^{(+)} I_1^{i+1} + \sum_{i=1}^{n-1} r_I^{(-)} I_{i+1}^1.$$

Here, $c = c_1 - c_2 = \mu_1 - \mu_s$ for arbitrary $s \geqq 2$.

1.5. Local laws modulo exact derivatives.

If we set $\chi_1 = \chi_x$, $\chi_2 = \chi_t$ for an arbitrary $\chi \in \mathbb{C}[\widetilde{U}]$ or $\mathbb{C}[\widetilde{Q}]$, we get a local conservation law $(\chi_1)_t = (\chi_2)_x$ which is trivial in the following sense. The corresponding integrals of motion $\int_{-\infty}^{\infty} \chi_1 dx = \chi(\infty) - \chi(-\infty)$ are zero under standard analytic assumptions (as in Exercise1.3). In this book, we do not discuss the classification of local conservation laws modulo trivial ones in full detail (see Ch.II for the corresponding analytical results). However we will compare all considered densities modulo exact derivatives.

Proposition1.4. *The power series ζ and α of §1.2 are connected by the relation*

$$\frac{\partial \zeta}{\partial k^{-1}} - \alpha = \theta_x,$$

where

$$\theta = \frac{\partial}{\partial k^{-1}} \log m_p(\Psi) - l_p\left(\Psi^{-1}\frac{\partial \Psi}{\partial k^{-1}}\right)$$

is a power series in k^{-1} with coefficients in $\mathbb{C}[\widetilde{Q}]$.

Proof. Using (1.7), (1.9) and the trivial equation

$$l_p(\Phi^{-1}\partial\Phi/\partial l^{-1}) \ = \ l_p(\Psi^{-1}\partial\Psi/\partial k^{-1}),$$

we can prove the desired relation. We see that θ is a power series in k^{-1} because the contribution of the exp-factors is zero. The independence of θ of the choice of Ψ follows from the formula

$$\theta = \mathrm{Sp}([\Psi]_p^{-1}\left[\frac{\partial \Psi}{\partial k^{-1}}\right]_p) - \mathrm{Sp}(\left[\Psi^{-1}\frac{\partial \Psi}{\partial k^{-1}}\right]_p),$$

which results from (1.7,9). Therefore, one can apply Theorem1.1, a). □

Pohlmeyer's law for PCF. We are going to discuss some specific points connected with the reality conditions. Let $U, V(x,t) \in \mathfrak{u}_n$. Then $U_0 \in \mathfrak{u}_n$. For the sake of simplicity, we impose the condition $p = 1$. The general case is left to the reader as an excercise. We can take $F_0(x,t) \in U_n$ and $Q(x,t) \in \mathfrak{u}_n = \mathrm{Lie}(U_n)$. Let us denote the first column of Ψ by ψ^1 (ϕ^1 corresponds to Φ). Here and below Φ and $\Psi = F_0^{-1}\Phi$ are solutions of (1.1) and (1.5) in the form (1.3). Using the hermitian form $(,)$ (anti-linear with respect to the second argument), we introduce the following χ and κ, which are more convenient for anti-hermitian U, V than ζ

and η. Let

$$\chi + \mu_1(k-\overline{k}) = (\log(\psi^1,\psi^1))_x =$$

$$(1.19)\qquad = (\log(\phi^1,\phi^1))_x = (k-\overline{k})(U\rho,\rho),$$

$$\kappa = (k'-\overline{k}')(V\rho,\rho),$$

(1.19A,B)

$$\kappa + c^2\mu_1(k^2-\overline{k}^2) = c^2(k^2-\overline{k}^2)(U\rho,\rho) + (k-\overline{k}^2)([U,U_x]\rho,\rho),$$

where $\overline{(\ )}$ denotes the complex conjugation and $\rho = \phi^1/|\phi^1|$.

If we use $\check{\Phi}$ satisfying the conditions of Proposition1.2, then $\kappa = (\log(\check{\phi}^1,\check{\phi}^1))_t$ for PCF(A), and $\kappa = (\log(\check{\phi}^1,\check{\phi}^1))_t - c^2\mu_1(k^2-\overline{k}^2)$ for GHM(B). We follow Theorem1.2 and Proposition1.4:

Theorem1.4. *The coefficients of the power series χ and κ expanded in k^{-1}, $\overline{k}^{-1}$ belong to $\mathbb{C}_0[\widetilde{U}]$ and $\mathcal{A},\mathcal{B}[\widetilde{U}]$, where $\mathcal{A}$ is linearly generated by the elements of V (for PCF), and $\mathcal{B} = \mathbb{C}$ (for GHM). The following conservation law holds:*

$$(1.20)\qquad \chi_t = \kappa_x.$$

The coefficients of the series $\chi - \zeta - \overline{\zeta}$ for ζ from Theorem1.2 are exact derivatives with respect to x of proper elements from $\mathbb{C}[\widetilde{U}]$. □

Let us denote the coefficient of χ of $k^{-r}\overline{k}^{-s}$ by $\chi_{r,s}$ and set $\chi_s = \chi_{s,0}$. Then $\chi_s = \zeta_s$ modulo exact derivatives. Since $\overline{\chi} = \chi$, then $\overline{\chi}_s = \chi_{0,s}$. Coefficients $\chi_{r,s}$ for $rs \neq 0$ are exact derivatives because of Theorem1.4. Hence $\{\chi_s\}$ are sufficient to consider. Moreover we have the following corollary.

Corollary1.2. *The real parts* $\mathrm{Re}\,\chi_s$ *and* $\mathrm{Re}\,\zeta_s$, $s = 0,1,\ldots$, *are exact derivatives of suitable elements from $\mathbb{C}[\widetilde{U}]$.*

Proof. Set $\overline{k} = k$ in (1.19) (i.e., take $k \in \mathbb{R}$). Then χ is identically zero. Consequently, $\chi_s + \overline{\chi}_s + \sum_{r=1}^{s}\chi_{r,s-r} = 0$. Taking the above observation into account, we obtain the claim of the corollary. □

If, in addition, $U, V(x,t) \in \mathfrak{o}_n$, then $\overline{\mu}_1 = \mu_{1'}$ for an appropriate index $1'$ ($\mu_{1'}$ has the same multiplicity $p = 1$). The series χ' constructed for $\mu_{1'}$ instead of μ_1 satisfies the relation $\chi'_{q,r} = \chi_{r,q} = \overline{\chi}_{q,r}$. In particular, $\chi'_s = \overline{\chi}_s$ (check it as an exercise, conjugating (1.1) and replacing Φ by $\overline{\Phi}(k) = \overline{\Phi(\overline{k})}$).

We see that conjugated eigenvalues ($\in \{\mu_i\}$) for orthogonal U, V correspond to the conjugated series among $\{\chi_s\}$, and $\mu_i = 0$ lead to the trivial series. These statements (as well as Theorem1.4 and Corollay1.2) are valid when the multiplicity p is arbitrary.

S^{n-1}-fields. Here we discuss the equation of the S^{n-1}-fields (0.5) for $q(x,t)$ with the normalization condition (0.6) (cf. the Introduction). Let us recall that the orthogonal PCF g with the condition $g^2 = I$ reduce to the S^{n-1} fields q. They have the form $g = I - 2P_q$, where $P_q z = (z,q)q$. The matrix $U = g_x g^{-1} = 2q \wedge q_x$ has three pairwise distinct eigenvalues $2i$, $-2i$, 0. The first two have multiplicity 1. The eigenvectors corresponding to $\pm 2i$ are proportional to $q_x \pm iq$.

Let us construct the series χ (in k^{-1}, $\overline{k}^{-1}$) for $\mu_1 = 2i$ and expand it as a power series in $\gamma = (1-2k)^{-1} = k^{-1}(k^{-1}-2)$ and its complex conjugate $\overline{\gamma}$. We denote the coefficient of χ of $\gamma^r\overline{\gamma}^s$ by $\chi^\gamma_{r,s}$ and set $\chi^\gamma_s = \chi^\gamma_{s,0}$. The latter are sufficient to consider since the sets $\{\chi_s\}$ and $\{\chi^\gamma_s\}$ are $\mathbb{R}$-linearly equivalent. Note that $\overline{\chi^\gamma_s} = \chi^\gamma_{0,s}$ and the densities $\operatorname{Re}\chi^\gamma_s$, $\operatorname{Re}\chi^\gamma_{r,s}$ are exact derivatives for $s \geqq 0$, $rs > 0$. This follows easily from the corresponding properties of $\chi_{r,s}$.

Corollary1.3. *For the S^{n-1}-fields q and $\mu_1 = 2i$, the densities χ^γ_s with even indices s represent exact derivatives in $\mathbb{C}[\widetilde{U}]$, where $U = 2q \wedge q_x$. If we set $\gamma = -\overline{\gamma}$ (i.e., take $\gamma \in i\mathbb{R}$) in the expansion of χ with respect to γ, $\overline{\gamma}$, then the even coefficients of the resulting series in γ are trivial modulo exact derivatives and all χ^γ_s and χ_s are expressed $\mathbb{R}$-linearly by the (pure imaginary) odd coefficients. Hence the latter are sufficient to consider in the case of S^{n-1}-fields.*

Proof. If Φ is a solution of (1.1), then (cf. [ZMi1, ZMi2, Ch2]), the function $g\Phi$ is a solution of the equation

$$(g\Phi)_x = (1-k)U(g\Phi). \tag{1.21}$$

Really, $(g\Phi)_x = g_x(gg)\Phi - kg(gg_x)\Phi = (1-k)g_x\Phi = (1-k)Ug\Phi$ ($gg_x + g_x g = 0$, since $g^2 = I$). For Φ in the form (1.3), the function $\dot{\Phi}(k) = g\Phi(1-k)$ satisfies (1.1) and has the form (1.3) with the exp-factor $\exp((1-k)U_0 x) = \exp(k\overline{U}_0 x)\exp(U_0 x)$. The correction factor $\exp(U_0 x)$ is not essential here, since it does not contain parameters. Hence, the series

$$\begin{aligned}\dot{\chi}(k) &= (\log(\dot{\phi}^1(k), \dot{\phi}^1(k)))_x \\ &= (\log(\phi^1(k), \phi^1(k)))_x = \chi(1-k)\end{aligned}$$

coincides with the series constructed from $\mu_{1'} = -2i$. Rewriting χ' in terms of χ (see above), we obtain the relation $\overline{\chi}(k) = \chi'(1-k) = \chi(1-k)$. Expressing k in terms of γ, we see that $\overline{\chi^\gamma_s} = (-1)^s\chi^\gamma_s$. In particular, the imaginary part of χ^γ_s is equal to zero for even s. The corollary follows from this. □

Exercise1.6. *Show that for the S^{n-1}-fields q in the normalized coordinates, the coefficients χ_s, χ_s^γ are polynomials of scalar products of derivatives of q with respect to x of order $\geqq 2$ (i.e., without q and q_x). Respectively, the ring $\mathbb{C}[\widetilde{U}]$ in Corollary1.3 can be replaced by the ring of "differential" polynomials of q, consisting of scalar products of the above derivatives.* (Integrate system (1.11), using derivatives of q and their scalar products; cf. also Ch.II §1, Exercise1.13). □

Exercise1.7. *Prove that the sum of the densities ζ of Theorem1.2 for all pairwise distinct eigenvalues μ_s (taken as μ_1) is expressed as an exact derivative with respect to x of a series with coefficients in $\mathbb{C}[\widetilde{Q}]$.* (Consider the exterior power $\bigwedge^n \mathbb{C}^n$ in (1.1)). *Check that zero eigenvalues μ_i lead to trivial conservation laws for the S^{n-1}-fields.* □

1.6. Comments.

This paragraph is based on the works by the author [Ch2,3]. The formal Jost functions Φ are closely related to the dressing transformation due to Zakharov-Shabat. Statements like Proposition1.1 (with or without the exp-factor) are in almost all papers on the "general" theory of soliton equations. The papers [ZMa1], [Kri3] appear closest to our exposition.

The local conservation laws for PCF were found for the first time in [Ch6] as a generalization of the local conservation laws by Pohlmeyer [Poh] for the S^{n-1} fields (cf. Theorem1.4 and below, Ch.II §1). These laws were obtained by the scattering theory (see Theorem1.2) in [Ch8]. The series of densities ζ is connected to those of the local conservation laws for the n-wave equation constructed by V. E. Zakharov and S. V. Manakov in [ZMa1]. Equations similar to (1.15) (the Riccati equations) were considered in quite a few works containing constructions of local conservation laws (cf., e.g., [F2], [ZMa2], [K1], [TF2] and below, Ch.II §3). The resolvent approach (Theorem1.3) develops the results by I. M. Gel'fand and L. A. Dickey (cf. [GD1–4], [Ch3]) and below §2). A certain interpretation of local conservation laws, their non-local, and higher dimensional generalizations are discussed in [VV].

A comparison of various constructions of local conservation laws of matrix soliton equations was started in [Ch8] (cf. also [TF2]). In papers [Ch2,3], the algebraic approach based on formal jets was worked out. The results established by the method of formal jets can be often obtained by direct methods (see §1.4) or with the aid of the scattering theory (cf. below, Ch.II). However the definition of formal jets is (as we try to show in this paragraph) a fruitful mathematical abstraction and give birth to quite a few new interesting questions in differential algebra. There exist certain classical results like Theorem1.1 (and Exercise1.4), but it seems that there were almost no natural problems to apply the "Galois theory" of formal jets before KdV and the appearance of the soliton theory.

§2. Generalized Lax equations

In §2.1 we introduce the higher equations of the GHM and VNS types, using the resolvents. In §§2.2, 2.3 we define the abstract fractional powers, associated with an indeterminate formal Ψ-function considered as an operator. In the next §2.4 we define generalized matrix Lax equations and Lax pairs and, in particular, establish their relation to the equations from §2.1. In §2.5 we study scalar Lax equations, their conservation laws, and τ-functions. We also discuss a connection of the Sin-Gordon equation with the higher KdV equations. A certain acquaintance with paper [GD2] and the dressing transformations (see [ZMa2], [ZS1], or [TF2]) could be useful and helpful.

2.1. Equations of the GHM and the VNS type.

Let $\widehat{\Phi} = \sum_{i=0}^{\infty} \Phi_i k^{-i}$, for indeterminates $\Phi_i \in \mathfrak{gl}_n$. Given a constant matrix $\mathcal{D} \in \mathfrak{gl}_n$ and $j > 0$, we introduce the *resolvent* $F^{\mathcal{D}}$ and the M-matrix :

$$F^{\mathcal{D}} = \widehat{\Phi}\mathcal{D}\widehat{\Phi}^{-1} = \sum_{i=1}^{\infty} F_i^{\mathcal{D}} k^{-i}, \quad M_{\mathcal{D}}^j = \sum_{i=0}^{j-1} F_i^{\mathcal{D}} k^{j-i}.$$

Supposing that $[\mathcal{D}_0, \mathcal{D}_1] = 0$, $j_0, j_1 > 0$, let $\check{\Phi} \stackrel{\text{def}}{=} \widehat{\Phi}\exp(t_0\mathcal{D}_0 k^{j_0} + t_1\mathcal{D}_1 k^{j_1})$. Since Φ_i are arbitrary (generic), we can define abstract differentiations ∂_l $(l = 0, 1)$ on the entries (matrix elements) of the coefficients $\{\Phi_i\}$, setting

$$\partial_l \check{\Phi} = M_{\mathcal{D}_l}^{j_l} \check{\Phi} \Leftrightarrow \partial_l \widehat{\Phi} = -\Big(\sum_{i=j_l}^{\infty} F_i^{\mathcal{D}_l} k^{j_l - i}\Big)\widehat{\Phi}. \tag{2.1}$$

Here ∂_l differentiate the exp-factor as $\partial/\partial t_l$:

$$\partial_s \exp(t_l \mathcal{D}_l k^{j_l}) = \mathcal{D}_l k^{j_l} \exp(t_l \mathcal{D}_l k^{j_l}) \delta_l^s$$

(we assume that $\partial_l k = 0$). By comparing the expansions with respect to k^{-1} on either side, we see that the derivatives of the entries of each $\partial_l \Phi_i$ are rationally expressed in terms of the entries of $\{\Phi_j\}$. This allows us to extend ∂_l to differentiations on the field of all rational functions of the entries of $\{\Phi_i\}$ via the standard Leibniz formula. In fact, (2.1) is a compact way to introduce these differentiations. Note that the corresponding ∂ is well-defined for any series $\widetilde{M}$ such that $\widetilde{M} - F^{\mathcal{D}} k^{-j} = O(1)$, where $O(1)$ is a series of non-negative powers of k^{-1}. But we need rather special $\widetilde{M}$ to ensure the following

Proposition2.1.

$$[\partial_0, \partial_1] = 0,$$

$$\partial_0 M_{\mathcal{D}_1}^{j_1} - \partial_1 M_{\mathcal{D}_0}^{j_0} = [M_{\mathcal{D}_0}^{j_0}, M_{\mathcal{D}_1}^{j_1}]. \tag{2.2}$$

Proof. It is sufficient to prove equation (2.2), since it is equivalent to the relation $[\partial_0, \partial_1]\check{\Phi} = 0$, which results in $[\partial_0, \partial_1]\phi = 0$ for any rational function ϕ of the entries of $\{\Phi_i\}$. We use henceforth the notation $(\sum_{i=-p}^{\infty} A_i k^{-i})_+ = \sum_{i=-p}^{-1} A_i k^{-i}$, where $p \in \mathbb{Z}_+$. In particular, $M_{\mathcal{D}}^j$ can be written in the form $(\check{\Phi}\mathcal{D}k^j\check{\Phi}^{-1})_+$ and

$$\begin{aligned}\partial_0 M_{\mathcal{D}_1}^{j_1} &= ((\partial_0\check{\Phi})\mathcal{D}_1 k^{j_1}\check{\Phi}^{-1} - \check{\Phi}\mathcal{D}_1 k^{j_1}\check{\Phi}^{-1}(\partial_0\check{\Phi})\check{\Phi}^{-1})_+ = \\ &= ([(\check{\Phi}\mathcal{D}_0 k^{j_0}\check{\Phi}^{-1})_+, \check{\Phi}\mathcal{D}_1 k^{j_1}\check{\Phi}^{-1}])_+.\end{aligned}$$

Consequently,

$$\partial_0 M_{\mathcal{D}_1}^{j_1} - \partial_1 M_{\mathcal{D}_0}^{j_0} = ([(F^{\mathcal{D}_0}k^{j_0})_+, F^{\mathcal{D}_1}k^{j_1}])_+ - ([(F^{\mathcal{D}_1}k^{j_1})_+, F^{\mathcal{D}_0}k^{j_0}])_+.$$

Using the trivial relation $([A - A_+, B - B_+])_+ = 0$, it is easy to see that the right hand side of the last equation is equal to

$$[(F^{\mathcal{D}_0}k^{j_0})_+, (F^{\mathcal{D}_1}k^{j_1})_+] + ([F^{\mathcal{D}_0}k^{j_0}, F^{\mathcal{D}_1}k^{j_1}])_+.$$

Taking into consideration the condition $[\mathcal{D}_0, \mathcal{D}_1] = 0$, we obtain the desired result. □

Since $[\partial_0, \partial_1] = 0$, we can consider $\{\partial_0, \partial_1\}$ as a commutative pair of vector fields on the manifolds of functions $\widehat{\Phi}$ (with the entries of $\{\Phi_i\}$ as the coordinates). Hence locally $\partial_l = \partial/\partial t_l$, $l = 0, 1$ for an appropriate dependence of the coefficients of the series Φ on (local) parameters $t_0, t_1 \in \mathbb{R}$. Respectively (2.2) becomes a system of differential equations for the coefficients of $M_{\mathcal{D}_l}^{j_l}$ and, in particular, is self-consistent. One can check the latter directly (without Proposition2.1).

Following §1, let us consider

$$\mathcal{D}_0 = U_0 \stackrel{\text{def}}{=} \operatorname{diag}(\mu_i), \qquad j_0 = 1,$$
$$\mathcal{D}_1 = \mathcal{D} \stackrel{\text{def}}{=} \operatorname{diag}(\delta_i), \qquad j_1 \stackrel{\text{def}}{=} m \geq 0,$$

where $\mu_i = \mu_j \Rightarrow \delta_i = \delta_j$. We also set $t_0 = x$, $t_1 = t$, $\partial_0 = \partial/\partial x$, $\partial_1 = \partial/\partial t_1$, $M_{U_0}^1 = kU$ where $U = \Phi_0 U_0 \Phi_0^{-1}$. Expanding (2.2) with respect to k, we see that the coefficients of $k^{m+1}, \dots, k^2$ are identically zero, and that the coefficient of k gives the only non-trivial equation

$$U_t = [U, F_m^{\mathcal{D}}]. \tag{2.3}$$

Indeed, $(F^{\mathcal{D}})_x = k[U, F^{\mathcal{D}}]$ and $(M_{\mathcal{D}}^m)_x - k[U, M_{\mathcal{D}}^m] = k[U, F_m^{\mathcal{D}}]$, which follows from the definition of $M_{\mathcal{D}}^m$. According to Theorem1.3, the entries of the coefficient $F_m^{\mathcal{D}}$ of the resolvent $F^{\mathcal{D}}$ of k^{-m} are differential polynomials of the entries of U with respect to x (moreover, $F_m^{\mathcal{D}}$ is represented as a differential polynomial of the whole matrix U – Exercise1.2). Equation (2.3) is a variant of (generalized) Lax equations. The above consideration gives that it is the compatibility condition of equations

$$\check{\Phi}_x = kU\check{\Phi}, \qquad \check{\Phi}_t = M_{\mathcal{D}}^m \check{\Phi}, \tag{2.1$'$}$$

i.e., has the zero curvature representation

$$(M_{\mathcal{D}}^m)_x - kU_t = k[U, M_{\mathcal{D}}^m]. \tag{2.2$'$}$$

If U_0 satisfies (0.10), $\mathcal{D} = c^2 U_0$, and $m = 2$, then we obtain the GHM equation (0.9).

We emphasize that $\check{\Phi}$ is not an abstract function (as it was at the beginning of this section) any more. It is regarded now as a solution of system (2.1$'$).

Now let us try to replace $F_m^{\mathcal{D}}$ in (2.3) by the "value" $F^{\mathcal{D}}(1/2)$ of the series $F^{\mathcal{D}}(k) = \sum_{m=0}^{\infty} F_m^{\mathcal{D}} k^{-m}$ at $k = 1/2$. Without analytical assumptions $F^{\mathcal{D}}(1/2)$ is not well-defined. However we can introduce the resolvent corresponding to $\mathcal{D}$ in a different way. Let $\widetilde{F}^{\mathcal{D}}$ be a solution of the equation $(\widetilde{F}^{\mathcal{D}})_x = k[U, \widetilde{F}^{\mathcal{D}}]$ which is equivalent (conjugated) to $\mathcal{D}$. This new definition generalizes the previous one because $F^{\mathcal{D}}$ satisfies this equation. We mention that $\widetilde{F}^{\mathcal{D}}$ is not uniquely determined.

For instance, let $\widetilde{F}^{\mathcal{D}}(1/2)$ be a solution of the equation $F_x = \frac{1}{2}[U, F]$ and $\widetilde{F}^{\mathcal{D}}(1/2) \sim \mathcal{D}$. Substituting $V = -2\widetilde{F}^{\mathcal{D}}(1/2)$, we can naturally interpret equation (2.3) as the system

$$U_x = \tfrac{1}{2}[V, U],$$
$$V_x = \tfrac{1}{2}[U, V],$$

which coincides with the system of equations for the currents of the PCF (0.2).

Thus we see that system (0.2) can be formally regarded as a Lax equation. Anyway if the analytic continuation of $\check{\Phi}(k)$ is well-defined up to the point $k = 1/2$, then the abstract differentiation $\partial_{1/2} U = [U, \widetilde{F}_{(1/2)}^{\mathcal{D}}]$ is well-defined and commutes with an arbitrary differentiation $\partial_m U = [U, F_m^{\mathcal{D}_m}]$ introduced for $[\mathcal{D}_m, \mathcal{D}] = 0$.

Equations of the VNS type. We will obtain another version of Lax type equations, using (1.5) instead of (1.1). Let $\widehat{\Psi} = I + \sum_{i=1}^{\infty} \Psi_i k^{-i}$ and

$$G^{\mathcal{D}} = \widehat{\Psi}\mathcal{D}\widehat{\Psi}^{-1} = \sum_{i=0}^{\infty} G_i^{\mathcal{D}} k^{-i},$$

$$L_{\mathcal{D}}^j = \sum_{i=0}^{j} G_i^{\mathcal{D}} k^{-j+1}, \qquad \mathcal{D} \in \mathfrak{gl}_n, j \geqq 0.$$

(cf. definition of $F^{\mathcal{D}}$, $M_{\mathcal{D}}^j$). As before, we define a series $\check{\Psi} = \widehat{\Psi}\exp(t_0\mathcal{D}_0 k^{j_0} + t_1\mathcal{D}_1 k^{j_1})$ for mutually commutative $\mathcal{D}_0$ and $\mathcal{D}_1$, and introduce differentiations by the formula

$$\partial_l \check{\Psi} = L_{\mathcal{D}_l}^{j_l} \check{\Psi} \Leftrightarrow \partial_l \widehat{\Psi} = -\Big(\sum_{i=j_l+1}^{\infty} G_i^{\mathcal{D}} k^{j_l-i}\Big)\widehat{\Psi}, \; l = 0, 1,$$

which determines $\partial_l \Psi_i$ uniquely after being expanded in powers of k^{-1}. Following Proposition2.1, we get

Proposition2.2. $[\partial_0, \partial_1] = 0$,

$$\partial_0 L_{\mathcal{D}_1}^{j_1} - \partial_1 L_{\mathcal{D}_0}^{j_0} = [L_{\mathcal{D}_0}^{j_0}, L_{\mathcal{D}_1}^{j_1}]. \tag{2.4}$$

□

Set $\mathcal{D}_0 = U_0$, $\mathcal{D}_1 = \mathcal{D}$, $\partial_0 = \partial/\partial x$, $\partial_1 = \partial/\partial t$. Let $j_0 = 1$, $L_{U_0}^1 = kU_0 - Q$, where $Q(x,t) \in \mathfrak{gl}_n' = [U_0, \mathfrak{gl}_n]$, $j_1 = m \geqq 0$. Then (2.4) gives the differential equation

$$Q_t = -[U_0, G_{m+1}^{\mathcal{D}}], \tag{2.5}$$

where the entries of $G_{m+1}^{\mathcal{D}}$ are differential polynomials of the entries of Q with respect to x (Theorem1.1).

Replacing $G_{m+1}^{\mathcal{D}}$ by $\widetilde{G}^{\mathcal{D}}(1/2)$ as above, and setting $P = -2\widetilde{G}^{\mathcal{D}}(1/2)$, we arrive at the system

$$Q_t = \tfrac{1}{2}[U_0, P], \qquad P_x = [\tfrac{1}{2}U_0 - Q, P]. \tag{2.5a}$$

Since $Q_t - P_x = [Q, P]$ because of (2.5a), we may assume that $Q = F_0^{-1}(F_0)_x$, $P = F_0^{-1}(F_0)_t$ for an appropriate matrix F_0. Then $U \stackrel{\text{def}}{=} F_0 U F_0^{-1}$ and $V \stackrel{\text{def}}{=} 2(F_0)_t F_0^{-1}$ give a solution of system (0.2), if (Q, P) is a solution of (2.5a) (cf. Corollary1.1). We see that (2.5) and (2.3) are equivalent "at the point $k = 1/2$". It holds true for arbitrary m as well:

Exercise2.1. *Let U be a solution of (2.3). If $Q = \Phi_0^{-1}(\Phi_0)_x$ (cf. Introduction) for the coefficient Φ_0 of the series Φ that satisfies system (2.1′) (such a series exists), then Q is a solution of equation (2.5).* (Check that $\check{\Psi} = \Phi_0^{-1}\check{\Psi}$ satisfies system

$$\check{\Psi}_x = kU_0\check{\Psi} - Q\check{\Psi}, \qquad \check{\Psi}_t = L_{\mathcal{D}}^m\check{\Psi},$$

using the relation $F^{\mathcal{D}} = \Phi_0 G^{\mathcal{D}} \Phi_0^{-1}$.) □

Exercise2.2. *Show that for U_0 with conditions (0.10) and $p = 1$ the construction of Exercise2.1 coincides with the reduction of GHM (0.9) to VNS (0.14). Write down the zero-curvature representation for VNS, using (2.4) (cf. [Mana2], [K2]).* □

2.2. Jost functions as operators.

Let $\widehat{\Psi} = I + \sum_{i=1}^{\infty} \Psi_i k^{-i}$ be a formal power series with coefficients in $\mathfrak{gl}_n$, depending on x inderterminately. Set $\Psi = \widehat{\Psi}\exp(kU_0x)$. The matrix $U_0 = \operatorname{diag}(\mu_i)$ (cf. §2.1) is supposed to be constant and invertible for the sake of simplicity; $\exp(kU_0x)$ should be considered as a symbol subject to usual laws of multiplication and differentiation. In contrast with the Jost function from §1 which satisfies certain differential equations with respect to x, the function Ψ is called the ***abstract Jost function***.

Set $\partial = \partial/\partial x$, $\xi = \partial^{-1}$, $\partial^p U = \partial^p U/\partial x^p$ for arbitrary matrix function U of x and $p \geqq 0$. We call $M = \sum_{i=0}^{r} V_i \circ \partial^i$ a differential operator of order r, where V_i are some matrix functions and "$\circ$" denotes the composition of operators. By the standard relation

$$\partial \circ U = U \circ \partial + \partial U, \tag{2.6a}$$

we can rewrite M in another ("right") form: $M = \sum_{i=0}^{r} \partial \circ W_i$ for suitable functions W_i (note that $W_r = V_r$). The expression of M with ∂ on the right is called the "left" form. The (associative) composition $M \circ N$ of differential operators M, N is defined in the usual way. For an arbitrary (matrix) function U of x, we set

$$MU = \sum_{i=0}^{r} V_i(\partial^i U), \qquad UM = \sum_{i=0}^{r} (-1)^i (\partial^i U) W_i.$$

Then the following relations hold:

$$(M \circ N)U = M(NU), \qquad U(M \circ N) = (UM)N.$$

Let us call MU and UM the left and right actions of M on U respectively.

In order to distinguish operators and their actions we write $\sum_{i=0}^{r} V_i \circ \partial^i$ instead of the traditional representation $\sum_{i=0}^{r} V_i\partial^i$. The latter means in this paragraph the function $\sum_{i=0}^{r}(-1)^i(\partial^i V_i/\partial x^i)$ and is not an operator.

The fundamental property (2.6a) of ∂ implies the following Leibniz rule:

$$\xi \circ U = \sum_{s=0}^{\infty}(-1)^s(\partial^s U) \circ \xi^{s-1}, \qquad U \circ \xi = \sum_{s=0}^{\infty} \xi^{s+1} \circ (\partial^s U), \tag{2.6b}$$

which can be checked by applying $\partial\circ$ or $\circ\partial$ respectively. The expression $P = \sum_{i=r}^{\infty} V_i \circ \xi^i$, $r \in \mathbb{Z}$, is called a formal *pseudo-differential operator* of order $(-r)$. By (2.6), $P = \sum_{i=r}^{\infty} \xi^i \circ W_i$ for suitable W_i ($W_r = V_r$) and the (associative) composition "$\circ$" of pseudo-differential operators is well-defined. We call an operator of the form $P = I + \sum_{i=1}^{\infty} V_i \circ \xi$ a (formal) *Volterra operator*.

Set $\langle P \rangle_p = V_p$, ${}_p\langle P \rangle = W_p$. If $\widehat{V}$ is a formal series $\widehat{V} = \sum_{i=-r}^{\infty} V_i k^{-i}$ we write $(\widehat{V})_p = \widehat{V}_p = V_p$. The forms $\langle\cdot\rangle_p$, ${}_p\langle\cdot\rangle$ are linear when the argument is multiplied by any matrix function on the left and on the right respectively. We will turn matrix formal (either "left" or "right") series in the form

$$\Phi^+ = \Big(\sum_{i=r}^{\infty} \Phi_i^+ k^{-i}\Big) \exp(kxU_0),$$

$$\Phi^- = \exp(-kxU_0) \sum_{i=r}^{\infty} \Phi_i^- k^{-i},$$

into formal pseudo-differential operators:

$$\widetilde{\Phi}^+ = \sum_{i=r}^{\infty} \Phi_i^+ \circ \xi^i \circ U_0^i, \tag{2.7a}$$

$$\widetilde{\Phi}^- = \sum_{i=r}^{\infty} U_0^i \circ \xi^i \circ \Phi_i^-. \tag{2.7b}$$

In particular, the above Ψ corresponds to the Volterra operator $\widetilde{\Psi} = I + \sum_{i=1}^{\infty} \Psi_i \circ \xi^i \circ U_0^i$. Using the well-known expansion $(1 - X)^{-1} = \sum_{i=0}^{\infty} X^i$, (2.6b), and the invertibility of U_0 we can represent the inverse (with respect to the composition law $\circ$) operator $\widetilde{\Psi}^{-1}$ as follows:

$$\widetilde{\Psi}^{-1} = I + \sum_{i=0}^{\infty} U_0^i \circ \xi^i \circ \Psi_i^\star.$$

The matrix functions Ψ_i^* are expressed in terms of differential polynomials of $\{\Psi_i\}$ and U_0, U_0^{-1}. Let us define the *Jost function dual to* Ψ:

$$\Psi^* = \exp(-kU_0 x)\Big(I + \sum_{i=1}^{\infty} \Psi_i^* k^{-i}\Big).$$

One can introduce Ψ and Ψ^* without any reference to the corresponding operators:

Proposition2.3. *The following three conditions on*

$$\Phi = \exp(-kU_0)(I + \sum_{i=1}^{\infty} \Phi_i k^{-i})$$

are equivalent:

(a) $\Phi = \Psi^*$, *i.e.*, $\widetilde{\Phi} \circ \widetilde{\Psi} = \widetilde{\Psi} \circ \widetilde{\Phi} = I$;
(b) $((M\Psi)U_0\Phi)_1 = 0$ *for any differential operator* M;
(c) $(\Psi U_0(\Phi M))_1 = 0$ *for any differential operator* M.

Proof. We use the following basic property of the operation $\Phi \to \widetilde{\Phi}$.

Lemma2.1. *For* $\Phi^{\pm}$ *of (2.7) and an arbitrary differential operator* M,

$$\widetilde{(M\Phi^+)} = M \circ \widetilde{\Phi}^+, \tag{2.8a}$$

$$\widetilde{(\Phi^- M)} = \widetilde{\Phi}^- \circ M. \tag{2.8b}$$

Proof. It is sufficient to check the statement only for $M = \partial$, since any M is represented in the form

$$M = \sum_{i=0}^{m} V_i \circ \partial^i = \sum_{i=0}^{m} \partial^i \circ W_i,$$

and relations (2.8a,b) are stable when M is multiplied by a function on the left or on the right respectively. Both formulae are checked in the same way. As for Φ^-:

$$\Big(\exp(-kU_0 x)\Big(\sum_{i=r}^{\infty} \Phi_i^- k^{-i}\Big)\Big)\partial = \exp(-kU_0 x)\sum_{i=r}^{\infty}(U_0\Phi_i^- k^{1-i} - (\partial\Phi_i^-)k^{-i}),$$

$$\widetilde{\Phi}^- \circ \partial = \sum_{i=r}^{\infty}(U_0^i \circ \xi^{i-1} \circ \Phi_i^- - U_0^i \circ \xi^i \circ (\partial\Phi_i^-)).$$

Replacing $(\cdot)k^{-i}$ in the first line by $U_0^i \circ \xi^i \circ (\cdot)$ and dropping the exp-factor, we obtain the second line. □

We need one more

Lemma2.2. *If we set* $X \circ Y = (\sum_{i=r}^{\infty} X_i \circ \xi^i) \circ (\sum_{i=s}^{\infty} \xi^i \circ Y_i) = \sum_{i=r+s}^{\infty} A_i \circ \xi^i = \sum_{i=r+s}^{\infty} \xi^i \circ B_i$, *then*

$$A_1 \stackrel{\text{def}}{=} \langle X \circ Y \rangle_1 = \sum_{i+j=1} X_i Y_j = {}_1\langle X \circ Y \rangle \stackrel{\text{def}}{=} B_1,$$

where $\{X_i, Y_i\}$ *are matrix functions and* $r, s \in \mathbb{Z}$.

Proof. Any operator $X_i \circ \xi^l \circ Y_{l-i}$ for $l > 1$ is written in terms of ξ^j, $j \geqq l > 1$ only either in the left or in the right form (use (2.6b)). If $l < 0$, this operator is differential and does not contain ξ at all. If $l = 1$, then

$$X_i \circ \xi \circ Y_{1-i} = X_i \circ Y_{1-i} \circ \xi + O(\xi^2) = \xi \circ X_i \circ Y_{1-i} + O(\xi^2),$$

where $O(\xi^2)$ contains ξ only in powers greater than 1. □

Now we come to the proof of Proposition2.3. First we will deduce statement (b) from (a). According to the definition of $\langle\cdot\rangle_1$, we can restrict ourselves to the case $M = \partial^m$, $m \geqq 0$. Using Lemma2.2 and (2.8a), we obtain:

$$\begin{aligned} ((\partial^m \Psi) U_0 \Phi)_1 &= \langle \widetilde{(\partial^m \Psi)} \widetilde{\Phi} \rangle_1 = \\ &= \langle \partial^m \widetilde{\Psi} \circ \widetilde{\Phi} \rangle_1 = \langle \partial^m \rangle_1 = 0. \end{aligned}$$

Conversely, if (b) is satisfied, then $\langle \partial^m \circ \widetilde{\Psi} \circ \widetilde{\Phi} \rangle_1 = 0$ for all $m \geqq 0$. Hence, $\widetilde{\Psi} \circ \widetilde{\Phi}$ cannot contain terms with ξ^s, $s > 0$, and $\widetilde{\Psi} \circ \widetilde{\Phi} = \boldsymbol{I}$. Similarly (applying (2.8b)) we show the equivalence of (a) and (c). □

2.3. Abstract fractional powers (generating operator).

Keeping the notations of the previous section, let us introduce *abstract fractional powers* $\{L_r\}$. For arbitrary $r \in \mathbb{Z}_+$ and a pseudo-differential operator $M = \sum_{i=-r}^{\infty} V_i \circ \xi^i$ of order r, we define its differential part as $M_+ = \sum_{i=r}^{0} V_{-i} \circ \partial^i$ and set $M_- = \sum_{i=1}^{\infty} V_i \circ \xi^i = M - M_+$. It is easy to see that this decomposition of M does not depend on the representation of M in the left or right form (i.e., with $\xi^{(\cdot)}$ on the right or $\xi^{(\cdot)}$ on the left). Let $\mathcal{D}$ be a constant matrix commutative with U_0.

Proposition2.4. *Given* $r \in \mathbb{Z}_+$, *the following conditions on differential operator* L_r *of order* r *with the highest coefficient* $U_0^{-r}\mathcal{D}$ *(of* ∂^r*) are equivalent:*

(a) $L_r = (\widetilde{\Psi} \circ U_0^{-r} \mathcal{D} \partial^r \circ \widetilde{\Psi}^{-1})_+$,
(b) $L_r \Psi = k^r \Psi \mathcal{D} + O(k)$,
(c) $\Psi^\star L_r = k^r \mathcal{D} \Psi^\star + O(k)$.

Hereafter $O(k)$ is a series of k^i, $i \leqq s$ (with or without the exp-factor).

Proof. First we deduce statement (b) from (a). The formula (2.8a) leads to

$$\begin{aligned}((\widetilde{\Psi} \circ U_0^{-r}\mathcal{D}\partial^r \circ \widetilde{\Psi}^{-1})_+ \Psi)^{\sim} &= (\widetilde{\Psi} \circ U_0^{-r}\mathcal{D}\partial^r \circ \widetilde{\Psi}^{-1})_+ \circ \widetilde{\Psi} \\ &= \widetilde{\Psi} \circ U_0^{-r}\mathcal{D}\partial^r + O(\xi),\end{aligned}$$

where $O(\xi)$ contains ξ^i for $i > 0$ only. Rewriting the above expression in the form (2.7a), we obtain (b). The statement (c) is derived analogously. This reasoning is easy to reverse. □

We can also prove the implications (b)⇒(a), (c)⇒(a), using the fact that the operators L_r are uniquely determined by (b) or (c). We can use these conditions to calculate such operators:

Exercise2.3. *For arbitrary operators $M = \sum_{s=0}^{r} V_s\partial^s = \sum_{s=0}^{r} \partial^s W_s$ and an invertible matrix function U, prove that*

$$(MU)U^{-1} = \sum_{s=0}^{r} V_s P_s((\partial U)U^{-1}),$$

$$U^{-1}(UM) = \sum_{s=0}^{r} (-1)^s R_s(U^{-1}(\partial U))W_s$$

for (non-commutative) polynomials P_r, R_r of $Z, \partial Z, \partial^2 Z, \dots$ such that

$$P_0 = R_0 = 1, \qquad P_1 = R_1 = Z,$$
$$P_s = \partial P_{s-1} + P_{s-1}P_1, \qquad R_s = \partial R_{s-1} + R_1 R_{s-1}.$$

□

Exercise2.4. *Using Exercise2.3, check that coefficients of the operator L_r (in either form) defined by condition (b) are polynomials of U_0, U_0^{-1}, the coefficients of the power series $(\partial\Psi)\Psi^{-1}$, their derivatives, and the coefficients of the series $\Psi\mathcal{D}\Psi^{-1}$ in k^{-1} (the latter enter linearly).* □

We call the operators constructed in Exercise2.4 the *abstract fractional powers*; the conjugation of the operators $U_0^{-r}\mathcal{D}\partial^r$ by the operator $\widetilde{\Psi}$ is naturally regarded as the abstract dressing transformation, following V. E. Zakharov and A. B. Shabat. We define the *abstract jets* $G^{p,q} = (\partial^p\Psi)U_0\mathcal{D}(\Psi^*\partial^q)$. In particular, we call $G = G^{0,0}$ the *resolvent*.

Proposition2.5. a) *Given* $r \in \mathbb{Z}, p \in \mathbb{Z}_+$,

$$\langle \widetilde{\Psi} \circ U_0^{-r} \mathcal{D} \partial^r \circ \widetilde{\Psi}^{-1} \rangle_{p+1} = (\Psi U_0 \mathcal{D}(\Psi^* \partial^p))_{r+1} = (G^{0,p})_{r+1},$$
$${}_{p+1}\langle \widetilde{\Psi} \circ U_0^{-r} \mathcal{D} \partial^r \circ \widetilde{\Psi}^{-1} \rangle = ((\partial^p \Psi) U_0 \mathcal{D} \Psi^*)_{r+1} = (G^{p,0})_{r+1}.$$

b) *Let* $G = \Psi U_0 \mathcal{D} \Psi^* \stackrel{\text{def}}{=} \sum_{i=0}^{\infty} G_i k^i$, $r \in \mathbb{Z}_+$. *Then*

$$L_r \Psi = k^r \Psi \mathcal{D} - G_{r+1} U_0^{-1} \Psi k^{-1} + O(k^{-2}),$$
$$\Psi^* L_r = k^r \mathcal{D} \Psi^* - \Psi^* U_0^{-1} G_{r+1} k^{-1} + O(k^{-2}).$$

Proof. Using Lemmas 2.1 and 2.2, we see that

$$\begin{aligned}
\langle \widetilde{\Psi} \circ U_0^{-r} \mathcal{D} \partial^r \circ \widetilde{\Psi}^{-1} \rangle_{p+1} &= \langle \widetilde{\Psi} \circ U_0^{-r} \mathcal{D} \partial^r \circ \widetilde{\Psi}^{-1} \circ \partial^p \rangle_1 \\
&= \langle \widetilde{\Psi} \circ U_0^{-r} \mathcal{D} \partial^r \circ (\widetilde{\Psi^* \partial^p}) \rangle_1 \\
&= \sum_{s=0}^{p+r+1} (\Psi_s U_0^s \mathcal{D} U_0^{-r} U_0^{r-s+1} (\Psi^* \partial^p)_{r-s+1}) \\
&= (\Psi U_0 \mathcal{D}(\Psi^* \partial^p))_{r+1}.
\end{aligned}$$

The second formula of a) is proved analogously. With the help of a), we can check the following general formula:

$$\begin{aligned}
\widetilde{L_r \Psi} &= (\widetilde{\Psi} \circ U_0^{-r} \mathcal{D} \partial^r \circ \widetilde{\Psi}^{-1})_+ \circ \widetilde{\Psi} \\
&= \widetilde{\Psi} \circ U_0^{-r} \mathcal{D} \partial^r - (\widetilde{\Psi} \circ U_0^{-r} \mathcal{D} \partial^r \circ \widetilde{\Psi}^{-1})_- \circ \widetilde{\Psi} \\
&= \widetilde{\Psi} \circ U_0^{-r} \mathcal{D} \partial^r - \sum_{p=0}^{\infty} \xi^{p+1} \circ (G^{p,0})_{r+1} \circ \widetilde{\Psi}.
\end{aligned}$$

Rewriting ${}_{s+1}\langle \cdot \rangle$ in terms of ${}_1\langle \cdot \rangle = \langle \cdot \rangle_1$ for $s \geqq 0$ (see above), we obtain the relation

$$\begin{aligned}
(2.9)\quad {}_{s+1}\langle \widetilde{\Psi} \circ U_0^{-r} \mathcal{D} \partial^r - (\widetilde{L_r \Psi}) \rangle &= (G^{s,0})_{r+1} + (G^{s-1,0})_{r+1} \Psi_1 U_0 + \\
&+ \sum_{p=0}^{s-2} ((\partial^{s-p-1} (G^{p,0}_{r+1} \Psi)) \exp(-kU_0 x))_1 U_0.
\end{aligned}$$

When $s = 0$, one has: $(L_r \Psi)_1 U_0 - (k^r \Psi \mathcal{D})_1 U_0 = -(G)_{r+1}$. The corresponding "right" formula of b) is proved analogously. □

Exercise2.5*. *Using the formula (2.9), show that*

$$(L_r\Psi - k^r\Psi\mathcal{D})_2 U_0^2 = (G^{0,1})_{r+1} - G_{r+1}\Psi_1 U_0.$$

Check statement b) of Proposition2.5, using (b) and (c) of Proposition2.4 (cf. [Ch5]). □

Generating operator. Let us define a formal power series Λ_+ of k^{-p-1}, $p \geqq 0$ which has the differential operators L_p as coefficients:

$$\Lambda_+ = \sum_{p=0}^{\infty} L_p k^{-p-1} = \Big(\widetilde{\Psi} \circ \Big(\sum_{p=0}^{\infty} U_0^{-p}\mathcal{D}\partial^p k^{-p-1}\Big) \circ \widetilde{\Psi}^{-1}\Big)_+.$$

We introduce the pseudo-differential operator

$$\Lambda_- = \Big(\widetilde{\Psi} \circ \Big(\sum_{p=-\infty}^{\infty} U_0^{-p}\mathcal{D}\partial^p k^{-p-1}\Big) \circ \widetilde{\Psi}^{-1}\Big)_-$$

of non-negative powers of ξ. The coefficient of this operator of ξ^s (in the left or right form) is a series in k^{-p-1} with $p \geqq -s$. We will use the power series $G = \Psi U_0 \mathcal{D}\Psi^\star$ in k^i, $i \geqq 0$ (Proposition2.5), and introduce $H = (\partial\Psi)\Psi^{-1}$ which contains k^i such that $i \geqq -1$ (Ψ^{-1} is the inverse of Ψ).

Theorem2.1. a) $(\partial - H) \circ \Lambda_+ = -G$,
b) $(\partial - H) \circ \Lambda_- = G$.

Proof. First we check relation b). Applying Proposition2.4, a), we find that

$$\begin{aligned}\langle\Lambda_-\rangle_{r+1} &= \sum_{p=-\infty}^{\infty} \Big(\Psi U_0 \mathcal{D}(\Psi^\star\partial^r)\Big)_{p+1} k^{-p-1}\\ &= \Psi U_0 \mathcal{D}(\Psi^\star\partial^r) = G^{0,r},\end{aligned}$$

where $r \geqq 0$. Consequently,

$$(\partial - (\partial\Psi)\Psi^{-1}) \circ \Lambda_- = (\partial - (\partial\Psi)\Psi^{-1}) \circ \sum_{r=0}^{\infty}\Big(\Psi U_0 \mathcal{D}(\Psi^\star\partial^r)\Big) \circ \xi^{r+1} =$$

$$\sum_{r=0}^{\infty}\{(\partial\Psi)U_0\mathcal{D}(\Psi^\star\partial^r) - \Psi U_0\mathcal{D}(\Psi^\star\partial^{r+1}) - (\partial\Psi)\Psi^{-1}\Psi U_0\mathcal{D}(\Psi^\star\partial^r)\} \circ \xi^{r+1}$$

$$+ \sum_{r=0}^{\infty}\Big(\Psi U_0(\Psi^\star\partial^r)\Big) \circ \xi^r = \Psi U_0 \mathcal{D}\Psi^\star.$$

Statement a) can now be derived from b), using the relation $(\partial - H)\circ\Lambda = 0$, for $\Lambda = \Lambda_+ + \Lambda_-$.

Let us give another (more direct) proof which will be useful later, and which does not rely on the operator interpretation of the Jost function.

It is necessary to check the formula

$$(\partial - (\partial\Psi)\Psi^{-1})\circ\sum_{p=0}^{\infty} L_p k^{-p-1} = -\Psi U_0 \mathcal{D}\Psi^{*}. \tag{2.10}$$

We introduce another parameter l and set $\Psi(l) = (\sum_{i=0}^{\infty}\Psi_i l^{-i})\exp(lU_0x)$. Since the left hand side of (2.10) is a differential operator, it is enough to verify the relation

$$\Big((\partial - (\partial\Psi)\Psi^{-1})\circ\sum_{p=0}^{\infty} L_p k^{-p-1}\Big)\Psi(l) = -\Psi U_0\mathcal{D}\Psi^{*}\Psi(l), \tag{2.11}$$

modulo $O(l^{-1})$. Hereafter we shall make the dependence on k explicit (write $\Psi(k)$ instead of Ψ, etc.). Propositions 2.4,(b) and 2.5,b) allow us to calculate the left hand side of (2.11) up to $O(l^{-1})$. The result is

$$(\partial - H(k))\Big(\Big(\sum_{p=0}^{\infty} l^p k^{-p-1}\Big)\Psi(l)\mathcal{D} - G(k)U_0^{-1}l^{-1}\Psi(l) + l^{-1}\mathcal{D}\Psi(l)\Big).$$

Differentiating and replacing $\sum_{p=0}^{\infty} l^p k^{-p-1}$ by $(k-l)^{-1}$, and using the relations $G(l) = U_0\mathcal{D} + O(l^{-1})$, $H(l) = U_0 l + O(l^0)$, we arrive at

$$((\partial - H)\circ\Lambda_+\Psi(l))\Psi(l)^{-1} = -\frac{H(k)}{k-l}\mathcal{D} + \frac{U_0 l}{k-l}\mathcal{D}$$
$$-G(k) + U_0\mathcal{D} \mod O(l^{-1}) = \frac{H(l)-H(k)}{k-l}\mathcal{D} - G(k) + U_0\mathcal{D} \mod O(l^{-1}).$$

The series $(H(l) - H(k))/(k-l)$ is equal to $-U_0$ up to $O(l^{-1})$. □

The series Λ_+ is called the *generating operator of fractional powers* L_r. According to the above theorem, it is written via H and G. Set

$$(\partial - H)_\partial^{-1} = U_0^{-1}k^{-1}(\partial U_0^{-1}k^{-1} - I + \Theta)^{-1} =$$
$$= -U_0^{-1}k^{-1}\Big(I + \sum_{i=1}^{\infty}(\partial U_0^{-1}k^{-1} + \Theta)^i\Big),$$

where $\Theta = I - HU_0^{-1}k^{-1} = O(k^{-1})$. Then $\Lambda_+ = (\partial - H)_\partial^{-1}\circ G$.

Exercise2.6. *Let* $(\partial - H)_\xi^{-1} \stackrel{\text{def}}{=} \xi \circ (I - H \circ \xi)^{-1} = \xi \circ (I + \sum_{i=1}^{\infty} (H \circ \xi)^i)$. *Then*

$$(\partial - H)_\xi^{-1} \circ G = \Psi \circ \xi \circ \Psi^* = \widetilde{\Psi} \circ \Big(\sum_{i=0}^{\infty} \xi^{i+1} k^i\Big) \circ \widetilde{\Psi}^{-1}. \quad \square$$

Exercise2.7. *Prove the "right" analogue of Theorem2.1:*

$$\Lambda_\pm \circ (\partial + (\Psi^*)^{-1}(\partial \Psi^*)) = \mp G. \quad \square$$

2.4. Generalized Lax equations (relations to VNS).

As in §2.1, let us suppose we are given two matrices $\mathcal{D}_1$ and $\mathcal{D}_2$ which commute with each other and with U_0. Let us introduce the corresponding formal differentiations $\partial_1 = \partial/\partial t_1$, $\partial_2 = \partial/\partial t_2$ of coefficients Ψ_k of the series Ψ. It is convenient to determine their action on the (coefficients of) Volterra operator $\widetilde{\Psi}$ associated to Ψ (equal to $U_0^k \Psi_k$ respectively - §2.2). For $j_r \in \mathbb{Z}_+$, $r = 1, 2$, we set

$$L_{(r)} = (\widetilde{\Psi} \circ \partial^{j_r} U_0^{-j_r} \mathcal{D}_r \circ \widetilde{\Psi}^{-1})$$

and

$$\partial_r \Psi \stackrel{\text{def}}{=} L_{(r)} \Psi - \Psi \mathcal{D}_r k^{j_r} \Leftrightarrow \partial_r \widetilde{\Psi} = L_{(r)} \circ \widetilde{\Psi} - \widetilde{\Psi} \circ U_0^{-j_r} \mathcal{D}_r \partial^{j_r}. \tag{2.12}$$

We assume here that the differentiations act trivially on k and the symbol ∂: $\partial_r k = \partial_r \partial = 0$. In particular, $\partial_r \exp(kxU_0) = 0$. One can check that $[\partial_r, \partial] = \partial_r \circ \partial - \partial \circ \partial_r = 0$. Expanding (2.12) with respect to k^{-1} with ξ on the right we obtain explicit formulae for $\partial_r \Psi_i$ in terms of $\{\Psi_i\}$ and their x-derivatives of appropriate orders (cf. §2.1). We can extend ∂_r uniquely to a differentiation of the ring of differential polynomials of the entries of $\{\Psi_i\}$. The action of ∂_r on any polynomial expressions in terms of k, k^{-1}, ∂, ξ with coefficients from this ring is also well-defined (∂_r are applied coefficient-wise).

Proposition2.6.

$$[\partial_1, \partial_2] = 0,$$
$$\partial_1 L_{(2)} - \partial_2 L_{(1)} = [L_{(1)}, L_{(2)}]. \tag{2.13}$$

Proof. For the sake of simplicity, let us use the formal operator

$$\widetilde{\Psi}^{\vee} = \widetilde{\Psi} \exp(A_1 t_1 + A_2 t_2)$$

instead of $\widetilde{\Psi}$ where $A_r = U_0^{-j_r}\mathcal{D}_r\partial^{j_r}$. Then, $\partial_r\widetilde{\Psi}^\vee = L_{(r)} \circ \widetilde{\Psi}^\vee$. We set $R_r = \widetilde{\Psi} \circ A_r \circ \widetilde{\Psi}^{-1}$, $r = 1, 2$. We have

$$\begin{aligned}\partial_1(\partial_2\widetilde{\Psi}^\vee) &= \partial_1(L_{(2)} \circ \widetilde{\Psi}^\vee)\\ &= ([L_{(1)}, R_2])_+ \circ \widetilde{\Psi}^\vee + (L_{(2)} \circ L_{(1)}) \circ \widetilde{\Psi}^\vee.\end{aligned}$$

Consequently,

$$([\partial_1, \partial_2]\widetilde{\Psi}^\vee) \circ (\widetilde{\Psi}^\vee)^{-1} = ([(R_1)_+, R_2])_+ + ([R_1, (R_2)_+])_+ + [(R_2)_+, (R_1)_+].$$

The right hand side is equal to $[R_1, R_2]_+$ (cf. Proposition2.1), which follows directly from the identity $[R_1, R_2]_+ = [(R_1)_+ + (R_1)_-, (R_2)_+ + (R_2)_-]_+$. Since $[\mathcal{D}_1, \mathcal{D}_2] = 0$, then $[R_1, R_2] = \widetilde{\Psi} \circ [A_1, A_2] \circ \widetilde{\Psi}^{-1} = 0$. □

In particular, we see that the differential operator $[L_{(1)}, L_{(2)}]$ is of order not more than $\max\{j_1, j_2\} - 1$. It can be checked directly:

Exercise2.8. *Let* $m \stackrel{\text{def}}{=} j_2$, $G^{(r)} = \Psi U_0 \mathcal{D}_r \Psi^*$. *If* $j_2 > j_1$, *then the differential operator*

$$L_{(3)} = (\widetilde{\Psi} \circ U_0^{-j_1-j_2}\mathcal{D}_1\mathcal{D}_2\partial^{j_1+j_2} \circ \widetilde{\Psi}^{-1})_+$$

modulo operators of type $O(\partial^{m-2})$ *(of order less than* $m-1$*) has the following form:*

$$\begin{aligned}L_{(3)} &= L_{(2)} \circ L_{(1)} + U_0^{-m}G^{(1)}_{j_1+1}\partial^{m-1}\mathcal{D}_2 + O(\partial^{m-2}) =\\ &= L_{(1)} \circ L_{(2)} + G^{(1)}_{j_1+1}U_0^{-m}\partial^{m-1}\mathcal{D}_2 + O(\partial^{m-2}).\end{aligned}$$

If $j_2 = j_1 = m$, *then*

$$L_{(3)} = L_{(2)} \circ L_{(1)} + (U_0^{-m}G^{(1)}_{m+1} + G^{(1)}_{m+1}U_0^{-m})\partial^{m-1}\mathcal{D}_2 + O(\partial^{m-2}).$$

(Calculate $L_{(3)}\Psi$ using Proposition2.5; cf. [Ch5]). □

If we replace ∂_r by $\partial/\partial t_r$, then (2.13), regarded as a differential equation for the coefficients $L_{(1)}$ and $L_{(2)}$ with respect to x, t_1, t_2, is called the *Zakharov-Shabat equation*. The *Kadomtsev-Petviashvili equation* (abbreviated as KP) is a special case (when $j_1 = 2$, $j_2 = 3$ and L is a scalar operator - see [ZMa2], [ZS1,2]). The *generalized Lax equations* (defined by I. M. Gelfand and L. A. Dickey) are obtained if $\partial/\partial t_1 = 0$ and Ψ is subject to the constraint $L_{(1)}\Psi - \Psi\mathcal{D}_1 k^{j_1} = 0$. They are written in the form

$$\frac{\partial}{\partial t}L_{(1)} = [L_{(2)}, L_{(1)}], \text{ where } t = t_2.$$

Now we will study these equations in more detail.

Let $\mathcal{D}_1 = I$, $\mathcal{D}_2 = \mathcal{D} = \operatorname{diag}(\varepsilon_i)$, $j_1 = p$, $j_2 = m$. We suppose that $\delta_i = \delta_j$ and $\mu_i = \mu_j$, if $(\mu_i)^p = (\mu_j)^p$ ($U_0 = \operatorname{diag}(\mu_i)$). We take the operator $L = U_0^{-p}\partial^p + \sum_{i=1}^p U_i\partial^{p-i}$ as $L_{(1)}$, where $U_1(x) \in [U_0^{-p}, \mathfrak{gl}_n]$. Note that we begin now with the operator L (the coefficients U_i are supposed to be unknown matrix functions of x); Ψ is not arbitrary anymore.

Proposition2.7. *a) There exists a solution* $\Psi = (I + \sum_{i=1}^{\infty} \Psi_i k^{-i}) \exp(kxU_0)$ *of the equation*

$$(2.14) \qquad L\Psi = \Psi k^p.$$

b) Any solution of (2.14) of the same type is the product ΨC *for proper* $C = I + \sum_{i=1}^{\infty} C_i k^{-i}$*, where* C_i *do not depend on* x *and commute with* U_0*.*

Proof. One can follow Proposition1.1 of §1 and examine the differential equation for Ψ_i directly (cf. the paper by Krichever [Kri3] and Proposition1.1, §1). We will sketch another way based on the Riccati equations (see [Ch5]):

Exercise2.9. *Rewrite equation (2.14) as an equation for* $H = (\partial\Psi/\partial x)\Psi^{-1} = U_0 k + \sum_{i=0}^{\infty} H_i k^{-i}$*. Show that the corresponding equation for* H_s *has the form*

$$u(H_s) = (\textit{differential polynomials of } U_0^{\pm}, H_0, \dots, H_{s-1}),$$

where $u(X) \overset{\text{def}}{=} \sum_{i=1}^{p} U_0^{-i} X U_0^{i-1}$*. Check that*

$$u([U_0^{-1}, U_0 X U_0^{1-p}]) = [U_0^{-p}, X]$$

for arbitrary matrix X*, and* $u(p^{-1}U_0 X) = X$ *for* X *commutative with* U_0*. Hence* $u(\mathfrak{gl}_n) = \mathfrak{gl}_n$ *and* H *is uniquely determined by (2.14) (via the above recurrent equations).* □

According to Proposition2.7, the formal power series $(\partial\Psi)\Psi^{-1}$, $\Psi\mathcal{D}\Psi^{-1}$, $G = \Psi U_0 \mathcal{D} \Psi^{\star}$ do not depend on the choice of the solution Ψ of equation (2.14). For the first two series this is obvious. As to G, use that $(\widetilde{\Psi C})^{-1} = C^{-1}\widetilde{\Psi}^{-1}$ and $(\Psi C)^{\star} = C^{-1}\Psi^{\star}$.

The proof of Theorem1.1, a) (§1) is easily adapted to the case of equation (2.14) instead of (1.5). We see that the coefficients of all three series above are written as differential polynomials of $U_0^{\pm}$ and $\{U_i, 1 \leqq i \leqq p\}$ (this also follows from Exercise2.9). In particular, the coefficients of the differential operator $L_m = \widetilde{\Psi} \circ (U_0^{-m} \partial^m \mathcal{D}) \circ \widetilde{\Psi}^{-1}$ are differential polynomials of the same type (cf. Exercise2.4), if Ψ satisfies (2.14). For this Ψ, we can make the statements of Exercise2.8 stronger:

Exercise2.10.

$$[L_m, L] = [U_0^{-p}, G_{m+1}]\partial^{p-1} + O(\partial^{p-2}),$$

where $G = \Psi U_0 \mathcal{D}\Psi^*$, $O(\partial^{p-2})$ is a differential operator of order less than $p-2$ (apply the operator $[L_m, L]$ to Ψ). □

Finally, the generalized Lax equation is written in the form

$$\frac{\partial}{\partial t}L = [L_m, L], \tag{2.15}$$

where the coefficients U_i $(1 \leqq i \leqq p)$ of the operator L are unknown. The right hand side is a differential operator of order less than $p-1$ determined uniquely by L, m, $\mathcal{D}$. Its coefficients are differential polynomials of U_0, U_0^{-1}, $\{U_i\}$.

First order operator. Let us establish the connection between (2.15) and the equations of §2.1 in the special case when $L = U_0^{-1}\partial/\partial x + U_0^{-1}Q$. It follows from Proposition2.3 b) that the dual Jost function Ψ^* corresponding to the Jost function Ψ from (2.14) is equal to $U_0^{-1}\Psi^{-1}U_0$. Let us check it. It is necessary to show that $((M\Psi)\Psi^{-1})_1 = 0$ for an arbitrary differential operator M. We can assume that $M = \sum_{i=0}^{r} V_i \circ L^i$ for appropriate matrix functions V_i, since L is of the first order. Then

$$((M\Psi)\Psi^{-1})_1 = \Big(\sum_{i=0}^{r} V_i(k^i\Psi\Psi^{-1})\Big)_1 = 0.$$

Using the obtained formula for Ψ^*, we see that $G = \Psi U_0 \mathcal{D}\Psi^*$ and, according to Exercise2.10, equation (2.15) has the form

$$\frac{\partial Q}{\partial t} = -[U_0, (\Psi\mathcal{D}\Psi^{-1})_{m+1}]. \tag{2.16}$$

The relation $L\Psi = k\Psi$ defining Ψ can be rewritten as: $\Psi_x + Q\Psi = kU_0\Psi$. Hence the series $\Psi\mathcal{D}\Psi^{-1}$ coincides with the resolvent $G^{\mathcal{D}}$ defined in §2.1 and equation (2.16) coincides with (2.5) (recall that equation (2.5) is, in its turn, equivalent to (2.3) - Exercise2.1):

For the operators L of first order, the technique of fractional powers leads to the same differential equations as the zero curvature representations from §2.1.

Exercise2.11. *For the operators L, L_m (see (2.15)) and the series Ψ from §2.2, show that*

$$L\Psi = k^p\Psi \Leftrightarrow \Psi^*L = k^p\Psi^*$$
$$\Psi_t = L_m\Psi - \Psi k^m\mathcal{D} \Leftrightarrow -\Psi_t^* = \Psi^*L_m - k^m\mathcal{D}\Psi^*$$

(cf. Proposition 2.4). □

Exercise2.12. *There exists a solution* $\check{\Psi} = (1 + \sum_{i=1}^{\infty} \Psi_i k^{-i}) \exp(kxU_0 + k^m \mathcal{D}t)$ *of the system*

$$L\check{\Psi} = k^p \check{\Psi}, \qquad L_m \check{\Psi} = \frac{\partial \Psi}{\partial t}$$

for L and L_m from (2.15). If we set $\check{\Psi}^\star = \exp(-kxU_0 - k^m \mathcal{D}t)(1 + \sum_{i=1}^{\infty} \Psi_i^\star k^{-i})$, *where $\{\Psi_i^\star\}$ are defined by $\{\Psi_i\}$ in §2.2, then*

$$\check{\Psi}^\star L = k^p \check{\Psi}^\star, \qquad \check{\Psi}^\star L_m = -\frac{\partial \check{\Psi}^\star}{\partial t}$$

(cf. Proposition1.2, §1, and Exercise2.11). □

Generating operator for Lax equations. Before ending this section, let us turn to the problem of calculating $[L_m, L]$. We will give the formula for $[\Lambda_+, L]$ where Λ_+ is defined in §2.3. We suppose that Ψ satisfies the assumption of Proposition2.7.

Exercise2.13. *Let us define differential operators K_i, M_i of order i $(0 \leqq i < n)$ by the relations $L - K_i \circ \partial^{p-i} = O(\partial^{p-i-1})$, $L - \partial^{p-i} \circ M_i = O(\partial^{p-i-1})$, where $O(\partial^{p-i-1})$ is a differential operator of order $\leqq p - i - 1$. Let $X = (\partial + A^{-1}(\partial A)) \circ C^{-1} = C^{-1} \circ (\partial - (\partial B)B^{-1})$, where A, B are invertible matrix functions, $C = BA$. Then*

$$L = X \circ R + A^{-1}(AL) = S \circ X + (LB)B^{-1},$$

where $R = \sum_{i=0}^{p-1} (CA^{-1}(AK_i)) \circ \partial^{p-i-1}$, $S = \sum_{i=0}^{p-1} \partial^{p-i-1} \circ ((M_i B)B^{-1}C)$ *(cf. [Ch5]).* □

Since $\Lambda_+^{-1} = -G^{-1} \circ (\partial - H) = -(\partial + H^\star) \circ G^{-1}$, where $H = (\partial\Psi)\Psi^{-1}$, $H^\star \stackrel{\text{def}}{=} (\Psi^\star)^{-1}(\partial\Psi^\star)$, $G = \Psi U_0 \mathcal{D} \Psi^\star$, it follows from the formulas of Exercise2.13 that

$$\begin{aligned}[\Lambda_+, L] &= \sum_{i=0}^{\infty} [L_i, L] k^{-i-1} = \\ &= \Lambda_+ \circ (L - k^p) - (L - k^p) \circ \Lambda_+ = \\ &= \sum_{i=0}^{p-1} \{\partial^{p-i-1} \circ ((M_i \Psi) U_0 \mathcal{D} \Psi^\star) - (\Psi U_0 \mathcal{D}(\Psi^\star K_i)) \circ \partial^{p-i-1}\}.\end{aligned}$$

In particular, this leads to the statement of Exercise2.10 (check this).

2.5. Scalar operators (τ-functions).

In contrast with the Lax type equations in §2.1, the equation (2.15) introduced above is quite interesting for the L-operator with the scalar coefficients U_i. We will take $U_0 = \mathcal{D} = 1$. Let $h = H$ and $g = G$, $\psi = (1 + \sum_{i=1}^{\infty} k^{-i}\psi_i)\exp(kx)$ be the abstract Jost function with undetermined scalars ψ_i, and L_r the corresponding abstract fractional power constructed in §2.3. The condition $U_0 = 1$ implies that

$$h = \psi_x\psi^{-1} = k + \sum_{i=1}^{\infty} k^{-i}h_i,$$

$$g = \psi\psi^{\star} = 1 + \sum_{i=2}^{\infty} k^{-i}g_i,$$

$$L_r = \partial^r + O(\partial^{r-2}),$$

i.e., the coefficients in h, g and L_r of k^0, k^{-1} and ∂^{r-1} respectively are zero. Following §2.4, we define pairwise commutative differentiations of the coefficients $\{\psi_j\}$, $\partial/\partial t_i$ $(i \geqq 0)$, setting

$$\frac{\partial\widehat{\psi}}{\partial t_i} = (L_i\psi)e^{-kx} - k^i\widehat{\psi}, \qquad \widehat{\psi} \stackrel{\text{def}}{=} \psi e^{-kx}. \tag{2.17}$$

In particular, $\partial\widehat{\psi}/\partial t_0 = 0$, $\partial\widehat{\psi}/\partial t_1 = \partial\widehat{\psi}/\partial x$, which allows us to drop $\partial/\partial t_0$ and identify $\partial/\partial x$ with $\partial/\partial t_1$.

With the help of $\partial/\partial t_i$, we can rewrite the formula (2.11) as follows:

$$\frac{\partial}{\partial x}\Big(\sum_{p=0}^{\infty}\frac{\partial\widehat{\psi}(l)}{\partial t_p}\widehat{\psi}(l)^{-1}\Big)k^{-p-1} +$$

$$(h(l) - h(k))\sum_{p=0}^{\infty}\frac{\partial\widehat{\psi}(l)}{\partial t_p}\widehat{\psi}(l)^{-1}k^{-p-1} + \frac{h(l) - h(k)}{k - l} = -g(k).$$

We obtain the relation

$$\frac{\partial}{\partial x}\sum_{p=0}^{\infty}\Big(\frac{\partial\widehat{\psi}(k)}{\partial t_p}\widehat{\psi}(k)^{-1}\Big)k^{-p-1} = -g(k) + \frac{\partial h(k)}{\partial k}, \tag{2.18}$$

when $l \to k$. It results in

$$g_{j+1} = \frac{\partial}{\partial x}f_{j+1}, \qquad f_{j+1} \stackrel{\text{def}}{=} j(\log\widehat{\psi})_j + \sum_{p=0}^{j-1}\frac{\partial}{\partial t_p}(\log\widehat{\psi})_{j-p}, \text{ for } j \geqq 0, \tag{2.19}$$

where $(\log\widehat{\psi})_i$ is defined by $\log\widehat{\psi} = \sum_{i=0}^{\infty}(\log\widehat{\psi})_i k^{-i}$.

Application to τ-functions. Let us add a formal element $\log\tau$ satisfying the equation $\partial\log\tau/\partial x = -\psi_1$ to the ring of the differential polynomials of $\{\psi_j\}$ with respect to x. According to Proposition2.5 b), $\partial\psi_1/\partial t_i = -g_{i+1}$. Therefore we can extend the differentiation $\partial/\partial t_i$ to this larger ring setting $\partial\log\tau/\partial t_i = -f_{i+1}$. It follows from Proposition2.6 that

$$\frac{\partial}{\partial x}\frac{\partial}{\partial t_j}\Big(\frac{\partial\log\tau}{\partial t_i}\Big) = \frac{\partial}{\partial x}\frac{\partial}{\partial t_i}\Big(\frac{\partial\log\tau}{\partial t_j}\Big),$$

which gives the pairwise commutativity of the extended $\partial/\partial t_i$. Substituting $g = 1 + \frac{\partial}{\partial x}\sum_{i=1}^{\infty}(\partial\log\tau/\partial t_i)k^{-i-1}$, we obtain the relation

$$\frac{\partial}{\partial x}\Big(\sum_{p=1}^{\infty}\frac{\partial}{\partial t_p}k^{-p-1} - \frac{\partial}{\partial k}\Big)(\log\widehat{\psi} + \log\tau) = 0.$$

We can omit $\partial/\partial x$ in the last formula, which follows from the construction of $\{f_j\}$. Finally, we have the equality

$$\widehat{\psi}\tau = e^{-\sum_{p=1}^{\infty}\frac{\partial}{\partial t_p}k^{-p}}\tau, \tag{2.20}$$

which connects $\widehat{\psi}$ with τ.

Local conservation laws. Now we take ψ of Proposition2.7 for

$$L = \partial^p\sum_{i=2}^{p}u_i\partial^{p-i}.$$

Let us suppose that $\{u_i\}$ satisfy the Lax equation (2.15). Set

$$\eta = (L_m\psi)\psi^{-1},$$
$$\beta = \frac{\partial}{\partial k}((L_m\psi)\psi^{-1}) + \sum_{p=0}^{\infty}((-L_m(L_p\psi))\psi^{-1} + (L_p\psi)\psi^{-1}(L_m\psi)\psi^{-1})k^{-p-1}.$$

Then we have the conservation laws

$$\frac{\partial h}{\partial t} = \frac{\partial\eta}{\partial x}, \tag{2.21a}$$
$$\frac{\partial g}{\partial t} = \frac{\partial\beta}{\partial x}. \tag{2.21b}$$

It is easy to derive equations (2.21) from (2.15) and relation (2.18). It is possible to use ψ that satisfies $\psi_t = L_m\psi - k^m\psi$, in the formulas for h, η, g, β (cf. Exercise2.12). Then relation (2.21a) becomes trivial ($h = (\log\psi)_x$), and (2.21b) follows directly from (2.18). The coefficients of h, η, g, β are x-differential polynomials of $u_2, \dots, u_p$ (cf. §2.4). Moreover, because of (2.18), the coefficients of $\partial h/\partial k - g$ are exact derivatives of suitable differential polynomials of $\{u_i\}$. Consequently, (2.21a) and (2.21b) give the same collections of local conservation laws modulo trivial conservation laws.

Exercise2.14. *Let $\check{\Psi}$ be the matrix valued series of Exercise2.12. Let $\zeta = (\log m_q(\check{\Psi}))_x$, $\alpha = l_q(\check{\Psi}^\star\check{\Psi})$, where $m_q(\cdot) = \det[\cdot]_q$, $l_q(\cdot) = \mathrm{Sp}[\cdot]_q$, and q is the multiplicity of the eigenvalue μ_1 of the matrix U_0 (see (1.7,9) §1). Generalizing (2.21), show that the coefficients of ζ and α are written in terms of differential polynomials of the entries of the coefficients $\{U_i\}$ of the operator L. Moreover the coefficients of $\partial\zeta/\partial t$, $\partial\alpha/\partial t$ are exact x-derivatives of suitable differential polynomials* (use the relation $\partial(\check{\Psi}^\star\check{\Psi})/\partial t = -(\check{\Psi}^\star L_m)\check{\Psi} + \check{\Psi}^\star(L_m\check{\Psi})$).

Reduction to the matrix operator of first order. We define a $(p\times p)$-matrix series $\Pi = (\pi^1, \dots, \pi^p)$ for scalar L and $\psi(k)$ of Proposition2.7, setting

$$\pi^j = \begin{pmatrix} \psi(\omega^{j-1}k) \\ \vdots \\ k^{-i+1}\frac{\partial^{i-1}\psi(\omega^{j-1}k)}{\partial x^{i-1}} \end{pmatrix},$$

where $\omega = \exp(2\pi i/p)$. Then

$$\Pi = (\sum_{s=0}^{\infty} \Pi_s k^{-s})\exp(kxU_0),$$

where the entries of Π_s are written in terms of $\{\psi_i\}$, $U_0 = \mathrm{diag}(1, \omega, \dots, \omega^{p-1})$, $\Pi_0 = (\omega^{(i-1)(j-1)})$ (i are for rows, j for columns). The operator L is stable under the substitutions $k \mapsto \omega^j k$. Differentiating ψ, we obtain the relation

$$\Pi_x = (kV_0 - V)\Pi.$$

Here $V_0 = (\delta^{j-1}_{i \bmod p})$. Moreover the last row of the matrix V is equal to $(u_p k^{-p+1}, \dots, u_2 k^{-1}, 0)$, and the remaining elements of V are equal to zero. It follows from the equation $V_0\Pi_0 = \Pi_0 U_0$, that

$$\Psi \stackrel{\mathrm{def}}{=} \Pi_0^{-1}\Pi = (I + \sum_{s=1}^{\infty}\Psi_s k^{-s})\exp(kxU_0)$$

is a solution of the equaton

$$\Psi_x + Q\Psi = kU_0\Psi \text{ for } Q = \Pi_0^{-1}V\Pi_0. \tag{2.22}$$

We can repeat this reasoning for the (abstract) differentiations $\partial\psi/\partial t_m = L_m\psi - k^m\psi$ instead of $\partial/\partial x$. Apply $\partial/\partial t_m$ to the vector $\pi = (\partial^{i-1}\psi/\partial x^{i-1})$ (recall that $\partial/\partial t_m$ commutes with $\partial/\partial x$). Then $\partial\pi = W'\pi - k^m\pi$ for a suitable (unique) matrix W' which depends polynomially on k^p. Replacing k by $\omega^j k$ $(0 \leqq j < n)$, and conjugating by the matrix $\operatorname{diag}(k^{1-j})$, we obtain the relation

$$\frac{\partial\Pi}{\partial t_m} = \widetilde{W}\Pi - \Pi U_0^m k^m,$$

where $\widetilde{W} = V_0^m k^m + \sum_{i=1}^m \widetilde{W}_i k^{m-i}$. Finally, turning to Ψ, we get to the identity

$$\frac{\partial\Psi}{\partial t_m} = W\Psi - \Psi U_0^m k^m, \tag{2.23}$$

where $W = \Pi_0^{-1}\widetilde{W}\Pi_0 = U_0^m k^m + \sum_{i=1}^m W_i k^{m-i}$. Since the left hand side of (2.23) does not contain terms with k^i, $i > 0$, and is a polynomial in k, then W coincides with the principal part of the series $\Psi U_0^m k^m \Psi^{-1}$. Using the notations from §2.1, we see that $W = L_{\mathcal{D}}^m$ for $\mathcal{D} = U_0^m$ and $\partial/\partial t_m = \partial_1 = \partial t$.

Now let us suppose that $\{u_i\}$ satisfy the Lax equation (2.15), which is the compatibility condition of (2.22) and (2.23) considered as an equation for Ψ for $t_m = t$. Then Q is a solution of equation (2.5) (which coincides with (2.16)). The converse is also true, if the matrix Q has the form (2.22) ($Q = \Pi_0^{-1}\widetilde{V}\Pi_0$ for a matrix $\widetilde{V}$ defined by $\{u_i\}$. We have proved that *scalar Lax equations have zero-curvature representation in matrices of order $p \times p$ where p is the order of the corresponding L operator.*

In fact, we calculated those representations in terms of the coefficients of L.

Exercise2.15. *Rewrite the Lax equation (2.15) for the general matrix (of order $n \times n$) operator L in the form of equation (2.5) in $(np \times np)$-matrices.*

Exercise2.16. *Using (2.22), check that*

$$\psi^* = \text{const} \cdot \widetilde{\pi}_1^p,$$

where $(\widetilde{\pi}_i^j) = \Pi^{-1}$.

Higher KdV equations and Sin-Gordon. Let us take $U_0 = i$ (i is the imaginary unit), $L = -\partial^2/\partial x^2 + u$,

$$\psi = \psi(k) = (1 + \sum_{s=1}^{\infty} \psi_s k^{-s}) \exp(ikx),$$
$$L\psi = k^2\psi.$$

We will show that $\psi^\star(k) = \psi(-k)w(k)^{-1}$, where

$$w(k) \stackrel{\text{def}}{=} \frac{i}{2k}(\psi(k)\psi_x(-k) - \psi_x(k)\psi(-k))$$

(cf. Exercise2.16). For any m, we can find a function $v_m = v_m(x)$ such that

$$(\partial^m\psi(k))\psi(-k) - \psi(k)(\partial^m\psi(-k)) = (\psi_x(k)\psi(-k) - \psi(k)\psi_x(-k))v_m,$$

using the relation $L\psi = k^2\psi$. Hence,

$$(\partial^m\psi(k))\psi(-k) - \psi(k)(\partial^m\psi(-k)) \;=\; 2ikw(k)v_m.$$

Since the left hand side of the last relation is odd with respect to k, the coefficient of $(\partial^m\psi(k))\psi(-k)w(k)^{-1}$ of k^{-1} is equal to zero for each $m \geqq 0$. Therefore $\psi^\star$ and $\psi(-k)w(k)^{-1}$ coincide by Proposition2.3.

Because of the last formula of the previous section §2.4 the Lax equation (2.15) for $\mathcal{D} = d \in \mathbb{C}^*$ looks as:

(2.24) $$u_t = -di\frac{\partial}{\partial x}(\psi\psi^\star)_{m+1}.$$

The resolvent $i\psi\psi^\star(k) = i\psi(k)\psi(-k)w(k)^{-1}$ is even with respect to k, hence the above equation is non-trivial only for odd m. For $d = 4i$, $m = 3$ this is nothing but *the KdV equation.*

We remind that system (0.2) for the PCF (0.2) was connected in §2.1 with the hierarchy of equations of the GHM type. Similarly, we will obtain the Sin-Gordon equation (0.8) as a variant of the higher KdV equations (2.24). Let $id = 2$. We replace $(\psi\psi^\star)_{m+1}$ by the "value" of $\psi\psi^\star$ at the point $k = 0$ in (2.24). To be more precise, we define $\psi(0)$ as a solution of the equation $\psi(0)_{xx} = u\psi(0)$. Then $a = \psi\psi^\star(0)$ means a solution of the equation $(\log a)_{xx} + \frac{1}{2}((\log a)_x)^2 = 2u$. Eliminating u from the equation $-u_t = 2a_x$ and setting $\alpha = i\log(-4a)$, we obtain the equation

(2.25) $$(i\alpha_{xx} + \tfrac{1}{2}\alpha_x^2)_t = (e^{-i\alpha})_x.$$

Note that the resolvent $i\psi\psi^*(k)$ for $k \in \mathbb{C}$, satisfies a differential equation of the third (not second) order (cf., e.g., [GD2],[Du1]). Therefore the above procedure at an arbitrary k leads to a system of two equations (u cannot be eliminated).

In a certain sense solutions of the Sin-Gordon equation form an open set dense everywhere in the set of all solutions of (2.25). Namely, if α satisfies (2.25), then $\alpha_{xt} + \sin\alpha = b(t)e^{i\alpha}$ for a function b of t (check this differentiating the function $(\alpha_{xt} + \sin\alpha)e^{-i\alpha}$ with respect to x). If $b \not\equiv (2i)^{-1}$, then the function α becomes a solution of the equation $\alpha_{xt} + \sin\alpha = 0$ after the substitution $\widetilde{\alpha}(x,t) \to \alpha(x, c(t)) + \log(c_t)$, which maps solutions of (2.25) to solutions of the same equation. Otherwise, $\alpha_{xt} = (2i)^{-1}e^{-i\alpha}$, i.e., α satisfies the *Liouville equation*.

2.6. Comments.

Equations (2.3), (2.5) and their zero curvature representations in various forms were introduced in a number of works (cf., e.g., [Du2, RSTS, Ch3] and books [ZMa2, TF2]). They are often called equations with a polynomial bundle. In [Ch11] we will discuss a more general class of equations with a rational bundle. Generalized Lax pairs and the corresponding equations of type (2.15) were studied by V. E. Zakharov, A. B. Shabat (cf. [ZS1,2]) and by I. M. Gelfand, L. A. Dickey [GD1–4] (cf. also [Kri3]).

The results from §2.3 on abstract fractional powers and their generating operators (including the introduction of the KP-hierarchy) were obtained in [Ch5]. They develop the theory by I. M. Gelfand, L. A. Dickey [GD2,3]. In fact, the generating operator Λ_+ for the Lax equations is, a different form of the generating function from [GD3] (which is a generalization, in its turn, of the generating function for the higher KdV equations introduced by B. A. Dubrovin [Du1]). The series $(\partial^i\Psi)U_0(\Psi^*\partial^j)$ (in particular, the resolvent $\Psi U_0 \Psi^*$) are from [Ch5]. They are directly connected with the jet of the resolvent of the operator L with respect to the diagonal from [GD3], [GD4] (if Ψ is constructed by L). Here we do not mention the earlier results on KdV, in particular, do not discuss an important contribution of P. Lax (cf., e.g., [GGKM, Lam, Lax1] and books [ZMa2, TF2]).

The equivalence of the language of formal pseudo-differential operators (for instance, in the construction of the dressing transformations) and the language of formal Jost-Baker functions (§2.2) was used by many specialists (D. P. Lebedev and G. Wilson studied this problem in a most explicit way - see e.g., [W1]). The notion of the dual formal Jost functions and their operator interpretation is based on [Ch5].

The commutativity of the flows corresponding to the fractional powers (Proposition2.1, 2.2, 2.6) was proved in several early works on Lax type equations via the Hamiltonian interpretation (cf. [BN], [GD2,3], [ZF], [Mani1] and books [ZMa2], [TF2]). It is hard to say who was the first to find the direct proof of such commutativity (presented in this paragraph). Anyway, for the scalar Lax equation, such proof is from [W1]. We note also the relations with the so-called Adler-Kostant-Manin-Lebedev scheme.

The construction of the τ-function in §2.5 with the help of Theorem2.1 and the proof of the equivalence of the series of local conservation laws (2.21) up to exact derivatives via (2.18) is mostly due to H. Flashka [Fl] (see also [FNR]). The theorem on the equivalence of $\partial h/\partial k$ and g was proved (directly) by G. Wilson [W2] (note also [Bo]). The formula (2.18) was obtained for the first time by E. Date, M. Jimbo, M. Kashiwara and T. Miwa (see, e.g., [DJM]). They used the infinite-dimensional Grassmanian and vacuum expectation values of free fermions to introduce τ-functions. See also [Kac]. We discuss these functions in [Ch11] in more detail.

The zero curvature representation for KdV was found by S. P. Novikov (cf. also [DuKN], [ZMa2]). This result was drastically generalized by V. G. Drinfeld, V. V. Sokolov [DrS] and by B. Kuperschmidt, G. Wilson [KW]. The transition from scalar to matrix operators in §2.5 is a special case of their construction. We established the relation of the Sin-Gordon equation and the higher KdV, following [Ch4] (where we developped an observation by V. A. Andreev). The interpretation of the system of equations for the currents of PCF as a higher Lax equation at a "finite" point is due to [Ch1,2].

§3. Algebraic-geometric solutions of basic equations

In §3.1 we give basic definitions related to algebraic curves (Riemann surfaces) and introduce Baker functions in two ways (algebraic and analytic). In §3.2 we construct algebraic-geometric solutions of the PCF equations (0.2), (0.1) and the GHM (0.9) and (0.10). §3.3 is devoted to the "reality" condition of constructed solutions and to the study of the manifold of the anti-hermitian solutions. We list several results on the reduction of the above construction in sections 3.2 and 3.3. As an illustration, we construct algebraic-geometric $O_{2m}(\mathbb{R})$-fields and S^{2m-1}-fields (§3.4). In §3.5 we discuss discrete analogues of the PCF equations. For more concrete examples and applications see §4 below.

The exposition is almost self-contained. We assume only (a special form of) the Riemann-Roch theorem and one corollary of the Riemann theorem (Corollary3.1), which is necessary for the proof of Theorem3.6. Some references for these theorems and basic properties of functions and divisors on algebraic curves are [La] and [Fay]. Another source is the monograph [Mu], the second part of which contains some applications of theta functions to the soliton theory.

3.1. Baker functions.

Let Γ be a smooth Riemann surface of genus $g \geqq 0$. Topologically, Γ is a sphere with g handles. The surface Γ can be realized as a connected nonsingular projective algebraic curve over $\mathbb{C}$. A *divisor* on Γ is a finite formal sum $\sum_i l_i P_i$ of points $P_i \in \Gamma$ with integral coefficients; l_i is called the multiplicity of P_i. Let supp $\mathcal{D} = \bigcup_i P_i$. The integer $\sum_i l_i$ is called the *degree* of $\mathcal{D}$ and is denoted by $\deg \mathcal{D}$. Addition and subtraction of divisors are defined in the obvious way. We can associate to each meromorphic function f on some open subset $A \subset \Gamma$ a divisor $(f) = \sum_P (\mathrm{ord}_P f)P$, where $P \in A$ runs over zeroes and poles of f and $\mathrm{ord}_P f$ is the (positive or negative) order of f at P. The sum of divisors of functions corresponds to their multiplication. Any meromorphic function on $A = \Gamma$ is rational and $\deg(f) = 0$. A divisor $\mathcal{D}$ is called effective if the multiplicity of each point is positive, and we write $\mathcal{D} \geqq 0$, $\mathcal{D}' \geqq \mathcal{D}$ if $\mathcal{D}' - \mathcal{D} \geqq 0$. The *linear equivalence* of divisors is defined as follows: $\mathcal{D}' \sim \mathcal{D}$ if $\mathcal{D}' - \mathcal{D} = (f)$ for some rational function f. We denote the equivalence class containing $\mathcal{D}$ by $\widetilde{\mathcal{D}}$. .

Let $H^0(\mathcal{D}) \stackrel{\text{def}}{=} \{f, (f)+\mathcal{D} \geqq 0\}$, where f is rational on Γ. $H^0(\mathcal{D})$ is a finite dimensional $\mathbb{C}$-linear space. By the Riemann-Roch theorem, which plays the fundamental role in the theory of algebraic curves, it follows that $\dim H^0(\mathcal{D}) \geqq \deg \mathcal{D} - g + 1$. If the inequality in this relation turns out to be equality (respectively, strict inequality), the divisor $\mathcal{D}$ is called *nonspecial* (respectively, *special*). The full statement of the Riemann-Roch theorem is the following:

$$\dim H^0(\mathcal{D}) - \dim H^0(K - \mathcal{D}) = \deg \mathcal{D} - g + 1,$$

where K is in the canonical class of Γ (the linear equivalence class $\widetilde{K}$ of divisors of arbitrary differentials on Γ). The dimension $\dim H^0(\mathcal{D})$ and the property of speciality depend only on the class $\widetilde{\mathcal{D}}$. For an effective divisor S of degree s, $\dim H^0(\mathcal{D}-S) \geqq \dim H^0(\mathcal{D}) - s$ since $H^0(\mathcal{D}-S)$ is cut out of $H^0(\mathcal{D})$ by s linear relations. The speciality of $\mathcal{D}$ implies that $\mathcal{D}'$ is special for any $\mathcal{D}' \geqq \mathcal{D}$.

Divisors of fixed degree d on Γ with respect to the linear equivalence form a connected smooth complete variety $\mathcal{J}^d$ of dimension g. For $d \geqq g-1$, classes of special divisors make a subvariety of dimension strictly less than g. The addition of divisors induces on $\mathcal{J}^0 \overset{\text{def}}{=} \operatorname{Jac}\Gamma$ the structure of an abelian variety (which is called the *Jacobian variety* of Γ), and on each $\mathcal{J}^d$ the structure of a principal homogeneous space over $\operatorname{Jac}\Gamma$ which means that for any $x \in \mathcal{J}^d$ the mapping $\mathcal{J}^0 \ni y \mapsto y+x \in \mathcal{J}^d$ is an isomorphism.

We call a formal sum $q = \sum_{i=1}^m q_i$ (i.e., a union) of principal parts q_i at pairwise distinct points $P_i \in \Gamma$ $(1 \leqq i \leqq m)$ a ***distribution*** on Γ. The set $\{P_i\}$ is called the support of q and is denoted by $\operatorname{supp} q$. If we choose local parameters k_i $(1 \leqq i \leqq m)$ at P_i, each q_i is written as a polynomial without constant term in the variable k_i^{-1}. Addition of distributions is understood formally for nonintersecting supports and as addition of corresponding principal parts at points that belong to both addends. Two distributions represented by series in local parameters are identified if their principal parts coincide.

Definition3.1. *a) For a distribution $q = \sum_{i=1}^m q_i$ with support $\operatorname{supp} q = \{P_i\}$, a function φ on Γ is called Baker (Baker-Akhiezer) function if it is meromorphic except at $\{P_i\}$ and for any i the function $\widehat{\varphi}_i \overset{\text{def}}{=} \varphi\exp(-q_i)$ is meromorphic in a neighbourhood of P_i.*

b) The divisor (φ) of a Baker function φ is defined to be the formal sum $\sum_{P\in\Gamma}(\operatorname{ord}_P \varphi)P$, where $\operatorname{ord}_P \varphi$ is the (positive or negative) order of φ at P for $P \notin \operatorname{supp} q$, and is the order of $\widehat{\varphi}_i$ at P_i $(1 \leqq i \leqq m)$.

c) For a divisor $\mathcal{D}$ on Γ we denote the space of Baker functions φ satisfying the condition $(\varphi) + \mathcal{D} \geqq 0$ by $H^0(\mathcal{D}; q)$. □

We can associate a linear equivalence class $\widetilde{q}$ of divisors Q of degree 0 to a distribution q. We outline an "analytic" construction for this at the end of this section. We now describe an "algebraic" construction.

Consider each polynomial q_i as a meromorphic function in an appropriate neighbourhood $A_i \subset \Gamma$ of the point P_i. Adding points P_i with trivial q_i, if necessary, we can assume that $\bigcup_{i=1}^m A_i = \Gamma$ and that q_i is holomorphic on $A_i \cap A_j$ for all i and j. We denote by $\widehat{Q}$ the analytic linear (invertible) sheaf on Γ, sections of which over arbitrary open sets $A \subset \Gamma$ are defined as collections $\{\widehat{f}_i,\ 1 \leqq i \leqq m\}$ of meromorphic functions $\widehat{f}_i$ on A_i with the compatibility condition $\widehat{f}_i\widehat{f}_j^{-1} = \exp(q_j - q_i)$ on

$A \cap A_i \cap A_j$ $(1 \leqq i \leqq m)$. Then for some divisor Q on Γ, the sheaf $\widehat{Q}$ is isomorphic to the sheaf $\mathcal{O}(Q)$. The sections of the latter sheaf over A are, by definition, meromorphic functions f on A with the condition $(f) + Q|_A \geqq 0$, where $Q|_A$ is obtained from the divisor Q by neglecting all points not included in the set A. Since q can be deformed continuously to zero (the zero distribution corresponds, by definition, to the divisor $Q = 0$), the degree of Q is equal to zero. We denote the class of the divisor Q by $\widetilde{q}$. If q coincides with the principal part of some rational function on Γ, then $\widetilde{q} = 0$ (check this as an exercise).

Theorem3.1. *a) The divisor (φ) of a Baker function φ corresponding to q is linearly equivalent to Q in the class $\widetilde{q}$.*
b) $\dim H^0(\mathcal{D}; q) = \dim H^0(\mathcal{D} + Q)$.

Proof. Represent φ by a collection of functions $\varphi_i = \exp(q_i)\widehat{f}_i$ $(1 \leqq i \leqq m)$ for suitable meromorphic functions $\widehat{f}_i$ on A_i. Then

$$\varphi_i = \varphi_j \Leftrightarrow \widehat{f}_i \widehat{f}_j^{-1} = \exp(q_j - q_i)$$

on $A_i \cap A_j$. Therefore, a Baker function φ with divisor (φ) corresponds to a rational section $\widehat{f} = \{\widehat{f}_i\}$ of the sheaf $\widehat{Q}$. Consequently, there is a suitable corresponding rational section of the sheaf $\mathcal{O}(Q)$ with divisor (φ). This section is represented by a rational function f on Γ with divisor $(f) = (\varphi) - Q$. Thus $0 \sim (f) \sim (\varphi) - Q$. Analogously, the Baker function φ with divisor $(\varphi) \geqq -\mathcal{D}$ corresponds to a rational function f on Γ with divisor $(f) = (\varphi) - Q \geqq -\mathcal{D} - Q$. □

This theorem reduces the computation of the dimenstion of the space $H^0(\mathcal{D}; q)$ to the classical problem of the computation of $\dim H^0(\mathcal{D} + Q)$. In particular, if $\deg \mathcal{D} \geqq g - 1$, then $\mathcal{D} + Q$ is not special for q in an open dense everywhere set (with respect to the natural coefficientwise topology on the space of sets of polynomials $\{q_i(k_i^{-1})\}$). Hence

$$\dim H^0(\mathcal{D}; q) = \deg \mathcal{D} - g + 1,$$

for such q (see above). This equality will be the fundamental theoretical result in the sequel. Various problems of construction of algebraic-geometric solutions of differential equations will be reduced to it.

The reader can find the Riemann-Roch theorem and the above cited results on algebraic curves, for example, in [La], [Fay], and [Mu]. To conclude this section, we turn to the more analytic (and constructive) point of view, connecting Baker functions to Riemann theta functions (see [Ba], [BC], [IM], [Kri3], [Mat3]).

Computations of Baker functions via theta functions. Let us choose a basis $a_1, \dots, a_g, b_1, \dots, b_g$ of 1-cycles on the Riemann surface Γ with the following intersection indices:

$$(a_i, a_j) = (b_i, b_j) = 0, \qquad (a_i, b_j) = \delta_i^j,$$

where δ_i^j is the Kronecker symbol. We normalize the basis of differentials of the first kind (holomorphic differentials) $\{\omega_j\}$ by the condition

$$\int_{a_i} \omega_j = \delta_i^j, \qquad (1 \leqq i, j \leqq g).$$

The matrix $B = (b_i^j)$, given by $b_i^j = \int_{b_i} \omega_j$ is called the period matrix. The *theta function* is defined by the following power series:

$$\theta(p) = \sum_{k \in \mathbb{Z}^g} \exp\{\pi i(Bk, k) + 2\pi i(k, p)\},$$

where $p \in \mathbb{C}^g$, $(x, y) = \sum_{j=1}^g x_j y_j$ for $x = {}^t(x_j)$, and $y = {}^t(y_j) \in \mathbb{C}^g$. The following relations are easily checked:

a) $\theta(-p) = \theta(p)$,
b) $\theta(p + k) = \theta(p)$ for $k \in \mathbb{Z}^g$,
c) $\theta(p + b^j) = \exp(-\pi i b_j^j - 2\pi i p_j)\theta(p)$, where b^j is the j-th column of B.

Let us fix a certain point $P_0 \in \Gamma$. Then the Jacobi map

$$\Gamma \ni P \mapsto \omega(P) = \begin{pmatrix} \int_{P_0}^P \omega_1 \\ \vdots \\ \int_{P_0}^P \omega_g \end{pmatrix} \in \mathbb{C}^g,$$

gives the vector $\omega(P)$ which is defined up to elements of the period lattice $\Lambda \overset{\text{def}}{=} \mathbb{Z}^g + \sum_{j=1}^g \mathbb{Z} b^j \subset \mathbb{C}^g$ and depends on the homotopy class of the path of integration from P_0 to P. With the help of ω, we can relate the g-dimensional complex torus $\mathbb{C}^g/\Lambda$ and the abelian variety $\operatorname{Jac}\Gamma$. If functions $\rho_r^p = \theta(\omega(P) - p - r)/\theta(\omega(P) - r)$ defined for $p \in \mathbb{C}^g$ and some $r = r_1, r_2 \in \mathbb{C}^g$ are not identically zero as functions of $P \in \Gamma$, then their ratio $\rho_{r_1}^p / \rho_{r_2}^p$ is a rational function on $\Gamma \ni P$ because of the properties b) and c) of theta functions. Consequently, under the assumption $\rho_r^p \not\equiv 0, \infty$ the class of the divisor (ρ_r^p) of the (multivalued) functions ρ_r^p on Γ does not depend on the choice of r. The corresponding r form an open subset of $\mathbb{C}^g$ depending on p - see below. The mapping

$$\alpha : \mathbb{C}^g/\Lambda \ni p \mapsto (\rho_r^p) \in \operatorname{Jac}\Gamma$$

induces an isomorphism between $\mathbb{C}^g/\Lambda$ and $\operatorname{Jac}\Gamma$, which obviously does not depend on the choice of the point P_0. Using α we can describe ω purely algebraically: $\alpha(\omega(P))$ is the class $\widetilde{P}-\widetilde{P}_0$ of the divisor $P-P_0$.

By the Riemann theorem (see, for example, [Fay] Theorem1.1), there exists a class of divisors $\widetilde{\Delta}\in\mathcal{J}^{g-1}$, for which $2\widetilde{\Delta}=\widetilde{K}$ ($\widetilde{K}$ is the canonical class of Γ) with the following properties. We define $w(\mathcal{D})\in\mathbb{C}^g \mod \Lambda$ for a divisor $\mathcal{D}$ of degree g by the relation $\alpha(w(\mathcal{D}))=\widetilde{\mathcal{D}}-\widetilde{\Delta}-\widetilde{P}_0$. Then

a) every nonspecial $\mathcal{D}\geqq 0$ is zero or the divisor of the (multivalued) function $\theta(\omega(P)-w(\mathcal{D}))$ of $P\in\Gamma$;

b) $\mathcal{D}$ is special $\Longrightarrow \theta(\omega(P)-w(\mathcal{D}))\equiv 0$ for all P;

c) $\mathcal{D}-P_0$ is special $\Longrightarrow \theta(w(\mathcal{D}))=0$.

Corollary3.1. *If a divisor S of degree $g-1$ with property $2\widetilde{S}=\widetilde{K}$ is not special, then*

$$w(S+P_0)=\zeta+\sum_{j=1}^{g}\xi_j b^j,$$

where $\zeta={}^t(\zeta_j)$, $\xi={}^t(\xi_j)\in\frac{1}{2}\mathbb{Z}^g\subset\mathbb{C}^g$. (see e.g. [Fay] Corollay 1.5).

Proof. Because of the property c) of the mapping w, it is enough to show that $\theta(z)\neq 0\Rightarrow {}^t\xi\zeta\in\frac{1}{2}\mathbb{Z}$ for any $z=\zeta+\sum_{j=1}^{g}\xi_j b^j$, $\zeta,\xi\in\frac{1}{2}\mathbb{Z}^g$, where ${}^t\xi\zeta\overset{\text{def}}{=}\sum_{j=1}^{g}\zeta_j\xi_j\in\frac{1}{2}\mathbb{Z}$. But by properties a), b), c) of theta functions, we have: $\theta(z)=\theta(-z)=\theta(z-2z)=\exp(4\pi i\,{}^t\xi\zeta)\theta(z)$. □

The definition of δ and w and the proof of Corollary3.1 can be given in a completely algebraic way (D. Mumford). In any case, it is easy to establish the following.

Exercise3.1*. *Without resorting to theta functions, and using the Riemann -Roch theorem, show that*

a) for arbitrary $\{P_j\}\subset\Gamma$, $1\leqq j\leqq g-1$, the divisor $K-\sum_{j=1}^{g}P_j$ is linearly equivalent to a divisor of the form $\sum_{j=1}^{g-1}P_j'$, $P_j'\in\Gamma$, where K is the canonical divisor of Γ;

b) for every $x\in\operatorname{Jac}\Gamma$ the intersection of the divisor $\{\sum_{j=1}^{g-1}P_j-(g-1)P_0\}\subset\operatorname{Jac}\Gamma$ shifted by x and $\Gamma=\alpha(\omega(\Gamma))\subset\operatorname{Jac}\Gamma$ either coincides with Γ or is represented in the form $\{Q_j-P_0,\quad 1\leqq j\leqq g\}$ for some nonspecial divisor $Q=\sum_{j=1}^{g}Q_j$, $Q_j\in\Gamma$ (use a)). □

For a distribution q we can make a differential ω_q on Γ of the second kind (with zero residues) that has principal part dq_j at the points P_j ($1\leqq j\leqq m$), is holomorphic outside $\operatorname{supp} q=\{P_j\}$, and is normalized by the conditions $\int_{a_i}\omega_q=0$. Set $v_q=\frac{i}{2\pi}{}^t(\int_{b_1}\omega_q,\dots,\int_{b_g}\omega_q)\in\mathbb{C}^g$. If q is the principal part of a rational function, then $v_q=0$.

Proposition3.1. *If $\mathcal{D} \geqq 0$ is not special of degree g then the function*

$$\varphi(P) = \exp\Big(\int_{P_0}^{P} \omega_q\Big) \frac{\theta(\omega(P) - w(\mathcal{D}) - v_q)}{\theta(\omega(P) - w(\mathcal{D}))}$$

belongs to $H^0(\mathcal{D}; q)$. Also $\alpha(v_q) = \widetilde{q}$ (see above). Here $\int_{P_0}^{P} \omega_q$ is calculated for the same path from P_0 to P as $\omega(P)$ for $P_0 \in \operatorname{supp} q$.

Proof. Add to the path from P_0 to P a 1-cycle homotopic to $\sum_{j-1}^{g}(\alpha_j a_j + \beta_j b_j)$ for some $\alpha_j, \beta_j \in \mathbb{Z}$. Then $\exp\Big(\int_{P_0}^{P} \omega_q\Big)$ is multiplied by $\exp\{-2\pi i(v_q, \beta)\}$, where $\beta = {}^t(\beta_j)$. It follows from the properties b), c) of theta functions that the ratio of theta functions in the formula for φ is multiplied by $\exp\{2\pi i(v_q, \beta)\}$. Hence φ is a single-valued function on Γ with exponential singularity of the form required in Definition3.1. Calculating the divisor of φ, we get the relation $\alpha(v_q) = \widetilde{q}$ (cf. §4 Ch.7 of the book by J.-P. Serre [Se1]). □

Exercise3.2. *Up to multiplication by a rational function, any Baker function can be obtained by the construction of Proposition3.1.* □

3.2. The main construction.

Let λ be a rational function on Γ of degree n. This means that the divisor of poles of λ is represented in the form $(\lambda)_\infty = \sum_{i=1}^{m} e_i^\infty R_i^\infty$, where $\sum_{i=1}^{m} e_i^\infty = n$ ($e_i^\infty \in \mathbb{Z}_+$ is the multiplicity of R_i^∞). The function λ defines a map $\Gamma \setminus \{R_i^\infty\} \to \mathbb{C}$, which can be extended to a regular covering $\widetilde{\lambda} : \Gamma \to \mathbb{P}^1$ of degree n. Here $\mathbb{P}^1 = \mathbb{C} \cup \infty$ is the projective line over $\mathbb{C}$ with λ considered as the coordinate. The fiber of $\widetilde{\lambda}$ over a point $z \in \mathbb{P}^1$ defined with the multiplicities is denoted by $(\lambda)_z$: $(\lambda)_z = \sum_{i=1}^{m_z} e_i^z R_i^z$, $\sum_{i=1}^{m_z} e_i^z = n$, where $\{R_i^z\} = \widetilde{\lambda}^{-1}(z)$ and e_i^z is the ramification index of $\widetilde{\lambda}$ at the point R_i^z. Points R_i^z are taken pairwise distinct. We fix a certain indexation of those points and choose a local parameter $k_{i,z}$ at each point R_i^z, $z \in \mathbb{C} \cup \infty$.

Let $q^1 = \{q_i^1, 1 \leqq i \leqq m_1\} = \sum_{i=1}^{m_1} q_i^1$ be a distribution, concentrated at the fiber $(\lambda)_1$ (at the zeros of $1 - \lambda$) and satisfying the regularity condition (i.e., trivial principal part) of the distribution $(1-\lambda)q^1$. Similarly, $q^{-1} = \{q_i^{-1}, 1 \leqq i \leqq m_{-1}\} = \sum_{i=1}^{m_{-1}} q_i^{-1}$, $\operatorname{supp} q_i^{-1} = R_i^{-1}$ and $(1+\lambda)q^{-1}$ is regular at the zeros of $1 + \lambda$. Now we write $k_{i,\pm 1}$ and $e_i^{\pm 1}$ instead of $k_{i_\pm}$ and $e_i^{\pm}$ respectively. Then $q_i^{\pm 1}$ is expressed in a neighbourhood of $R_i^{\pm 1}$ as a polynomial of $k_{i_\pm}$ of degree $\leqq e_i^{\pm}$ without constant term. Set $q' = \frac{1}{1-\lambda} q^1$. We will further investigate the following two cases:

$$\text{(3.1A,B)} \qquad q = q_0 + xq^1 + tq^{-1}, \qquad q = q_0 + xq^1 + tq'c^2,$$

where $x, t \in \mathbb{R}$, $c \in \mathbb{C}^*$, q_0 is an arbitrary distribution on Γ.

Let us fix a divisor $\mathcal{D}$ on Γ of degree $g+n-1$, the support of which $\operatorname{supp}\mathcal{D}$ does not intersect $\widetilde{\lambda}^{-1}(\infty) = \{R_i^\infty\}$. Further we always assume that x and t are taken from the subset in $\mathbb{R}^2$, for which the divisor $\mathcal{D}+Q-(\lambda)_z$ of degree $g-1$ is not special, where $Q \in \widetilde{q}$ in the notations of the previous section (see Theorem3.1). This subset is open and dense everywhere. The concrete choice of z is not essential here, because all fibers of $\widetilde{\lambda}$ are linearly equivalent.

We use the abbriviations $k_i = k_{i,\infty}$, $e_i = e_i^\infty$, $m = m^\infty$, $R_i = R_i^\infty$. Let us introduce the functions φ_{rs} $(1 \leqq r \leqq m,\ 1 \leqq s \leqq e_r)$ in $H^0(\mathcal{D};q)$ normalized by the conditions

$$(k_j^{1-e_j}\varphi_{rs} - k_j^{s-e_j}\delta_j^r)(R_j) = 0 \tag{3.2}$$

for $1 \leqq j \leqq m$. Because of the nonspeciality of the divisors $\mathcal{D}+Q-(\lambda)_\infty$ of the Baker functions, φ_{rs} are uniquely determined by (3.2) and form a basis in $H^0(\mathcal{D};q)$. We change the indices as follows:

$$\varphi_i = \varphi_{rs} \text{ for } i = s + e_{r-1} + \cdots + e_1 + e_0 \qquad (1 \leqq i \leqq n, e_0 = 0),$$

and put $\varphi = {}^t(\varphi_1, \ldots, \varphi_n)$.

Theorem3.2. *There exist unique $\mathfrak{gl}_n$-valued functions $U, V(x,t)$ (q of (3.1A)) and $U, W(x,t)$ (q of (3.1B)) such that*

$$\varphi_x = \frac{U}{1-\lambda}\varphi, \tag{3.3}$$

$$\varphi_t = \frac{V}{1+\lambda}\varphi, \quad \varphi_t = \Big\{\frac{c^2U}{(1-\lambda)^2} + \frac{W}{1-\lambda}\Big\}\varphi. \tag{3.3A,B}$$

Proof. It results from the nonspeciality of $\mathcal{D}+Q-(\lambda)_\infty$ that the functions $\varphi_1, \ldots, \varphi_n$ are linearly independent not only over $\mathbb{C}$, but also over the field $\mathbb{C}(\lambda)$ of rational functions of λ. Indeed, any $\mathbb{C}(\lambda)$-linear relation on $\{\varphi_i\}$ can be written in the form $\sum_{i=1}^n c_i\varphi_i = \lambda^{-1}(\sum_{i=1}^n r_i(\lambda^{-1})\varphi_i)$, where not all $c_i \in \mathbb{C}$ are zero and r_i are some polynomials. But then $\sum_{i=1}^n c_i\varphi_i \in H^0(\mathcal{D}-(\lambda)_\infty;q) = \{0\}$.

Using the nonspeciality of $\mathcal{D}+Q+l(\lambda)_z)$ when $l \geqq -1$ and the linear independence of $\{\varphi_i\}$, we obtain that $(z-\lambda)^i\varphi_j$ for $1 \leqq i \leqq l$, $1 \leqq j \leqq n$ form a basis of the space $H^0(\mathcal{D}+l(\lambda)_z);q)$. Functions $(\varphi_i)_x$ and $(\varphi_i)_t$ in the case A, B belong respectively to $H^0(\mathcal{D}+(\lambda)_1;q)$ and $H^0(\mathcal{D}+(\lambda)_{-1};q)$ (A), $H^0(\mathcal{D}+2(\lambda)_{+1};q)$ (B) (additional poles arise only because of the differentiating the exp factor). $\square$

Note that if we replace the divisor $\mathcal{D}$ with an equivalent $\mathcal{D}'$, the matrix functions corresponding to $\mathcal{D}'$, U', V', W' are obtained from U, V, W by conjugating by a constant invertible matrix (check this as an exercise). In particular, we can also relate the matrices U, V, W, defined up to a conjugation, to a divisor $\mathcal{D}$ containing points of $\lambda^{-1}(\infty) = \{R_i\}$. In order to do this, it is necessary to replace $\mathcal{D}$ by equivalent $\mathcal{D}'$, the support of which does not intersect $\lambda^{-1}(\infty)$, and to apply the costruction of Theorem3.2. It gives the following.

Exercise3.3. *Let l be a distribution on Γ such that $v_l \in \Lambda$ (under the notations at the end of §3.1), $U = U(q_0)$ be that of Theorem3.2. Here we show explicitly the dependence of q_0 from the definition of q (see (3.1A,B)). Then for a constant invertible matrix $\mathcal{E} = \mathcal{E}_l$*

a) $U(q_0 + l) = \mathcal{E}U(q_0)\mathcal{E}^{-1}$ (analogous statements hold for V, W with the same $\mathcal{E}$);

b) $\mathrm{Sp}(UV)(q_0+l) = \mathrm{Sp}(UV)$, $\mathrm{Sp}(U_x^2)(q_0+l) = \mathrm{Sp}(U_x^2)(q_0)$, i.e., $\mathrm{Sp}(UV)$, $\mathrm{Sp}(U_x^2)$ are quasi periodic functions of x, t. □

Jordan form of U, V. We suppose that for $z \in \mathbb{C} \cup \infty$ the fiber $(\lambda)_z$ does not intersect $\mathrm{supp}\,\mathcal{D}$. We make an $n \times n$ matrix $\Phi_{(z)}$ of "values" of φ at the points of $(\lambda)_z$ from the vector function φ of the previous section. As the j-th column of $\Phi_{(z)}$ ($j = e_0^z + \cdots + e_{r-1}^z + s$, $e_0^z = 0, 1 \leqq r \leqq m^z$, $1 \leqq s \leqq e_r^z$), we take the coefficient of $k_{r,z}^{s-1}$ in the expansion of the function φ for $z \notin \lambda(\mathrm{supp}\,q)$ and of the function $\varphi\exp(-q)$ for $z \in \lambda(\mathrm{supp}\,q)$ in a neighbourhood of points of R_r^z. If the fiber $(\lambda)_z$ is simple (i.e., $m^z = n$, $e_1^z = \cdots = e_m^z = 1$) and $\lambda^{-1}(z) \cap \mathrm{supp}\,q = \emptyset$, then the j-th column of $\Phi_{(z)}$ is the value of φ at the point $R_j^z \in \lambda^{-1}(z)$:

$$\Phi_{(z)} = (\varphi(R_1^z), \ldots, \varphi(R_n^z)).$$

For arbitrary z the matrix $\Phi_{(z)}$ is invertible, since $\mathcal{D} + Q - (\lambda)_z$ is not special and $\mathrm{supp}\,\mathcal{D} \cap \lambda^{-1}(z) = \emptyset$. For $z = \infty$, $\Phi_{(\infty)} = \boldsymbol{I}$ because of the normalization condition. Note that the columns of $\Phi_{(z)}$ are permuted when we change the indices of the points in $\widetilde{\lambda}^{-1}(z)$.

Until the end of this section we suppose that the local parameters $k_{r_\pm}$ satisfy the relation $k_{r_\pm}^{e_r^\pm} = 1 \mp \lambda$ for $1 \leqq r \leqq m^\pm \overset{\mathrm{def}}{=} m^{\pm 1}$. In a neighbourhood of each point $R_r^\pm \overset{\mathrm{def}}{=} R_r^{\pm 1} \in \widetilde{\lambda}^{-1}(\pm 1)$ we write the distribution $q^{\pm 1}$ in the form

$$q^{+1} = \sum_{s=1}^{e_r^+} \mu_{r,s}(k_{r_+}^{-1})^{e_r^+ - s+1} + \cdots, \tag{3.4$^+$}$$

$$q^{-1} = \sum_{s=1}^{e_r^-} \nu_{r,s}(k_{r_-}^{-1})^{e_r^- - s+1} + \cdots, \tag{3.4$^-$}$$

modulo holomorphic part for appropriate $\mu_{r,s}, \nu_{r,s} \in \mathbb{C}$. The latter expansion is only for the case (3.1A). We make a $n \times n$ matrix from m^+ square blocks $U_0^1, \dots, U_0^r = (m_i^{r,j}), \dots, U_0^{m^+}$ located along the diagonal constructed as follows. For $i > j$ set $m_i^{r,j} = 0$, for $i \leqq j$ set $m_i^{r,j} = \mu_{r,j-i+1}$, where $1 \leqq i, j \leqq e_r^+$ are the indices of rows and columns. Analogously we make V_0^r and V_0 replacing $+1$, μ by -1, ν.

Proposition3.2. *The (constant) matrices U_0, V_0 defined for the distributions $q^{\pm 1}$ (see (3.1), (3.4)) are equivalent to the matrix functions U, V in Theorem3.2 with respect to conjugation by invertible matrices.*

Proof. Put $\psi^{(\pm)} = (\Phi_{(\pm 1)})^{-1}\varphi$. Then $\psi^{(\pm)}\exp(-q)$ is normalized in a neighbourhood of the fiber $(\lambda)_{\pm 1}$ by the relation (3.2) for $k_{j_\pm}$ instead of k_j, and

$$(3.5) \qquad (\psi^{(+)})_x = (1-\lambda)(\Phi_{(+1)}^{-1} U \Phi_{(+1)})\psi^{(+)} - \Phi_{(+1)}^{-1}(\Phi_{(+1)})_x \psi^{(+)}.$$

Differentiating the exp-factor of the expansion of $\psi^{(+)}$ in a neighbourhood of the points of $\widetilde{\lambda}^{-1}(\pm 1)$, we can easily show that $(\psi^{(+)})_x$ is written in the form $(1-\lambda)U_0\psi^{(+)}$ modulo regular terms. Consequently $\Phi_{(+1)}^{-1} U \Phi_{(+1)} = U_0$. Analogously, replacing $+$ with $-$ we can check the relation $\Phi_{(-1)}^{-1} V \Phi_{(-1)} = V_0$ (for the case (3.1A)). □

Exercise3.4. *Put $\mu_r = \mu_{r,1}$ $(1 \leqq r \leqq m \stackrel{\text{def}}{=} m^{=1})$. We denote the order of zero of $(1-\lambda)q^{+1} - \mu_r$ at the point R_r^{+1} by κ_r $(1 \leqq \kappa_r \leqq e_r \stackrel{\text{def}}{=} e_r^{+1})$. Let $[x]$ be the integer part of $x \in \mathbb{R}$ and $M_\alpha(s)$ be the Jordan block of order $(s+1) \times (s+1)$ with $\alpha \in \mathbb{C}$ on the principal diagonal and with the unity (if $s > 0$) on the (right) neighbouring diagonal. Then the Jordan standard form of U_0 and, hence, of U consists of κ_1 blocks $M_{\mu_1}([e_1 - 1/\kappa_1]), \dots, M_{\mu_1}([e_1 - \kappa_1/\kappa_1])$, κ_2 blocks $M_{\mu_2}([e_2 - 1/\kappa_2]), \dots, M_{\mu_2}([e_2 - \kappa_2/\kappa_2])$,..., κ_m blocks $M_{\mu_m}([e_m - 1/\kappa_m]), \dots, M_{\mu_m}([e_m - \kappa_m/\kappa_m])$.* □

Theorem3.3. *a) The functions U, V considered in Theorem3.2 satisfy system (0.2) in the case (A); U, V are equivalent respectively to the matrices U_0, V_0 in Proposition3.2. If $\operatorname{supp} \mathcal{D} \cap (\lambda)_0 = \emptyset$, then for arbitrary invertible constant matrices G_1, G_2 the function $g \stackrel{\text{def}}{=} G_1 \Phi_{(0)} G_2$ is a solution of equation (0.1).*

b) In the case of (3.1B), let $q^1 = q_r^1 = (1-\lambda)^{-1}c_1$ in a neighbourhood of the points $R_r^+ \in \widetilde{\lambda}^{-1}(+1)$ for $1 \leqq r \leqq r_0$, $q^1 = q_r^1 = (1-\lambda)^{-1}c_2$ in a neibourhood of the R_r^+ for $r_0 < r \leqq m_+$ and $e_1^+ + \dots + e_{r_0}^+ = p$, where $c_1, c_2 \in \mathbb{C}$, $c = c_1 - c_2$. Then $W = [U, U_x]$ and U is a solution of the GHM equation (0.9) with the condition (0.10).

Proof. For an open set $\emptyset \neq A \subset \mathbb{C}$ the fibers $(\lambda)_z$ $z \in A$ are simple and $\widetilde{\lambda}^{-1}(z) \cap \operatorname{supp} \mathcal{D} = \emptyset$. We can assume that the points R_i^z depend on $z \in A$ continuously for a fixed index $1 \leqq i \leqq n$. It results from (3.3) and (3.4) that U, V, and $\Phi(\lambda) \stackrel{\text{def}}{=} \Phi_{(\lambda)}$ satisfy relation (0.3) (see Introduction), where $\lambda \in A$. Therefore $U, V(x,t)$ are solutions of (0.2) because $\lambda \in A$ is arbitrary and $\Phi(\lambda)$ is invertible. (We use $\Phi_{(\lambda)}$ for the sake of simplicity. One can obtain (0.2) directly from (3.3,3.3A)). In a neighbourhood of the points R_r^0 $(1 \leqq r \leqq m^0)$ we have: $1-\lambda = 1 + O(k_{r,0}^{e_0^r})$. Hence, (3.3, 3.3A) imply that $(\Phi_{(0)})_x = U\Phi_{(0)}$, $(\Phi_{(0)})_t = V\Phi_{(0)}$ and $g = G_1\Phi_{(0)}G_2$ satisfies (0.1) (see Introduction).

Let us turn to statement b) and consider the functions $\breve{\varphi}^{(l)}(k_{r_+}) \stackrel{\text{def}}{=} \varphi(\rho^l k_{r_+})$ in a sufficiently small punctured neighbourhood of each point $R_r^+ \in \widetilde{\lambda}^{-1}(+1)$ for $\rho = \exp(2\pi i/e_r^+)$, $0 \leqq l \leqq e_r^+ - 1$. Then the functions of $1-\lambda$,

$$\breve{\varphi}^{rs} \stackrel{\text{def}}{=} \frac{1}{e_r^+} k_{r_+}^{-(s-1)} \sum_{l=0}^{e_r^+-1} \rho^{(s-1)l} \breve{\varphi}^{(l)}, \qquad (1 \leqq s \leqq e_r^+)$$

satisfy (3.3, 3.3B) in a suitable punctured neighbourhood of $+1 \in \mathbb{P}^1$. After the obvious change of indices (cf. the definition of φ) let us construct the matrix $\breve{\Phi}$, the columns of which are the vector functions $\breve{\varphi}^{rs}$. Then the expansion of $\breve{\Phi}$ in $k^{-1} = 1-\lambda$ is of the form (1.3B) (see §1) with constant term Φ_0 equal to $\Phi_{(+1)}$, and $\breve{\Phi}$ satisfies the same relation (3.3, 3.3B) as φ. Hence $W = [U, U_x]$ (used in the calculation in §1.4) also fulfills (0.11). Consequently, U is a solution of (0.9) in the desired equivalence class (see Proposition3.2). □

Note that in the case of general position the fibers $(\lambda)_{\pm 1}$ are simple. Then $q^{\pm 1}$ are uniquely determined by the numbers $\mu_r = \mu_{r,1}$, $\nu_r = \nu_{r,1}$ $(1 \leqq r \leqq m^{\pm 1} = n)$, and U, V are equivalent to $\operatorname{diag}(\mu_i)$, $\operatorname{diag}(\nu_i)$ respectively. (For simple $(\lambda)_{+1}$ the above $\breve{\Phi}(k)$ coincides with $\Phi(1-k^{-1})$). The trivial conservation laws $\operatorname{Sp}(U^s)_t = (\operatorname{Sp} V^s)_x$ for $s = 1, 2, \ldots$ (see Introduction) make the reduction of (0.2) or (0.9) possible from the very beginning. It is necessary to consider the solutions U, V with prescribed constant eigenvalues. We constructed precisely solutions of this kind. Moreover in the case of simple fibers $(\lambda)_{\pm 1}$ the dependence of the distribution q on x, t is uniquely determined by the eigenvalues of U, V.

Exercise3.5. *In a neighbourhood of R_r^+ for $1 \leqq r \leqq m^+$ $((\lambda)_{+1} = \sum_{r=1}^{m^+} e_r^+ R_r^+)$ put*

$$q_r^1 = \frac{\mu_r}{1-\lambda}, \qquad q_r^2 = \frac{\delta_r}{(1-\lambda)^m},$$

where $m \geqq 1$, $\{\mu_r, \delta_r\} \subset \mathbb{C}$, $\delta_r = \delta_s$ if $\mu_r = \mu_s$. Construct the vector Baker functions φ for the distribution $q = q_0 + q^1 x + q^2 t$ (see (3.2)). Show that the function U defined by (3.3) satisfies equation (2.3), §2. Going from φ to $\breve{\Phi}$ (see the *proof of Theorem3.3, b) and (3.5)), check that* $Q = \Phi^{-1}_{(+1)}(\Phi_{(+1)})_x$ *is a solution of equation (2.5), §2 (cf. [Du2], [Kri3]).*

3.3. Reality conditions.

Let T be a ramification divisor of the covering $\widetilde{\lambda} : \Gamma \to \mathbb{P}^1$ of §3.2. By definition $T = \sum_{P\in\Gamma}(e(P)-1)P$, where $e(P)$ is the index of the branch of $\widetilde{\lambda}$ at the point P. If $P = R^z_r$ for $z \in \mathbb{C} \cup \infty$, $1 \leqq r \leqq m^z$ (see above), then $e(P) = e^z_r$. The divisor $(d\lambda)$ of the differential (the differential 1-form) $d\lambda$ on Γ is easily calculated via T: $(d\lambda) = T - 2(\lambda)_\infty$. Since $(d\lambda)$ belongs to the canonical class, and, hence, its degree is equal to $2g-2$, then $\deg T = 2(g+n-1)$. The last equation is a special case of the Hurwitz formula (see, for example, [La]).

For a rational function f on Γ we denote the *trace* of f with respect to λ by $\operatorname{Tr} f$. It is a rational function of λ defined as the sum of the images of f by n distinct embeddings of the field $\mathbb{C}(\Gamma)$ of rational functions on Γ into the algebraic closure of the field $\mathbb{C}(\mathbb{P}^1) = \mathbb{C}(\lambda)$ of rational functions of λ. If the fiber $(\lambda)_z = \sum_{r=1}^{m^z} e^z_r R^z_r$ does not intersect the poles of f,

$$\operatorname{Tr} f(z) = \sum_{r=1}^{m^z} e^z_r f(R^z_r) \in \mathbb{C}, \qquad z \in \mathbb{P}^1.$$

An important (and defining) property of T is that for any function $f \in H^0(T)$ the function $\operatorname{Tr} f$ does not depend on λ (i.e., is constant).

Let us suppose further that Γ is equipped with a *real structure* introduced as an *anti-involution* σ of the field $\mathbb{C}(\Gamma)$ (an automorphism of order two, acting on $z \in \mathbb{C} \subset \mathbb{C}(\Gamma)$ by the usual conjugation $z \mapsto \overline{z}$). the action of σ on points $P \in \Gamma$ is extended by the usual formula

$$\overline{f(P)} = f^\sigma(P^\sigma). \tag{3.6}$$

The point P^σ is uniquely determined by (3.6), since we can compute the value of every rational function in $\mathbb{C}(\Gamma)$ at P^σ by this relation. The set $\Gamma_\mathbb{R} = \{P \in \Gamma, P^\sigma = P\}$ of real points of Γ is either empty or a disjoint union of the components diffeomorphic to the circle S^1 (called the real ovals of Γ). The action of σ is naturally transferred to the divisors (their linear equivalence classes) and distributions by extending the pointwise action of σ on their supports.

If Γ is supplied with projective coordinates (i.e., Γ is detrmined by polynomial equations in an appropriate projective space), then the complex conjugation of the coordinates gives an anti-involution on Γ, if the ideal of the defining equations of Γ is invariant under the coefficientwise conjugation. In this case real points of Γ are simply points with real coordinates. This more traditional definition is less convenient for us because even in the simplest cases (for example, for hyperelliptic curves) it is easier to introduce σ directly on rational functions generating $\mathbb{C}(\Gamma)$, without projective embeddings.

Let us assume that $\lambda^\sigma = \lambda$ hereafter. Then we can choose the system of local coordinates $k_{i,z}$ at points $R_i^z \in (\lambda)_z$, $z \in \mathbb{C} \cup \infty$ in such a way that $k_{i,z}^{e_i^z} = \pm(z-\lambda)$ or $\pm\lambda^{-1}$ for $z \in \mathbb{C}$, $z = \infty$ and that $k_{i,z}^\sigma = k_{j,\bar{z}}$ for all pairs $(R_i^z)^\sigma = R_j^z$. Let us check it. In fact, $k_{i,z} = (\pm(z-\lambda))^{1/e_i^z}$ for $z \in \mathbb{C}$ and $k_i = \lambda^{-1/e_i}$ for $z = \infty$ are defined up to a root of unity of order $2e_i^z$. Let us denote the sign of $(z-\lambda)$ or λ^{-1} by $\varepsilon_{i,z}$. We can put $\varepsilon_{i,z} = +$ either for $(R_i^z)^\sigma \neq R_i^z$ or for odd e_i^z. Then it is easy to find the desired k. If $(R_i^z)^\sigma = R_i^z$, $e_i^z = 2l$, $l \in \mathbb{Z}_+$, and $\widetilde{k}_{i,z}^\sigma = \rho\widetilde{k}_{i,z}$ for $\rho^l \neq 1$ and some $\widetilde{k}_{i,z} = (z-\lambda)^{1/e_i^z}, \lambda^{-1/e_i}$, then it is necessary to take $\varepsilon_{i,z} = -$. In this case $k_{i,z} = \zeta\widetilde{k}_{i,z}$ for $\zeta^2 = \rho$ will be a real parameter at R_i^z of the desired type.

Let us take the distribution q and the divisor $\mathcal{D}$ of Theorem3.2, satisfying the relations

$$q^\sigma = -q, \tag{3.7a}$$

$$\mathcal{D}^\sigma + \mathcal{D} \sim T, \tag{3.7b}$$

which are consistent in the following sense. If q and $\mathcal{D}$ are subject to (3.7), then $Q^\sigma + Q \sim 0$ for $Q \in \widetilde{q}$ and $\mathcal{D} + Q$ has proberty (3.7b). Here (3.7a) consists of three relations of the same kind for $q^{\pm 1}$ and q_0 (recall that x, t are real parameters). The equivalence of $\mathcal{D}^\sigma + \mathcal{D}$ and T means that $\mathcal{D}^\sigma + \mathcal{D} - T = (w)$ for a rational function w on Γ. Being the ramification divisor, T is obviously σ-invariant (though the points of its support can be non-real). Hence we can choose real w (with the property $w^\sigma = w$).

Let us suppose that for some $z \in \mathbb{P}^1_{\mathbb{R}} \stackrel{\text{def}}{=} \mathbb{R} \cup \infty$ the fiber $(\lambda)_z$ does not intersect $\operatorname{supp}\mathcal{D}$ (hence $\operatorname{supp}\mathcal{D}^\sigma$ either). The permutation of the points of $\widetilde{\lambda}^{-1}(z) = \{R_r^z, 1 \leqq r \leqq m^z\}$ under the action of σ will be denoted by $c = c^z : (R_r^z)^\sigma = R_{c(r)}^z$. Define the permutation d of indices from 1 to n by the relation

$$d(e_0^z + \cdots + e_{r-1}^z + s) = \sum_{i=0}^{c(r)-1} e_i^z + s,$$

where $e_0^z = 0$, $1 \leqq r \leqq m^z$, $1 \leqq s \leqq e_r^z$, $e_r^z = e(R_r^z)$. Put

$$C = C_z = (c_i^j), \quad c_i^j = \delta_i^{d(j)}, \qquad 1 \leqq i, j \leqq n.$$

In a neighbourhood of R_r^z $(1 \leqq r \leqq m)$ we expand

$$e_r^z(k_{r,z}^{-1})w = \sum_{s=1}^{e_r^z} w_{r,s}^z (k_{r,z}^{-1})^s + \cdots,$$

modulo holomorphic part. For $1 \leqq r \leqq m^z$ consider a $e_r^z \times e_r^z$ matrix W^r, setting $W^r = (w_i^{r,j})$, where $w_i^{r,j} = w_{r,i+j-1}^z$ $(1 \leqq i,j \leqq e_r^z)$ if $i+j-1 \leqq e_r^z$ and $w_i^{r,j} = 0$ if $i+j > e_r^z + 1$. Let us recall that $k_{r,z}^{e_r^z} = \varepsilon_{r,z}(z-\lambda), \varepsilon_{r,z}\lambda^{-1}$ for $z \neq \infty$, $z = \infty$, $\varepsilon_{r,z} = \pm 1$. Arranging $W^1, \ldots, W^r, \ldots, W^{m^z}$ along the principal diagonal, we get a matrix W_z of order $n \times n$. Finally, let $\Omega_z \stackrel{\text{def}}{=} W_z C_z$. Since $w_{r,e}^z \neq 0$ for $e = e_r^z$, $1 \leqq r \leqq m^z$, Ω_z is invertible. The matrix Ω_z is hermitian: $\Omega_z^+ = \Omega_z$, where the sign "+" means hereafter the hermitian conjugation (composition of the transposition and the complex conjugation). In fact, $(W^r)^+ = W^{c(r)}$ because of the reality of w and the relation $k_{r,z}^\sigma = k_{c(r),z}$ for $1 \leqq r \leqq m^r$. The conjugation $W_z \to C_z W_z C_z^{-1}$ induces the permutation $c = c^z$ of the blocks $\{W_r\}$. We set $\Omega \stackrel{\text{def}}{=} \Omega_\infty$. Let us define $\Phi_{(z)}$ by the above system of "real" parameters $\{k_{i,z}\}$.

Theorem3.4. *For φ, U, V of Theorems 3.2,3 and λ, q, $\mathcal{D}$, satisfying the condition $\lambda^\sigma = \lambda$ and (3.7) with respect to some anti-involution σ on Γ, the following relations hold:*

a) $\Omega = (\omega_i^j)$, $\omega_i^j \stackrel{\text{def}}{=} \operatorname{Tr}(\varphi_i w \varphi_j^\sigma)$, $1 \leqq i,j \leqq n$;

b) $\Phi_{(z)}\Omega_z\Phi_{(z)}^+ = \Omega$, $z \in \mathbb{R} \cup \infty$, $\operatorname{supp}\mathcal{D} \cap \widetilde{\lambda}^{-1}(z) = \emptyset$;

c) $U\Omega + \Omega U^+ = 0 = V\Omega + \Omega V^+$.

Proof. First of all, the trace of $\omega_i^j = \operatorname{Tr}(\varphi_i w \varphi_j^\sigma)$, where $\varphi = {}^t(\varphi_1, \ldots, \varphi_n)$, is well defined, since $\varphi_i\varphi_j^\sigma$ belong to $\mathbb{C}(\Gamma)$ by virture of (3.7a). Moreover, $\varphi_i\varphi_j^\sigma \in H^0(T)$ and ω_i^j are constant (do not depend on λ) because of the fundamental property of the ramification divisor T. Computing ω_i^j, we will use the relation of Tr and the residue Res for arbitrary $z \in \mathbb{P}^1$, $f \in \mathbb{C}(\Gamma)$:

$$\operatorname{Res}_z(\operatorname{Tr}(f)d\lambda) = \sum_{r=1}^{m^z} \operatorname{Res}_{R_r^z}(f\,d\lambda).$$

For $z \neq \infty$ we have

$$\begin{aligned}\omega_i^j &= \operatorname{Res}_z(\operatorname{Tr}(\varphi_i w \varphi_j^\sigma)(\lambda - z)^{-1} d(\lambda - z)) = \\ &= \sum_{r=1}^{m^z} e_r^z \operatorname{Res}_{R_r^z}(\varphi_i w \varphi_j^\sigma k_{r,z}^{-1} dk_{r,z}^{-1}),\end{aligned}$$

because $z - \lambda = \varepsilon_{r,z} k_{r,z}^{e_r^z}$ in a neighbourhood of R_r^z. Strictly speaking, in this computation we can take arbitrary constant $\varepsilon_{r,z}$ instead of $\varepsilon_{r,z} = \pm 1$. Expressing $e_r^z k_{r,z}^{-1} w$ via $w_{r,s}^z$ (see above), we find that the matrix (ω_i^j) coincides with $\Phi_{(z)}\Omega_z\Phi_{(z)}^+$

(this is also true for $z = \infty$). In particular, the latter matrix does not depend on the choice of z (and on the parameters $x, t \in \mathbb{R}$ either) and $(\omega_i^j) = \Phi_{(\infty)}\Omega\Phi^+_{(\infty)} = \Omega$.

To prove b) we use the relation $(\Phi_{(z)})_x = (1-z)^{-1}U\Phi_{(z)}$ for a simple fiber $(\lambda)_z$, $z \neq \pm 1$ which does not intersect $\operatorname{supp}\mathcal{D}$ (see Theorem3.3). The relation $\Phi^+_{(z)} = \Omega_z^{-1}\Phi^{-1}_{(z)}\Omega$ gives c) for U. Analogously we can check c) for V. □

The matrix Ω^{-1} can be regarded as an hermitian form: $\Omega^{-1}(\boldsymbol{a},\boldsymbol{b}) \stackrel{\text{def}}{=} (\Omega^{-1}\boldsymbol{a},\boldsymbol{b})$, where $(\boldsymbol{a},\boldsymbol{b}) = \boldsymbol{b}^+\boldsymbol{a}$ is the standard hermitian form, $\boldsymbol{a},\boldsymbol{b} \in \mathbb{C}^n$. With respect to the anti-involution defined by this form (sending A to $A^* \stackrel{\text{def}}{=} \Omega A^+\Omega^{-1}$) the matrices U, V are $*$-anti-hermitian because of Theorem3.4,b). If $\operatorname{supp}\mathcal{D} \cap (\lambda)_0 = \emptyset$, then $g = \Phi_{(0)}\dot{\Phi}^{-1}_{(0)}$ is a $*$-unitary solution of (0.1) for $\dot{\Phi}_{(0)} = \Phi_0(\dot{x},\dot{t})$, where $\dot{x}$, $\dot{t}$ are some fixed (admissible) values of x, t. This is a result of claim a) and Theorem3.3a). Note that ω and Ω^{-1} are defined up to multiplication by real constants.

We will show that linear equivalent divisors $\mathcal{D}$ lead to the equivalent forms Ω^{-1} (with respect to base changes in $\mathbb{C}^n$ and multiplication by constants from $\mathbb{R}^*$). We can examine the form $\Omega(\boldsymbol{a},\boldsymbol{b}) = (\Omega\boldsymbol{a},\boldsymbol{b})$ instead of Ω^{-1}, since they are equivalent $(\Omega^+\Omega^{-1}\Omega = \Omega)$. If $\mathcal{D}' = \mathcal{D} + (f)$, where $f \in \mathbb{C}(\Gamma)$, then the fuction w' such that $(w') = \mathcal{D}' + (\mathcal{D}')^\sigma - T$, $(w')^\sigma = w'$ is of the form $w' = \rho f w f^\sigma$ for suitable $\rho \in \mathbb{R}^*$. Here, as for $\mathcal{D}$, $\operatorname{supp}\mathcal{D}' \cap \widetilde{\lambda}^{-1}(\infty) = \emptyset$. Hence we can introduce an invertible matrix $\Omega' = \Omega'_\infty$ for w'. Set $f = \sum_{s=1}^{e_r-1} f_{r,s}k_r^s + O(k_r^{e_r})$ in a neighbourhood of $R_r \in \widetilde{\lambda}^{-1}(\infty)$ and define the matrix F following the construction $\{w^z_{r,s}\} \mapsto W_z$ with $f_{r,s}$ instead of $\{w^z_{r,s}\}$. Then the matrix $C_\infty F^+ C_\infty$ corresponds to $\{f^\sigma_{r,s}\}$ and $\Omega' = W'_\infty C_\infty = \rho F W_\infty C_\infty F^+ C_\infty C_\infty = \rho F \Omega F^+$.

The equivalence of Ω and Ω' can be established in a different way. Let us take a fiber $(\lambda)_z$ not intersecting $\operatorname{supp}\mathcal{D},\mathcal{D}'$. Then Ω is equivalent to Ω_z and Ω' is equivalent to Ω'_z by Theorem3.4b). If $(\lambda)_z$ is simple, then Ω_z and Ω'_z are diagonal and

$$\Omega'_z = \rho F \Omega_z F^+$$

for $F = \operatorname{diag}(f(R_1^z), \dots, f(R_n^z))$.

Let $\widehat{\mathcal{J}}^{g+n-1}_\sigma$ be the variety of linear equivalence classes of divisors of degree $g+n-1$ satisfying condition (3.7b). In the notation of §3.1,

$$\widehat{\mathcal{J}}^{g+n-1}_\sigma = \{x \in \mathcal{J}^{g+n-1}, x^\sigma + x = \widetilde{T}\}.$$

The above construction makes it possible to assign an equivalence class of non-degenerate hermitian forms Ω (up to proportionality) to each class $\widetilde{\mathcal{D}} \in \widehat{\mathcal{J}}^{g+n-1}_\sigma$. For this purpose one can either: a) choose $\mathcal{D}' \in \widetilde{\mathcal{D}}$ not intersecting $(\lambda)_\infty$ and construct Ω' from $\mathcal{D}'$ or b) given any $\mathcal{D} \in \widetilde{\mathcal{D}}$, define Ω_z at a general enough fiber

$(\lambda)_z$. The equivalence class of Ω' or Ω_z is the class of hermitian forms corresponding to $\widetilde{\mathcal{D}}$. If (p_+, p_-) is the signature (type) of the form Ω, the map $\mathcal{D} \mapsto \kappa \stackrel{\text{def}}{=} |p_+ - p_-|$ is extended to a continuous map

$$\kappa : \widehat{\mathcal{J}}_\sigma^{g+n-1} \to \{0, \dots, n\}.$$

Really, any continuous deformation of $\widetilde{\mathcal{D}}$ in $\widehat{\mathcal{J}}_\sigma^{g+n-1}$ can be lifted to a pointwise continuous deformation of divisors $\mathcal{D} \in \widetilde{\mathcal{D}}$ not intersecting $(\lambda)_\infty$. Theorem3.4b) gives that κ is constant on connected components of $\widehat{\mathcal{J}}_\sigma^{g+n-1}$.

Exercise3.6*. $\widehat{\mathcal{J}}_\sigma^{g+n-1} \neq \emptyset$. *(see [Fay]; a purely algebraic proof is in the author's paper "On reality conditions in finite zoned integration", Doklady of the Academy of Sciences USSR, 1980,* **252**-*5, pp.1104–1108).* □

In the particular case of $\kappa = n$, which will be mostly considered further, the statement of the exercise is checked below.

Proposition3.3. a) *The variety $\widehat{\mathcal{J}}_\sigma^{g+n-1}$ is a principal homogeneous space over a commutative (real) Lie group $\widehat{\mathcal{J}}_\sigma^0 \stackrel{\text{def}}{=} \{x \in \mathcal{J}^0 = \operatorname{Jac}\Gamma, x^\sigma = -x\}$. In particular, $\widehat{\mathcal{J}}_\sigma^{g+n-1}$ and $\widehat{\mathcal{J}}_\sigma^0$ are isomorphic as smooth manifolds.*

b) The connected component of zero $(\widehat{\mathcal{J}}_\sigma^0)_0$ of the group $\widehat{\mathcal{J}}_\sigma^0$ is isomorphic to the torus $(S^1)^g$, $S^1 \stackrel{\text{def}}{=} \{z \in \mathbb{C}, |z| = 1\}$. The factor group $\widehat{\mathcal{J}}_\sigma^0/(\widehat{\mathcal{J}}_\sigma^0)_0$ is isomorphic to $\mathbb{Z}_2^{d-g}$, where 2^d is the number of real (σ-invariant) points of order 2 in $\mathcal{J}^0$. In particular, $\widehat{\mathcal{J}}_\sigma^0$ is isomorphic to the disjoint union of 2^{d-g} copies of $(S^1)^g$.

Proof. If $\widetilde{\mathcal{D}} \in \widehat{\mathcal{J}}_\sigma^{g+n-1}$, then the map $\widehat{\mathcal{J}}_\sigma^0 \ni x \mapsto \widetilde{\mathcal{D}} + x \in \widehat{\mathcal{J}}_\sigma^{g+n-1}$ is an isomorphism ($\mathcal{D}_1, \mathcal{D}_2 \in \widehat{\mathcal{J}}_\sigma^{g+n-1} \Rightarrow \mathcal{D}_1 - \mathcal{D}_2 \in \widehat{\mathcal{J}}_\sigma^0$). But the existence of at least one point $\widetilde{\mathcal{D}} \in \widehat{\mathcal{J}}_\sigma^{g+n-1}$ is guaranteed by the statement of Exercise3.6 (see also Proposition3.4,b)).

Let us denote the kernel of the map $\varepsilon : \mathcal{J}^0 \ni x \mapsto x - x^\sigma \in \widehat{\mathcal{J}}_\sigma^0$ by $\mathcal{J}_\sigma^0$. Since every connected Lie group is divisible and $\varepsilon(y) = 2y$ for $y \in \widehat{\mathcal{J}}_\sigma^0$, then $\varepsilon(\mathcal{J}^0) \supset (\widehat{\mathcal{J}}_\sigma^0)_0$. Moreover, $\varepsilon(\mathcal{J}^0)$ is connected because $\mathcal{J}^0$ is, and $\varepsilon(\mathcal{J}^0) = (\widehat{\mathcal{J}}_\sigma^0)_0$. Therefore all connected components of $\widehat{\mathcal{J}}_\sigma^0$ are identified under the action of ε. This means that each connected component of $\widehat{\mathcal{J}}_\sigma^0$ contains at least one point of the group $\widehat{\mathcal{J}}_\sigma^0 \cap \mathcal{J}_\sigma^0 = \{x \in \mathcal{J}^0, 2x = 0, x^\sigma = x\}$, which is isomorphic to $\mathbb{Z}_2^d$ for some $d \leqq 2g$. A proper real neighbourhood of an arbitrary nonsingular real point of an algebraic g-dimensional variety with anti-involution is isomorphic to a domain in $\mathbb{R}^g$. In particular, the group $\widehat{\mathcal{J}}_\sigma^0$ is g-dimensional and locally isomorphic to the group $\mathbb{R}^g$. Hence the compact group $(\widehat{\mathcal{J}}_\sigma^0)_0 \subset \mathcal{J}^0$ must be isomorphic to the torus $(S^1)^g$. Consequently, the subgroup S of the points of the second order of the group $(\widehat{\mathcal{J}}_\sigma^0)_0$

is isomorphic to $\mathbb{Z}_2^g$. Since $S \subset \widehat{\mathcal{J}}_\sigma^0 \cap \mathcal{J}_\sigma^0$ and $d \geqq g$ (see above), the factor group $(\widehat{\mathcal{J}}_\sigma^0 \cap \mathcal{J}_\sigma^0)/S$ is isomorphic to $\widehat{\mathcal{J}}_\sigma^0/(\widehat{\mathcal{J}}_\sigma^0)_0$. □

Anti-hermiticity. Here we discuss the condition which implies the (positive or negative) definiteness of the form Ω considered above (i.e., the equality $\kappa = |p_+ - p_-| = n$).

Proposition3.4. *a) If there exists $\widetilde{\mathcal{D}} \in \widehat{\mathcal{J}}_\sigma^{g+n-1}$ for some $\kappa = n$, then all fibers $(\lambda)_z$, $z \in \mathbb{R} \cup \infty$ are simple and real.*

b) Conversely, the reality and simplicity of all $(\lambda)_z$, where $z \in \mathbb{R} \cup \infty$, lead to a representation $T = \mathcal{D}_0 + \mathcal{D}_0^\sigma$ for some $\mathcal{D}_0 \geqq 0$. Such a divisor $\mathcal{D}_0$ corresponds to the form Ω^0 with $\kappa = n$.

Proof. First, let us suppose that a certain fiber $(\lambda)_z$, where $z \in \mathbb{R} \cup \infty = \mathbb{P}_\mathbb{R}^1$, has non-real points. Changing the point z a little (in the topology of $\mathbb{P}_\mathbb{R}^1$), we may assume that $(\lambda)_z$ is a simple fiber. Take a divisor $\mathcal{D} \in \widetilde{\mathcal{D}}$ such that $\widetilde{\lambda}^{-1}(z) \cap \operatorname{supp} \mathcal{D} = \emptyset$. Then the form $\Omega_z = \operatorname{diag}(w(R_r^z))C_z$ for the permutation matrix C_z of σ acting on $\{R_r^z\}$, should be definite. However $C_z \neq \boldsymbol{I}$ and $(C_z\boldsymbol{a}, \boldsymbol{a}) = 0$ for a vector $\boldsymbol{a}$ with only one nonzero component not stable under the action of C_z. This contradiction means that $C_z = \boldsymbol{I}$.

Second, Ω_z is equal to $\operatorname{diag}(W^r)$ because of the "absence" of C_z at an arbitrary (not necessarily simple) fiber (see the definition of Ω_z). If some index $e_r^z = e$ is not equal to the unity, then $(W^r\boldsymbol{a}, \boldsymbol{a}) = 0$ for $\boldsymbol{a} = {}^t(0, \ldots, 0, 1) \in \mathbb{C}^e$. Hence $e_r^z = 1$ and the fiber is simple.

Conversely, if T does not contain real points then $\operatorname{supp} T$ is represented as follows: $T = \mathcal{D}_0 + \mathcal{D}_0^\sigma$. For such $\mathcal{D}_0$ we can take $w_0 = 1$. The reality of the fibers $(\lambda)_z$, $z \in \mathbb{P}_\mathbb{R}^1$, results in $\Omega^0 = \boldsymbol{I}$. □

Till the end of this section we assume that Γ, σ, λ satisfy the conditions of Proposition3.4. Then Γ is *separable*: $\Gamma \setminus \Gamma_\mathbb{R}$ consists of two connected components (by definition). Indeed, $\mathbb{P}^1$ is separated by the circle $\mathbb{P}_\mathbb{R}^1$ into two components. Therefore the image (with respect to $\widetilde{\lambda}$) of any path connecting two points $P_1, P_2 \in \Gamma$ with projections $\widetilde{\lambda}(P_1)$, $\widetilde{\lambda}(P_2)$ in different components of $\mathbb{P}^1 \setminus \mathbb{P}_\mathbb{R}^1$ intersects $\mathbb{P}_\mathbb{R}^1$ and, hence, the path itself intersects $\widetilde{\lambda}^{-1}(\mathbb{P}_\mathbb{R}^1) = \Gamma_\mathbb{R}$. The number of components does not exceed 2. This is a general fact. In the given situation it is easily derived from the existence of λ.

Theorem3.5. *a) If $\kappa = n$ for $\widetilde{\mathcal{D}} \in \widehat{\mathcal{J}}_\sigma^{g+n-1}$, then there exists a divisor $\mathcal{D} \in \widetilde{\mathcal{D}}$ such that $\mathcal{D} - (\lambda)_\infty \geqq 0$.*

b) In the case $\kappa = n$, the function φ is regular for all $x, t \in \mathbb{R}$, and the solutions $U, V(x,t)$ of system (0.2) and equation (0.9) of Theorem3.4 are non-singular and bounded (see Exercise3.3).

Proof. Constructing the function w by $\mathcal{D}$, put $\omega = w\,d\lambda$. Since Ω_z is definite for any $z \in \mathbb{P}^1_{\mathbb{R}}$, w has a constant sign over any oval of $\Gamma_{\mathbb{R}}$ (the poles of w belong to $\operatorname{supp} T \subset \Gamma \setminus \Gamma_{\mathbb{R}}$). The form ω is regular on Γ. Choosing the orientation on $\mathbb{P}^1_{\mathbb{R}}$ and using $\widetilde{\lambda}$, we can pull it back to an orientation on ovals of $\Gamma_{\mathbb{R}}$. With respect to this orientation, ω does not change its sign on $\Gamma_{\mathbb{R}}$, which contradicts its regularity because of the Cauchy theorem ($\int_{\Gamma_{\mathbb{R}}} \omega$ is zero for regular ω).

Let us recall that φ, U, V were constructed under the condition that $\mathcal{D}+Q-(\lambda)_\infty$ is non-special for $\mathcal{D}$, q satisfying (3.7), $Q \in \widetilde{q}$. However, because of statement a), any divisor of type (3.7b) with $\kappa = n$ (in particular, $\mathcal{D} + Q$) becomes non-special after $(\lambda)_\infty$ is subtracted. □

Exercise3.7. *Prove directly that any solution g of equation (0.1) constructed in Theorem3.3 is regular for all $x, t \in \mathbb{R}$, if it is unitary.* (Show that g can have only poles with respect to x or t. Such a singularity necessarily causes a singular point of the function $gg^+ \equiv \boldsymbol{I}$). □

Theorem3.6. *The form Ω is definite only for one component of $\widehat{\mathcal{J}}^{g+n-1}_\sigma$. Every other component contains at least one class with a representative $\mathcal{D}$ such that $\mathcal{D} - (\lambda)_\infty \geqq 0$, which leads to the existence of singular solutions $U, V(x,t)$ for a suitable choice of the initial data $(\mathcal{D}, q_0)$.*

Proof. The set of classes of divisors $S \in \mathcal{J}^{g+n-1}$ for which $2S \sim T$ is identified with $\mathbb{Z}_2^{2g}$ in the following way (see Corollary3.1 §3.1). Introduce a quadratic form on $\mathbb{Z}_2^{2g} = \{(x;y),\ x, y \in \mathbb{Z}_2^g\}$,

$$\langle (x;y) \rangle = {}^t y x = \sum_{i=1}^{g} x_i y_i \mod 2,$$

where $x = {}^t(x_i)$, $y = {}^t(y_i)$. Then the class S corresponds to an element $s \in \mathbb{Z}_2^{2g}$ such that

$$H^0(S - (\lambda)_\infty) = \{0\} \Rightarrow \langle s \rangle = 0.$$

As follows from Proposition3.3, the classes S belonging to different connected components $(\widehat{\mathcal{J}}^{g+n-1}_\sigma)_\alpha \subset \widehat{\mathcal{J}}^{g+n-1}_\sigma$ correspond to different affine $\mathbb{Z}_2$-subspaces $\Sigma_\alpha \subset \mathbb{Z}_2^{2g}$ by the mapping $S \mapsto s$. These affine subspaces Σ_α do not intersect and are translations of a certain linear $\mathbb{Z}_2$-subspace $\mathbb{Z}_2^g \cong \Sigma \subset \mathbb{Z}_2^{2g}$. By virtue of Theorem3.5a), the component $(\widehat{\mathcal{J}}^{g+n-1}_\sigma)_0$ containing the divisor $\mathcal{D}_0$ of Proposition3.4,b) corresponds to an isotropic affine subspace Σ_0 ($s \in \Sigma_0 \Rightarrow \langle s \rangle = 0$). The dimension Σ_0 over $\mathbb{Z}_2$ ($= \dim_{\mathbb{Z}_2} \Sigma$) is equal to g, i.e., the half of the dimension of $\mathbb{Z}_2^{2g}$. Therefore Σ_0 is a maximal isotropic affine subspace in $\mathbb{Z}_2^{2g}$ (check this as an exercise or see [Fay]). Any other connected component $(\widehat{\mathcal{J}}^{g+n-1}_\sigma)_1$ not containing divisors S such that

$S - (\lambda)_\infty \geqq 0$ corresponds to $\kappa = n$ and is mapped to $\Sigma_1 \neq \Sigma_0$. But then $\Sigma_0 \cup \Sigma_1$ becomes an isotropic affine subspace of dimension $g+1$, which is impossible. □

We can assign not only κ, but also a sharper discrete invariant $\Pi = \pm\{\mathrm{sgn}_j\, w\}$ to each component of $\widehat{\mathcal{J}}_\sigma^{g+n-1}$. Here w is a function constructed by $\mathcal{D}$ from the corresponding component, $\mathrm{sgn}_j\, w = w/|w|_{\Gamma^j_{\mathbb{R}}}$ is the sign of w on the j-th real oval $\Gamma^j_{\mathbb{R}}$ of the curve Γ ($1 \leqq j \leqq p$, $p \leqq n$). The set $\pm\{\mathrm{sgn}_j\, w\}$ does not depend on the choice of concrete $\mathcal{D}$ and w and is a function of the component. The sign of w on each oval $\Gamma^j_{\mathbb{R}}$ is constant, since $(w) = \mathcal{D} + \mathcal{D}^\sigma = T$ and, hence, the poles and zeros of w on $\Gamma_{\mathbb{R}}$ have even multiplicities. Note that $\kappa = |\sum_{j=1}^p l_j\, \mathrm{sgn}_j\, w|$, where l_j is the degree of the covering $S^1 \cong \Gamma^j_{\mathbb{R}} \to \mathbb{P}^1 \cong S^1$ induced by $\widetilde{\lambda}$.

In a more general situation of an arbitrary real curve Γ that is a union of p real ovals $\Gamma^j_{\mathbb{R}}$ ($\Gamma_{\mathbb{R}} = \bigcup_{j=1}^p \Gamma^j_{\mathbb{R}}$) we can construct an analogous invariant Π' as well. Instead of $\widehat{\mathcal{J}}_\sigma^{g+n-1}$ we take $\widehat{\mathcal{J}}_\sigma^{g-1} = \{x \in \mathcal{J}^{g-1}, x + x^\sigma = \widetilde{K}\}$, where $\widetilde{K}$ is the canonical class of Γ. There exists a real differential form ω on Γ with divisor $(\omega) = \mathcal{D} + \mathcal{D}^\sigma$ for each $\mathcal{D} \in x \in \widehat{\mathcal{J}}_\sigma^{g-1}$. The zeros and poles of ω on $\Gamma_{\mathbb{R}}$ have even multiplicities. Therefore, fixing the orientation on one of components of $\Gamma \setminus \Gamma_{\mathbb{R}}$ and the corresponding orientation on $\Gamma_{\mathbb{R}}$, we obtain that ω is of constant sign on each oval $\Gamma^j_{\mathbb{R}}$. Hence we can set $\Pi' \stackrel{\mathrm{def}}{=} \pm\{\mathrm{sgn}_j\, \omega\}$. In the case considered above the mapping $\mathcal{D} \mapsto \mathcal{D} - (\lambda)_\infty$ results in an isomorphism $\widehat{\mathcal{J}}_\sigma^{g+n-1} \to \widehat{\mathcal{J}}_\sigma^{g-1}$ and the coincidence $\Pi' = \Pi$, since $\omega = w\, d\lambda$. Theorem3.5 can be generalized as follows:

Exercise3.8*. *The invariant Π' establishes an isomorphism of the set of connected components of $\widehat{\mathcal{J}}_\sigma^{g-1}$ and the set* $\{(\pm 1)_j\} \mod (\pm 1)$. *In particular, the number of connected components is equal to 2^{p-1}. The component of $\widehat{\mathcal{J}}_\sigma^{g-1}$ corresponding to* $\Pi' = \{+1, \ldots, +1\} \mod (\pm 1)$ *does not contain $x \ni \mathcal{D}$ for $\mathcal{D} \geqq 0$ (see [Fay], Proposition 6.8).* □

3.4. Curves with an involution.

The previous discussion gave us algebraic-geometric solutions with values in $\mathfrak{gl}_n$ and its real forms ($\mathfrak{u}_n$ etc.). In order to obtain solutions with values in other Lie algebras (or Lie groups) or in homogeneous spaces of type S^{n-1}, we need more special curves. We will describe below methods of constructing of algebraic-geometric $Sp_{2m}(\mathbb{C})$-fields, $O_{2m}(\mathbb{C})$-fields, and S^{2m-1}-field. All notations from §3.2 remain the same.

Let us assume that there is an involution τ on Γ (its action is written on the left) such that ${}^\tau\lambda = \lambda$. By definition, τ is a $\mathbb{C}$-automorphism of the second order of the field $\mathbb{C}(\Gamma)$ extended to an action on points $P \in \Gamma$ by the formula $f({}^\tau P) = ({}^\tau f)(P)$ for arbitrary $f \in \mathbb{C}(\Gamma)$ (see (3.6)). We impose the following conditions on a divisor

$\mathcal{D}$ of degree $g+n-1$ on Γ and a distribution q from (3.1A):

$$q + {}^{\tau}q = 0, \qquad \mathcal{D} + {}^{\tau}\mathcal{D} \sim T, \tag{3.8a,b}$$

where the ramification divisor T is defined in §3.3. These relations are self-consistent (see (3.7a,b)): ${}^{\tau}(\mathcal{D}+Q)+(\mathcal{D}+Q) \sim T$, if $Q \in \widetilde{q}$. For a suitable function $v \in \mathbb{C}(\Gamma)$ we can set $(v) = \mathcal{D} + {}^{\tau}\mathcal{D} - T$. Moreover ${}^{\tau}v = \pm v$, since ${}^{\tau}vv^{-1} \in \mathbb{C}$. Let us determine which data lead to the sign $+$.

Let us denote the smooth curve with the function field $\mathbb{C}(\Gamma') = \{f \in \mathbb{C}(\Gamma),\ {}^{\tau}f = f\}$ by Γ'. Then the covering $\widetilde{\lambda} : \Gamma \to \mathbb{P}^1$ is factorized as $\widetilde{\lambda} = \widetilde{\lambda}' \circ \pi$, where $\Gamma \to \Gamma'$ is induced by the embedding $\mathbb{C}(\Gamma') \subset \mathbb{C}(\Gamma)$ and $\widetilde{\lambda}' : \Gamma' \to \mathbb{P}^1$ corresponds to the function λ restricted to Γ'. In particular, $n = 2m$ for $m = \deg \widetilde{\lambda}'$ and $T = \pi^*T' + T_\pi$, where $\pi^*T' \subset \Gamma$ is the pullback (the inverse image) of the ramification divisor T' of the covering $\widetilde{\lambda}'$, and T_π is the ramification divisor of the covering π. If $T_\pi \neq 0$, then v cannot belong to $\mathbb{C}(\Gamma')$, since the divisors $\mathcal{D} + {}^{\tau}\mathcal{D} - T$ and T_π are obtained from some divisors on Γ'. When $T_\pi = 0$, the covering π is non-ramified (by definition) and the divisor (v) can be regarded as a divisor on Γ'. However v does not necessarily belong to $\mathbb{C}(\Gamma')$ (it holds only when the sign is $+$).

Proposition3.5. *The variety $\widehat{\mathcal{J}}_{\tau}^{g+n-1}$ of the classes of divisors $\mathcal{D}$ with properties (3.8b) is connected in the case $T_\pi \neq 0$ (in this case ${}^{\tau}vv^{-1} = -1$). If $T_\pi = 0$ then it consists of two isomorphic components $(\widehat{\mathcal{J}}_{\tau}^{g+n-1})_{\pm}$ for $T_\pi \neq 0$ where the signs are the signs of ${}^{\tau}vv^{-1}$. The connected components of $\widehat{\mathcal{J}}_{\tau}^{g+n-1}$ are isomorphic to the abelian variety $P = \{{}^{\tau}x - x, x \in \mathcal{J}^0\}$ (called Prym variety of the covering π) of dimension $g - g' = g' + \frac{\deg T}{2} - 1$, where g' is the genus of Γ'.*

Proof. (see [Fay] Ch4; cf. Proposition3.3). The variety $\widehat{\mathcal{J}}_{\tau}^{g+n-1}$ is isomorphic to the algebraic group $\widehat{\mathcal{J}}_{\tau}^{0} = \{x \in \mathcal{J}^0,\ {}^{\tau}x + x = 0\}$. Let $(\widehat{\mathcal{J}}_{\tau}^{0})_0$ be the connected component of zero of $\widehat{\mathcal{J}}_{\tau}^{0}$ and $(\mathcal{J}_{\tau}^{0})_0$ be the component of zero of the group $\mathcal{J}_{\tau}^{0} = \{x \in \mathcal{J}^0,\ {}^{\tau}x = x\}$. Then $(\mathcal{J}_{\tau}^{0})_0 + (\widehat{\mathcal{J}}_{\tau}^{0})_0 \supset 2\mathcal{J}^0$ and, consequently, $(\mathcal{J}_{\tau}^{0})_0 + (\widehat{\mathcal{J}}_{\tau}^{0})_0 = \mathcal{J}^0$ because of the connectivity of $\mathcal{J}^0$. In particular, each component of $\widehat{\mathcal{J}}_{\tau}^{0}$ contains at least one point of the group $\widehat{\mathcal{J}}_{\tau}^{0} \cap (\mathcal{J}_{\tau}^{0})_0$, which consists of the points of the second order. Therefore the components of $\widehat{\mathcal{J}}_{\tau}^{0}$ are in one-to-one correspondence with the points of the second order in $(\mathcal{J}_{\tau}^{0})_0$ (all of them lie in $\widehat{\mathcal{J}}_{\tau}^{0}$) modulo $(\mathcal{J}_{\tau}^{0})_0 \cap (\widehat{\mathcal{J}}_{\tau}^{0})_0 \stackrel{\text{def}}{=} I_\tau = \{{}^{\tau}x + x = {}^{\tau}y - y, x, y \in \mathcal{J}^0\}$. Since

$$(g) = {}^{\tau}\mathcal{D} + \mathcal{D} + {}^{\tau}\mathcal{D}^* - \mathcal{D}^* \Leftrightarrow ({}^{\tau}gg) = 2\pi(\mathcal{D}),$$

and $\mathbb{C}(\Gamma')^* = \{{}^{\tau}gg\}$ for $g \in \mathbb{C}(\Gamma)^*$, $\mathcal{D}, \mathcal{D}^* \in \Gamma$ (see [Se2] Ch.2 §3), then the points of the second order of $\mathcal{J}'^0 = \operatorname{Jac}\Gamma'$ generate I_τ with respect to the natural map $\pi^* : \mathcal{J}'^0 \to \mathcal{J}^0$.

When $T_\pi \neq 0$, π^* does not have the kernel (divisors non-equivalent to zero remain non-equivalent after being lifted to Γ). Indeed, if $(f) = \pi^*\mathcal{D}'$ for a divisor $\mathcal{D}'$ on Γ', then ${}^\tau f = \pm f$, $f^2 \in \mathbb{C}(\Gamma')$, and $(f^2) = 2\mathcal{D}'$. But then, adding the function f to $\mathbb{C}(\Gamma')$, we can construct a curve between Γ and Γ' which is an unramified covering of Γ'. Hence, $f \in \mathbb{C}(\Gamma')$. An analogous argument shows that the kernel of π^* is isomorphic to $\mathbb{Z}_2$ when $T_\pi = 0$. Really, if $f \notin \mathbb{C}(\Gamma')$ and $(f) = \pi^*\mathcal{D}'$ for some $\mathcal{D}'$, then f is uniquely determined up to the multiplication by elements of $\mathbb{C}(\Gamma')^*$.

Since the dimension of $\mathcal{J}'^0$ is g, the number of points of second order in $\mathcal{J}'^0$ is equal to $2^{2g'}$. Therefore the group I_τ contains either $\mathbb{Z}_2^{2g'}$ $(T_\pi \neq 0)$ or $\mathbb{Z}_2^{2g'-1}$ (when $T_\pi = 0$). The dimension of $(\mathcal{J}_\tau^0)_0 = \pi^*\mathcal{J}'^0$ is equal to g'. Consequently, the group of points of the second order $(\mathcal{J}_\tau^0)_0$ is isomorphic to $\mathbb{Z}_2^{2g'}$. This proves the proposition. □

Following §3.2, let us define the vector Baker function $\varphi = {}^t(\varphi_j)$ for $\mathcal{D}$, q of type (3.3), assuming the non-speciality of $\mathcal{D} + Q - (\lambda)_\infty$. Note that for $Q \in \widetilde{q}$ the class $\widetilde{\mathcal{D}+Q}$ of the divisor $\mathcal{D} + Q$ lies in the same connected component as $\widetilde{\mathcal{D}}$. Theorems 3.2 and 3.3 allow us to construct a solution U, V of system (0.2).

Let us discuss the restriction on U, V following from condition (3.3). We define a matrix Λ_z for $z \in \mathbb{P}^1 = \mathbb{C} \cup \infty$ exactly in the same way as Ω_z in §3.3 for the function v instead of w. Here we assume that $\operatorname{supp}\mathcal{D} \cap \widetilde{\lambda}^{-1}(z) = \emptyset$ and take the matrix of the permutation τ of the points of the fiber $(\lambda)_z$ instead of the matrix C_z of the permutation σ in the same fiber. However (in contrast with §3.3) ${}^t\Lambda_z = \varepsilon\Lambda_z$, where $\varepsilon = 1$ only if $T_\pi = 0$ and $\widetilde{\mathcal{D}} \in (\widehat{\mathcal{J}}_\tau^{g+n-1})_+$; otherwise $\varepsilon = -1$. Just as in Theorem3.4, we can check

Proposition3.6. *a)* $\Lambda \stackrel{\text{def}}{=} \Lambda_\infty = (l_i^j)$, *where* $l_i^j = \operatorname{Tr}(\varphi_i v\, {}^\tau\varphi_j)$;
b) $\Phi_{(z)}\Lambda_z\, {}^t\Phi_{(z)} = \Lambda$,
c) $U\Lambda + \Lambda\, {}^tU = 0 = V\Lambda + \Lambda\, {}^tV$. □

Depending on the sign $\varepsilon = \pm 1$, statement c) allows us to obtain solutions of (0.2) with values respectively in the Lie algebra either $\mathfrak{o}_n(\mathbb{C})$ or $\mathfrak{sp}_n(\mathbb{C})$ conjugating U, V by proper constant matrices. Let us recall that here $n = 2m$. In particular, if the fiber $(\lambda)_\infty$ is simple, then $\Lambda = \begin{pmatrix} 0 & \Delta \\ \varepsilon\Delta & 0 \end{pmatrix}$, where $\Delta = \operatorname{diag}(v(R_j))$, $1 \leqq j \leqq m$, $(\lambda)_\infty = \sum_{j=1}^m R_j + {}^\tau(\sum_{j=1}^m R_j)$.

Replace $\mathcal{D}$ with an equivalent divisor $\mathcal{D}' = \mathcal{D} + (f)$, $f \in \mathbb{C}(\Gamma)^*$ (which conjugates U and V by a constant matrix). Then v is multiplied by ${}^\tau f\, f$ and the matrix Δ is replaced by $\Delta\operatorname{diag}(f(R_j)f({}^\tau R_j))$. Changing the function f, we can make Δ equal to the unit matrix I_m.

$O_{2m}(\mathbb{R})$-fields. In addition to (3.8), now let the assumptions of §3.4 be fulfilled, $\mathcal{D}$, q satisfy the relations (3.7), and $\tau\sigma = \sigma\tau$. To make things concrete, we shall assume that $\varepsilon = 1$ and Ω is definite (see Proposition3.4). The general case is left to the reader as an exercise. Under the above assumptions, the fiber $(\lambda)_\infty$ is simple, $\Lambda = \begin{pmatrix} 0 & \Delta \\ \Delta & 0 \end{pmatrix}$, $\Delta = \mathrm{diag}(v(R_j))$, $1 \leqq j \leqq m$ (see above), $\Omega = \mathrm{diag}(w(R_i), 1 \leqq i \leqq 2m$. Replacing $\mathcal{D}$ by a linear equivalent divisor, we may assume that $\Omega = \boldsymbol{I}_{2m}$. Since $({}^\tau w\, w) = (v\, v^\sigma)$, then ${}^\tau w\, w = \rho v v^\sigma$ for $\rho \in \mathbb{R}^*$ (here the commutativity of σ and τ and the reality of w were used). Considering the latter equation at the points of the fiber $(\lambda)_\infty$, we obtain that $\rho > 0$. Substituting v for $\rho^{-1/2}v$, we come to the equality ${}^\tau w\, w = v\, v^\sigma$, which gives the relation $v(R_j)\overline{v(R_j)} = 1$, $1 \leqq j \leqq m$. There exists a diagonal matrix A of order m such that $A\overline{A} = \boldsymbol{I}_m$ and $\Delta = A^2$. Let F be the following matrix of order $2m$: $F = \frac{1}{\sqrt{2}}\begin{pmatrix} A & iA \\ A & -iA \end{pmatrix}$. Then we have

$$F\,{}^tF = \Lambda, \qquad FF^+ = \boldsymbol{I}_{2m}.$$

Consequently, the pair $F^{-1}UF$, $F^{-1}VF$ is a solution of (0.2) with values in $\mathfrak{o}_{2m}(\mathbb{R})$ (due to Proposition3.6c), Theorem3.4c)).

The variety $\widehat{\mathcal{J}}^{g+n-1}_{\sigma,\tau}$ of the classes of divisors $\mathcal{D}$ with conditions (3.7, 8b) is isomorphic to the disjoint union of the components $(S^1)^{g'-1}$, where $g = 2g' - 1$. We recommend calculating the number of the connected components of $\widehat{\mathcal{J}}^{g+n-1}_{\sigma,\tau}$ as an exercise. Anyway it is easy to see that $\widehat{\mathcal{J}}^{g+n-1}_{\sigma,\tau}$ does not contain the classes of divisors $\mathcal{D}$ for which $\mathcal{D} - (\lambda)_\infty \geqq 0$ (Theorem3.5). Hence U, V are well defined for all x, t (and bounded). Analogous statements hold for $\mathfrak{sp}_{2m}(\mathbb{R})$ (instead of $\mathfrak{o}_{2m}(\mathbb{R})$) in the case of $\varepsilon = -1$.

S^{2m-1}-fields. First, we are not suppose n to be even and adjust the construction of §3.2 in order to get a solution g of the PCF equation (0.1) with the condition $g^2 = \boldsymbol{I}$. Let us assume that there exists an involution γ on Γ such that ${}^\gamma\lambda = \lambda^{-1}$. Then γ permutes $(\lambda)_0$ and $(\lambda)_\infty$ and preserves $(\lambda)_{\pm 1}$. We numerate the points $\{R^0_j\} = \widetilde{\lambda}^{-1}(0)$, $\{R^\infty_j\} = \widetilde{\lambda}^{-1}(\infty)$ with respect to γ: ${}^\gamma R^\infty_j = R^0_j$. In §3.2 we constructed a matrix $g = \Phi_{(0)}$ of "values" of the vector function φ at the points of $(\lambda)_0$. By virtue of Therem3.3, g is a solution of (0.1).

We also assume that $\mathcal{D}$, q (see above and §3.2) satisfy the relations

$$(3.9) \qquad {}^\gamma q = q, \qquad {}^\gamma\mathcal{D} = \mathcal{D}.$$

Replacing $\mathcal{D}$ by a linear equivalent divisor, we may assume that ${}^\gamma\mathcal{D} = \mathcal{D}$ (for the sake of simplicity). This is a special case of the Hilbert Theorem 90. We will sketch the proof here. Put $(f) = {}^{\gamma}\mathcal{D} - \mathcal{D}$. Then ${}^{\gamma}f\,f = c \in \mathbb{C}^*$ and (dividing f by $c^{-1/2}$) we can take f such that ${}^{\gamma}f\,f = 1$. Consequently, $f^{-1} = {}^{\gamma}y\,y^{-1}$, where $y = (x + {}^{\gamma}x\,f)$ for a sufficiently general $x \in \mathbb{C}(\Gamma)$. The divisor $\mathcal{D} + (y)$ becomes γ-invariant. Since we replaced $\mathcal{D}$ by an equivalent γ-invariant divisor, the new matrix g is conjugated with the old one by a constant matrix.

The invariance of $\mathcal{D}$, q gives that γ acts on the space $H^0(\mathcal{D}; q)$. Since γ is an involution, we can choose a basis $\{\widehat{\varphi}_j, 1 \leqq j \leqq n\}$ of $H^0(\mathcal{D}; q)$ such that ${}^{\gamma}\widehat{\varphi}_j = \varepsilon_j\widehat{\varphi}_j$, $\varepsilon_j = \pm 1$. As in §3.2, we define the "values" $\widehat{\Phi}_{(0)}$, $\widehat{\Phi}_{(\infty)}$ of the vector function $\widehat{\varphi} = {}^t(\widehat{\varphi}_j)$ at the fibers $(\lambda)_0$, $(\lambda)_\infty$. If $(\lambda)_0$, $(\lambda)_\infty$ are simple, $\widehat{\Phi}_{(0,\infty)} = (\widehat{\varphi}_i(R_j^{0,\infty}))$, where i is the index of rows. Put $\mathcal{E} = \operatorname{diag}(\varepsilon_j)$. Then $\widehat{\Phi}_{(\infty)} = \mathcal{E}\widehat{\Phi}_{(0)}$. On the other hand, $g = \Phi_{(0)} = \widehat{\Phi}_{(\infty)}^{-1}\widehat{\Phi}_{(0)}$. Thus, $g = \widehat{\Phi}_{(0)}^{-1}\mathcal{E}\widehat{\Phi}_{(0)}$ and $g^2 = \boldsymbol{I}$.

Now let us assume that there is another anti-involution σ on Γ (see §3.3) which commutes with γ, and $\mathcal{D}$, q satisfy (3.7), (3.9). We also assume that Ω is definite, in particular, all fibers $(\lambda)_z$, $z \in \mathbb{R} \cup \infty$ are purely real and simple. As above, we take $\mathcal{D}$ such that ${}^{\gamma}\mathcal{D} = \mathcal{D}$. Then ${}^{\gamma}w = \pm w$. However w does not change the sign on $\Gamma_{\mathbb{R}}$. Hence ${}^{\gamma}w = w$. In particular, $\Omega_0 = \Omega_\infty \stackrel{\text{def}}{=} \Omega$. Replacing $\mathcal{D}$ by an equivalent γ-invariant divisor, we may get $\Omega_0 = \Omega = \boldsymbol{I}$. Then (see Theorem3.4) $gg^+ = \boldsymbol{I}$ for g constructed above.

We need one more involution τ on Γ which commutes with σ, γ. Let $\mathcal{D}$ and q satisfy (3.8). Since ${}^{\gamma}\mathcal{D} = \mathcal{D}$, then ${}^{\gamma}v = \pm v$. Let ${}^{\gamma}v = v$. Then $\Lambda_0 = \Lambda$ and $g\Lambda\,{}^t g = \Lambda$ (Proposition3.6). Hence, $F^{-1}gF$ is an S^{2m-1}-field.

The analysis of the variety of the classes of divisors $\mathcal{D} = {}^{\gamma}\mathcal{D}$ with conditions (3.7–3.8b) for which Ω is definite and ${}^{\gamma}v = v$ is left to the reader as an exercise. In the following section we will discuss examples which illustrate the constructions of this section. Note that the above results are directly related to the Euler equation on $\mathfrak{o}_4$ and other Lie algebras (see [A1], [Du3], [Manal], and [Ch11]).

3.5. Application: discrete PCF equation.

In this section, following [Ch9], we briefly describe a technique of integrating certain difference analogues of equation (0.1) and system (0.2). This discretization is based on the process of constructing algebraic-geometric solutions. A more systematic approach to difference analogues of integrable soliton equations makes use of the r-matrix Hamiltonian formalism and the general theory of the τ-functions and Hirota equations (cf., for example, [TF2], [DJM], [UT]). We do not discuss these directions here. The integration of difference analogues of GHM and other equations studied above can be done by similar technique and is left to the reader as an exercise (for the HM and the NS, see [TF2], [DJM]; we also refer to the papers of M. Ablowitz with coauthors). As to the theory of the Toda lattice and related topics, see the book of M. Toda "Theory of nonlinear lattices" (Springer), and books [CD], [TF2] (the last book contains systematic exposition of various problems connected with models on lattices).

The *discrete PCF system* is the system of equations

$$\frac{dU_k}{dt} = \frac{1}{\alpha}(V_k U_k - U_k V_{k-1}), \tag{3.10a}$$

$$V_k - V_{k-1} = \frac{\kappa}{\alpha}(V_k U_k - U_k V_{k-1}), \tag{3.10b}$$

on the functions $\{U_k, V_k, V_{M-1}, M \leqq k \leqq N\}$ of $t \in \mathbb{R}$ with values in $n \times n$-matrices. Let $M \leqq N \in \mathbb{Z}$, $\kappa, \alpha \in \mathbb{C}$ be certain fixed parameters, $\kappa \neq \alpha$, $\kappa \neq 0 \neq \alpha$. If we take $\alpha = 2$ and formally replace U_k, V_k by $U(x)$, $V(x)$ and $\kappa^{-1}(V_k - V_{k+1})$ by $\partial V/\partial x$ we will arrive at (0.2). Similar to (0.2), system (3.10) is the compatibility condition of the equaitons

$$\frac{d\Phi_k}{dt} = \frac{1}{\alpha - \lambda} V_k \Phi_k,$$

$$\Phi_k - \Phi_{k-1} = \frac{\kappa}{\kappa - \lambda}(U_k - 1)\Phi_{k-1},$$

for arbitrary λ, $M - 1 \leqq k \leqq N$ (I is identified with 1 hereafter).

It follows from (3.10b) that

$$V_k^s\Big(1 - \frac{\kappa}{\alpha}U_k\Big) = \Big(1 - \frac{\kappa}{\alpha}U_k\Big)V_{k-1}^s$$

for the powers V_k^s, $s = 1, 2, \ldots$. Let us assume further that U_k, V_k are diagonizable and that their eigenvalues are in a general position. Then the traces $\mathrm{Sp}(V_k^s)$ and (equivalently) eigenvalues $\{\nu^i(t)\}$ of the matrices V_k for $1 \leqq i \leqq n$ do not depend on k. Using (3.11a), we obtain that $\mathrm{Sp}\, U_k$, $\mathrm{Sp}\, U_k^2$, etc. do not depend on t. Hence the eigenvalues $\{\mu_k^i\}$ of the matrices U_k for $M \leqq k \leqq N$ do not depend on t either (cf. the analogous statement for the PCF).

Our goal is to integrate system (3.10) under the boundary condition

$$V_{M-1} C = C V_N \tag{3.10c}$$

for a diagonal constant invertible matrix C and fixed $\{\nu^i(t)\}$, $\{\mu_k^i\}$ in the case of a general position.

Let Γ, λ, $(\lambda)_z = \sum_{i=1}^n R_i^z$ denote the same as in §3.3. We take a divisor $\mathcal{D}$ on Γ of degree $g+n-1$ ($g = \mathrm{genus}(\Gamma)$, $n = \deg\lambda$) and $N - M + 1$ divisors $S_k = \sum_{i=1}^n Q_k^i$ of degree n. Let us suppose that $\sum_{k=M}^N (S_k - (\lambda)_\infty) = (h)$ for a certain rational function h on Γ. Put $\mathcal{D}_k = \mathcal{D} + \sum_{i=M}^k ((\lambda)_\kappa - S_i)$, $M - 1 \leqq k \leqq N$. We will assume that neither all S_k nor $(\lambda)_\infty$, nor $(\lambda)_\alpha$ have points of multiplicities more than 1. Moreover let the support of $\mathcal{D}_k$ not intersect the fibers $\widetilde{\lambda}^{-1}(\infty)$, $\widetilde{\lambda}^{-1}(\alpha)$.

We can introduce the set of Baker functions $\{\varphi_i^k\}$ which are meromorphic on $\Gamma\backslash(\lambda)_\alpha$ with divisors (φ_i^k) such that $(\varphi_i^k)+\mathcal{D}_k \geqq 0$. They have essential singularities in the form $\exp((\alpha-\lambda)^{-1}\int_0^t \nu^i(\tau)d\tau)$ at the points R_i^α, and are normalized by the condition $\varphi_i^k(R_j^\infty) = \delta_i^j$ (cf. §§3.1,2). These functions are well-defined for almost all $t \in \mathbb{R}$ and the indices $M-1 \leqq k \leqq N$, $1 \leqq i,j \leqq n$. Let us form the vector functions

$$\varphi^k = \begin{pmatrix} \varphi_1^k \\ \vdots \\ \varphi_n^k \end{pmatrix}.$$

Theorem3.7. *a) The matrix functions U_k, V_k of t are uniquely determined by*

$$\frac{d\varphi^k}{dt} = \frac{1}{\alpha-\lambda}V_k\varphi^k, \qquad M-1 \leqq k \leqq N, \tag{3.11a}$$

$$\varphi^k - \varphi^{k-1} = \frac{\kappa}{\kappa-\lambda}(U_k - 1)\varphi^{k-1}, \qquad M \leqq k \leqq N. \tag{3.11b}$$

b) The functions $\{U_k, V_k\}$ satisfy system (3.10) with the boundary matrix

$$C = (\delta_i^j h(R_i^\infty)).$$

Moreover $\mu_k^i = \kappa^{-1}\lambda(Q_k^i)$ and the eigenvalues of $\{V_k\}$ coincide with $\nu^i(t)$ from the definition of φ^k.

Proof. It is parallel to the proofs of Theorems 3.2, 3.3. (and goes even when the supports of the divisors and μ, ν are not generic). The formulas for μ result from relation (3.11b) at the points Q_k^i as φ^k tend to zero. The expression for C follows from the equation $\varphi^N = C^{-1}\varphi^{M-1}$. □

It can be shown that the construction of the theorem covers all "non-degenerate" solutions of system (3.10) modulo conjugation by constant matrices (with respect to t and k). The proof is based on the technique of the work [vMM] (cf. also [Kri2]).

Now, assuming that the fiber $(\lambda)_{\alpha/2}$ is simple and that it does not intersect all $\mathcal{D}_k$, let us consider the matrices $g_k = (\varphi_i^k(R_j^{\alpha/2})) \stackrel{\text{def}}{=} \Phi_{(\alpha/2)}^k$ (see Theorem3.3,a)). Then g_k is a solution of the *discrete principal chiral field* equation

$$\frac{dg_k}{dt}g_k^{-1}(g_k + g_{k-1}) = (g_{k-1} + g_k)g_{k-1}^{-1}\frac{dg_{k-1}}{dt}, \tag{3.12}$$

where $M \leqq k \leqq N$, $Cg_N = g_{M-1}C'$, $C' = (\delta_i^j h(R_i^{\alpha/2}))$. Relations (3.11), (3.10) imply that:

$$\frac{dg_k}{dt} = \frac{2}{\alpha}V_k, \qquad (g_k + g_{k-1})g_{k-1}^{-1} = \frac{\kappa}{\kappa - \alpha/2}\Big(U_k - \frac{\alpha}{\kappa}\Big).$$

We will also describe the procedure of integration of the "twice discrete" analogue of the PCF equation in the generic case. Given $\widetilde{\kappa} \neq \kappa$, $0 \neq \widetilde{\kappa} \neq \alpha$, $\widetilde{M} \leqq \widetilde{N} \in \mathbb{Z}$, and a set of divisors $\widetilde{S}_l = \sum_{i=1}^n \widetilde{Q}_l^i$ of degree n ($\widetilde{M} \leqq l \leqq \widetilde{N}$), let us define the divisors $\mathcal{D}_{k,l} = \mathcal{D}_k + \sum_{i=M}^{l}((\lambda)_{\widetilde{\kappa}} - \widetilde{S}_i)$, where $\mathcal{D}_k$ were introduced above, $\widetilde{M} - 1 \leqq l \leqq \widetilde{N}$, and $\sum_{l=\widetilde{M}}^{\widetilde{N}}(\widetilde{S}_l - (\lambda)_{\widetilde{\kappa}}) = (\widetilde{h})$ for a certain rational function $\widetilde{h}$ on Γ. We define a set of rational functions $\varphi_i^{k,l}$ and a vector function $\varphi^{k,l}$ by the relations $(\varphi_i^{k,l}) + \mathcal{D}_{k,l} \geqq 0$ and the same normalization condition as above.

Exercise3.9. a) *Matrices $U_{k,l}$, $\widetilde{U}_{k,l}$ that are uniquely defined by the relations*

$$\varphi^{k,l} - \varphi^{k-1,l} = \frac{\kappa}{\kappa - \lambda}(U_{k,l} - 1)\varphi^{k-1,l}, \qquad M \leqq k \leqq N,$$
$$\varphi^{k,l} - \varphi^{k,l-1} = \frac{\widetilde{\kappa}}{\widetilde{\kappa} - \lambda}(\widetilde{U}_{k,l} - 1)\varphi^{k-1,l}, \qquad \widetilde{M} \leqq k \leqq \widetilde{N},$$

satisfy the equations

$$U_{k,l}\widetilde{U}_{k-1,l} - \widetilde{U}_{k,l}U_{k,l-1} = U_{k,l} - U_{k,l-1} - \frac{\widetilde{\kappa}}{\kappa}(\widetilde{U}_{k,l} - \widetilde{U}_{k-1,l}),$$
$$\kappa(U_{k,l} - U_{k,l-1}) + \widetilde{\kappa}(\widetilde{U}_{k,l} - \widetilde{U}_{k-1,l}) = 0,$$

with the boundary conditions

$$C\widetilde{U}_{N,l} = \widetilde{U}_{M-1,l}C, \qquad \widetilde{C}U_{k,\widetilde{N}} = U_{k,\widetilde{M}-1}\widetilde{C},$$

for C, $\widetilde{C}$ defined with respect to h, $\widetilde{h}$ as in Theorem3.7.

b) *The function $g_{k,l} = \Phi_{(\xi)}^{k,l} = (\varphi_i^{k,l}(R_j^\xi))$ for $\xi = (\kappa + \widetilde{\kappa})/(\kappa^2 + \widetilde{\kappa}^2)$, $M-1 \leqq k \leqq N$, $\widetilde{M} - 1 \leqq l \leqq \widetilde{N}$ satisfies the equation*

$$g_{k,l}(g_{k,l-1}^{-1} + g_{k-1,l}^{-1}) = (g_{k,l-1} + g_{k-1,l})g_{k-1,l-1}^{-1} \tag{3.12}$$

with the boundary conditions $Cg_{N,l} = g_{M-1,l}C'$, $\widetilde{C}g_{k,\widetilde{N}} = g_{k,\widetilde{M}-1}\widetilde{C}'$ for $C' = (\delta_i^j h(R_i^\xi))$, $\widetilde{C}' = (\delta_i^j \widetilde{h}(R_i^\xi)$. The eigenvalues μ_k^i, $\widetilde{\mu}_l^i$ of the matrices $U_{k,l}$, $\widetilde{U}_{k,l}$ do not depend respectively on l and on k and are equal to $\kappa^{-1}\lambda(Q_k^i)$, $\widetilde{\kappa}^{-1}\lambda(\widetilde{Q}_k^i)$. □

The equation (3.12) is an analogue of the one suggested by A. M. Polyakov [Pol]. Another variant of the Polyakov lattice (the periodic nonabelian Toda lattice) was integrated by I. M. Krichever (see [Du2]). In conclusion we mention that there are quite a few papers devoted to the two-dimensional nonabelian Toda lattice including [LeS], [MOP], [DrS], [KW], [UT]. There are also a lot of results on integrating various concrete difference equations including those due to H. Flashka, D. MacLaughlin, S. V. Manakov, O. I. Bogoyavlensky, V. B. Matveev, M. A. Sall. E. Date.

Our aim here is to demonstrate the fundamental principle of integrating periodic lattices, which is the passage from the continuous parameters x, t connected with distributions on Γ to discrete parameters k, l corresponding to translations in the group $\operatorname{Jac}\Gamma$ by divisors of degree zero.

3.6. Comments.

The formula for the functions on Γ with exponential singularities and g poles from Proposition3.1 is, in fact, from [Ba] (similar functions also appeared in ealiear works). Those functions were intoroduced by G. Baker to simplify the results by J. Burchnall, T. Chaundy [BC], who studied pairs of commuting differential operators of one variable x of coprime orders. Later this problem appeared to be very important for the so-called "finite zoned integration". If one of these operator is of order 2 (and the order of the second is odd) the description of such pairs is equivalent to integrating the stationary ($\partial/\partial t \equiv 0$) higher KdV equations (§2.5). S. P. Novikov reduced the construction of the finite zoned solutions of the KdV equation to the last problem (see [N1], [Du1], [IM], [Mar], [Lax] and [McKvM]). In the context of the KdV and similar equations, the Baker (Baker-Akhiezer) functions played the main role in the works by S. P. Novikov, B. A. Dubrovin. A. R. Its, V. B. Matveev (see [DuM]). I. M. Krichever developped the technique of Baker functions and made it more universal. He applied it (see [Kri3]) to obtain algebraic-geometric solutions of the matrix Zakharov-Shabat equations (see §2.4 above). The results of J. Burchnall, T. Chaundy were generalized to differential operators whose orders are not coprime with each other ([Dr], [Kri4]), which gave new types of solutions of soliton equations (see [KN], [Mani], and the papers by S. P. Novikov and P. G. Grinevich).

In paper [Ch7] the construction of algebraic-geometric ("finite zoned") solutions of the PCF equations and of general equations with rational zero curvature representation (the Zakharov-Mikhailov systems) was made (including anti-hermitian solutions). The variety of anti-hermitian algebraic-geometric solutions of the PCF (and the general Zakharov-Mikhailov systems) was proved to be connected in [Ch10]. These solutions are always regular (which was demonstrated in the same paper). Concerning the results of §3.4, we note that the curves with complex involution were studied by E. Date, S. P. Novikov, I. M. Krichever, B. A. Dubrovin and other authors for various problems. One can find typical examples applications of the "finite zoned integration" in works [Du3], [Kri6], [N3].

The present exposition is based mostly on works [Ch1,7]. To study the reality conditions we reproved several results from the book by J. Fay [Fay] (to make the book self-contained and because Baker functions are very convenient for many problems concerning real curves and their Jacobians). The reality and regularity of algebraic-geometric solutions are connected with rather delicate (and sometimes insufficiently studied) problems of real algebraic geometry. In particular, it is important to describe the varieties of special divisors $\mathcal{D}$ of type (3.7–9) in the most general situation (when the forms are not definite).

In this section we do not touch upon the "completeness" of the constructed families of algebraic-geometric solutions. For this purpose one should define the latter as solutions of higher stationary equations of Lax type (i.e., as generalized "finite zoned" solutions). Actually this "abstract approach" is directly related to the classification of maximal subtori of proper linear groups over projective line $\mathbb{P}^1$ (see [Ch11]).

There are quite a few more specific questions. For example, we can show that the construction of §3.2 gives anti-hermitian U, V in §3.2 if and only if there exists a certain anti-involution σ on the corresponding curve and $\mathcal{D}$ and q satisfy property (3.7). Checking such statements is beyond the framework of this book.

§4. Algebraic-geometric solutions of Sin-Gordon, NS, etc.

In §4.1 we construct algebraic-geometric solutions of the Sin-Gordon equation, based on the results of §2.5 (that is, without using the results of §3). These solutions are then lifted to S^2-fields. In §4.2, without consulting the results of §3, we describe algebraic-geometric solutions of the VNS equation, and analyze the NS equation in detail. §4.3 is devoted to comparing the constructions in §3 and §§4.1,2. In §4.4, as an application, we present a method of constructing algebraic-geometric solutions of the four-dimensional duality equation (the self dual Yang-Mills fields).

In the first two sections we use the definition of the Baker function and Proposition3.3 from the material of §3. The rest of the section is independent of §3 except for the proof of Theorem4.4, b) which is based on the results of §3.3.

4.1. Sin-Gordon equation, S^2-fields.

Let Γ be a hyperelliptic curve given by the equation

$$\mu^2 = \lambda \prod_{j=1}^{2g} (\lambda - \alpha_j), \tag{4.1}$$

for pairwise distinct $0 \neq \alpha_j \in \mathbb{C}$. More precisely, the curve Γ is the totality of points $(\lambda, \mu) \in \mathbb{C}^2$ satisfying (4.1) as a set, completed by the point $O_\infty \stackrel{\text{def}}{=} (\lambda = \infty, \mu/\lambda = \infty)$. The embedding $\Gamma \setminus O_\infty \subset \mathbb{C}^2$ gives the structure of a smooth (check it) submanifold on $\Gamma \setminus O_\infty$. The analytic structure at O_∞ is introduced, given a local parameter k_∞ in a (punctured) neighbourhood of O_∞, by the relations $\lambda^{-1} = k_\infty^2$, $\mu^{-1} = k_\infty^{2g+1} + O(k_\infty^{2g+2})$. Here $O(k_\infty^{2g+2})$ is a power series in k_∞^j, $j \geqq 2g+2$ and the second relation is necessary only to fix the sign $k_\infty = \lambda^{-1/2}$. The field of rational functions $\mathbb{C}(\Gamma)$ on Γ is generated by λ, μ.

By the formulae ${}^\tau\lambda = \lambda$, ${}^\tau\mu = -\mu$ we can define an involutive automorphism τ on $\mathbb{C}(\Gamma)$, which extends to the involution of points of Γ by the relations (3.6). (The coordinate λ of an arbitrary point $P \in \Gamma$ does not change under the action of τ, while the μ-coordinate, $\mu(P)$, is multiplied by -1.) The action of τ on functions, points, divisors and distributions of Γ is denoted by ${}^\tau(\cdot)$. The branching points of the mapping $\widetilde{\lambda} : \Gamma \to \mathbb{P}^1$ induced by λ (i.e., fixed points of τ) are the points O_∞, $O_1, \dots, O_{2g}, O_0$, defined by $\lambda(O_j) = \alpha_j$, $\lambda(O_0) = 0$. In a neighbourhood of O_0 we choose the local coordinate $k_0 = \mu/\lambda$. The involution τ acts on the parameters k_∞, k_0 by multiplication of -1. Following the notation of §3.3, let $\widetilde{\mathrm{Tr}}$ be the trace of $\widetilde{\lambda}$ and let $T = O_\infty + O_0 + \sum_{j=1}^{2g} O_j$ be the branching divisor of $\widetilde{\lambda}$. By the Hurwitz formula the genus of Γ is equal to $\deg T/2 - \deg \widetilde{\lambda} + 1$, and thus equal to g.

Let us assume that the coefficients of the polynomial $\prod_{j=1}^{2g}(\lambda - \alpha_j)$ are real. Then we can introduce an anti-involution σ on Γ, putting

$$\lambda^\sigma = \lambda, \qquad \mu^\sigma = \mu.$$

Under the action of σ each point $P \in \Gamma$ with the coordinates (λ, μ) maps to the point with the coordinates $(\overline{\lambda}, \overline{\mu})$. The points O_∞, O_0 are real (i.e., σ-invariant) and $k_\infty^\sigma = k_\infty$, $k_0^\sigma = k_0$. As a consequence of λ being real, the set of points $\{O_j\}$ is mapped to itself under the action of σ.

We call divisors $\mathcal{D} \geqq 0$ on Γ of degree g *admissible* if they satisfy the condition

$$\mathcal{D}^\sigma + \mathcal{D} \sim \sum_{j=1}^{2g} O_j. \tag{4.2}$$

For each such $\mathcal{D}$ we can set $\mathcal{D}^\sigma + \mathcal{D} - \sum_{j=1}^{2g} O_j = (v)$, where v is a suitable real ($v^\sigma = v$) rational function on Γ uniquely written in the form

$$v = c_\infty + \mu^{-1} \sum_{j=1}^{g} c_j \lambda^j, \tag{4.3}$$

where $c_\infty = v(O_\infty) \in \mathbb{R}$, $c_j \in \mathbb{R}$.

Theorem4.1. a) *A necessary and sufficient condition for the existence of admissible divisors on Γ is that all real numbers among $\{\alpha_1, \dots, \alpha_{2g}\}$ are negative.*

b) *Let us suppose that there exist exactly $2m$ real numbers among $\{\alpha_j\}$, $1 \leqq j \leqq g$ (their number should be even, since remaining α_j must be pairwise conjugate), and all of them are negative. Then the variety of classes of admissible divisors $\mathcal{D}$ is isomorphic to a disjoint union of 2^m copies of $(S^1)^g$.*

Proof. We suppose that there exists an admissible divisor $\mathcal{D}$. Then for a function v corresponding to $\mathcal{D}$, applying τ to (4.3), we obtain the relation

$$c_\infty^2 - {}^\tau v\, v = \lambda\Big(\sum_{j=1}^{g} c_j \lambda^{j-1}\Big)^2 \prod_{j=1}^{2g} (\lambda - \alpha_j)^{-1}. \tag{4.4}$$

From Proposition3.3b) it follows that classes of admissible divisors form a g-dimensional real variety. There exist only a finite number ($\leqq 2^{g-m}$) of admissible divisors with the same function v. Thus, there exists an admissible divisor $\mathcal{D}$ for which $c_\infty \neq 0$ and zeros of v do not have multiplicities (in particular, $\alpha_j \in \mathbb{R} \Rightarrow O_j \notin \operatorname{supp} \mathcal{D}, \mathcal{D}^\sigma$). Define P_j by $\mathcal{D} = \sum_{j=1}^{g} P_j$ and put $\lambda(P_j) = \delta_j$. Then

$$^\tau v\, v = c_\infty^2 \prod_{j=1}^{g} (\lambda - \delta_j)(\lambda - \overline{\delta}_j) \mu^{-2} \lambda. \tag{4.5}$$

Comparing this with (4.4), we have the identity

$$\prod_{j=1}^{2g}(\lambda-\alpha_j)-\lambda(\sum_{j=1}^{g}c_j\lambda^{g-1})^2c_\infty^{-2}=\prod_{j=1}^{g}(\lambda-\delta_j)(\lambda-\bar{\delta}_j),$$

and calculation of the constant term of both sides gives us the relation

$$\prod_{j=1}^{2g}\alpha_j=\prod_{j=1}^{g}\delta_j\bar{\delta}_j. \tag{4.6}$$

In particular $\prod_{j=1}^{2g}\alpha_j>0$, and consequently in a small enough neighbourhood of $\lambda=0$, the right hand side of (4.4) is positive under the condition $\lambda>0$. If $\alpha_1=\min\{\alpha_j,\mathbb{R}\ni\alpha_j>0\}$, $1\leqq j\leqq g$, then in the interval $0<\lambda<\alpha_1$ the right hand side of (4.4) takes all values from 0 to ∞ and, among others, c_∞^2. Then ${}^\tau v\,v(\lambda')=0$ for some $0<\lambda'<\alpha_1$ and both points on Γ with coordinate $\lambda=\lambda'$ must be real $(\lambda'\prod_{j=1}^{2g}(\lambda'-\alpha_j)>0)$. One of the points is necessarily in the zero divisor of v and since this has the form $\mathcal{D}+\mathcal{D}^\sigma$, this point must have multiplicity $\geqq 2$. But we chose $\mathcal{D}$ such that all zeros of v are simple. This is a contradiction.

Now, conversely, let us suppose that, if $\alpha_j\in\mathbb{R}$, then $\alpha_j<0$, $1\leqq j\leqq 2g$. In order to prove the existence of an admissible $\mathcal{D}$ it is enough to check that the class of the divisor $O_\infty-O_0$ belongs to the connected component of zero $(\mathcal{J}_\sigma^0)_0$ of the group $\mathcal{J}_\sigma^0$ of real (σ-invariant) points of $\operatorname{Jac}\Gamma$. In fact, the group $\mathcal{J}_\sigma^0\simeq(S^1)^g$ is divisible. Therefore we can put $O_\infty-O_0\sim 2Q$, where the class Q belongs to $\mathcal{J}_\sigma^0$ and then the divisor $gO_\infty+Q=\mathcal{D}$ is equivalent to an admissible divisor. The fact that the class $O_\infty-O_0$ belongs to the group $(\mathcal{J}_\sigma^0)_0$ follows from the fact that the points O_∞ and O_0 are contained in one real oval $\{(\lambda,\mu)\in\Gamma_{\mathbb{R}},0\leqq\lambda\leqq+\infty\}$ of the curve Γ.

To calculate the number of components of $\widehat{\mathcal{J}}_\sigma^0$ (cf. Proposition3.3) we count the number of real points of second order of $\operatorname{Jac}\Gamma$. That α_j is negative is not used here. An arbitrary point of second order of $\operatorname{Jac}\Gamma$ is uniquely represented by a class of divisors $o_L\overset{\text{def}}{=}\sum_{j=1}^{s}O_{l_j}-sO_\infty$, where $L=\{l_1,\dots,l_s\}\subset\{1,\dots,2g\}$, $1\leqq s\leqq 2g$, $l_i\neq l_j$ for $i\neq j$ (check this as an exercise). Conversely, the classes of divisors o_L of the sets $L=\{l_1,\dots,l_s\}$ containing $l_{j'}$ and l_j such that $O_{l_j}^\sigma=O_{l_{j'}}$ correspond to the real points. Hence, the number of such points is equal to $2^{2m+(g-m)}=2^{m+g}$, where $2m$ is the number of real points O_j, $1\leqq j\leqq 2g$. Consequently, $\widehat{\mathcal{J}}_\sigma^0$ consists of 2^m connected components. □

Corollary4.1. *a) For an arbitrary admissible divisor $\mathcal{D}$, divisors $\mathcal{D}-O_0$ and $\mathcal{D}-O_\infty$ are not special, i.e., $\mathcal{D}=\sum_{j=1}^{g}P_j$, where all $\lambda(P_j)$ are pairwise distinct and not equal to 0, ∞.*

b) Let $\mathcal{D}$ be admissible, $\lambda(P) = z \in \mathbb{R}$ for some point $P \in \Gamma$ and $\{\alpha_j\}$ be numbered so that $\{\alpha_j, 1 \leqq j \leqq 2m\} = \{\alpha_j\} \cap \mathbb{R}$ and $0 > \alpha_1 > \alpha_2 > \ldots > \alpha_{2m}$. Then for $0 \leqq i < m$,

$$\alpha_{2i+1} > z > \alpha_{2i+2} \Rightarrow P^\sigma = P, \quad v(P)v({}^\tau P) \leqq 0,$$
$$z < \alpha_{2m} \text{ or } \alpha_{2i} > z > \alpha_{2i+1} \Rightarrow P^\sigma = {}^\tau P, \quad v(P)v({}^\tau P) \geqq 0,$$
$$z > \alpha_1 \Rightarrow P^\sigma = P, \quad v(P)v({}^\tau P) \geqq 0.$$

Proof. It is necessary to prove that $v(O_\infty) = c_\infty \neq 0$ for an admissible divisor $\mathcal{D}$ ($v(O_0) = v(O_\infty)$ by virtue of (4.3)). If $c_\infty = -$, then, as follows from (4.4), $\mathcal{D}+\mathcal{D}^\sigma+{}^\tau\mathcal{D}+{}^\tau\mathcal{D}^\sigma$ contains the point O_∞ with multiplicity $2(2l+1)$ for a certain integer $l \geqq 0$. This is, however, impossible, since the point O_∞ appears in each divisor $\mathcal{D}$, $\mathcal{D}^\sigma$, ${}^\tau\mathcal{D}$, ${}^\tau\mathcal{D}^\sigma$ with the same multiplicity. Statement b) is obtained by direct analysis of formula (4.4). As $\lambda \in \mathbb{R}$ varies from $+\infty$ to $-\infty$, the functions ${}^\tau v\, v(P)$ and $\mu^2(P)$ change sign at the moment when λ goes over each point $\alpha_1, \ldots, \alpha_{2m}$ (μ^2 changes sign also when λ goes over O_0 and O_∞). The zeros of ${}^\tau v\, v$ for real λ may have only even multiplicities. Therefore, for $z < \alpha_{2m}$, $z > \alpha_1$, and also in an arbitrary interval $\alpha_{j+1} < z < \alpha_j$, the function ${}^\tau v\, v(P)$ does not change sign. □

Solutions of the Sin-Gordon equation. Let $\mathcal{D} \geqq 0$ be admissible, v the rational function on Γ corresponding to the divisor $\mathcal{D}$ (cf. above). Turning to the construction of algebraic-geometric solutions, we fix a distribution $q = ik_\infty x + ik_0 t$, $x, t \in \mathbb{R}$. As $q^\sigma = -q$, $\mathcal{D}+Q$ for $Q \in \widetilde{q}$ (cf. §3.1) is equivalent to a certain admissible divisor (which depends on x, t). In particular, divisors $\mathcal{D}+Q-O_0$ and $\mathcal{D}+Q-O_\infty$ are not special for any x, t (Collorary4.1) and we can always choose a generator ψ in the one-dimensional space $H^0(\mathcal{D}; q)$ which is normalized by $\psi \exp(-q)(O_\infty) = 1$. In addition, necessarily, $\psi \exp(-q)(O_0) \neq 0$.

Proposition4.1. *For $P \notin \{O_j, 1 \leqq j \leqq 2g\}$*

$$v\psi\psi^\sigma(P) + v\psi\psi^\sigma({}^\tau P) = v(P) + v({}^\tau P) = 2v(O_\infty), \tag{4.7}$$
$$\psi\psi^\sigma(O_0) = 1. \tag{4.8}$$

Proof. Since $q^\sigma = -q$, the function $\psi\psi^\sigma$ is rational on Γ and

$$v, v\psi\psi^\sigma \in H^0(\sum_{j=1}^{2g} O_j) \subset H^0(T).$$

By virtue of the fundamental property of the branching divisor T, the functions $\mathrm{Tr}\, v$, $\mathrm{Tr}(v\psi\psi^\sigma)$ do not depend on λ. This implies (4.7) (moreover, the equality

$v + {}^{\tau}v = 2v(O_\infty)$ is an obvious consequence of (4.3)). Substituting $P = O_0$ in (4.7), we obtain (4.8). □

Put, as above, $\mathcal{D} = \sum_{j=1}^{g} P_j$, $\lambda(P_j) = \delta_j$. Set $\prod_{j=1}^{2g} \alpha_j = c^2$, $c \in \mathbb{R}$, $\prod_{j=1}^{g} \delta_j = \Delta$. Then $c^2 = \Delta\overline{\Delta}$ (cf. (4.6)). We introduce a function

$$a(x,t) = (-1)^g \Delta\, {}^{\tau}\psi\psi(O_0),$$

where ${}^{\tau}\psi\psi(O_0) = \exp(-2q)\psi^2(O_0)$.

From (4.8) it follows that $a\bar{a} = c^2$.

Theorem4.2. *The quasi-periodic real-valued function*

$$\alpha(x,t) = i\log(c^{-1}a(x,(c^{-1}/4)t)) \tag{4.9}$$

defined for all $x, t \in \mathbb{R}$ (modulo 2π) satisfies the Sin-Gordon equation (0.8): $\alpha_{xt} + \sin\alpha = 0$.

Proof. If we put $\psi\exp(-q) = 1 + \psi_1 k_\infty + O(k_\infty^2)$ in a neighbourhood of O_∞, then $\psi_{xx} = (u-\lambda)\psi$ for $u = 2i(\psi_1)_x$. In fact, $\psi_{xx} - (u-\lambda)\psi \in H^0(\mathcal{D} - O_\infty; q) = \{0\}$. The function ${}^{\tau}\psi\psi_t w$ for $w = \lambda\mu^{-1}\prod_{j=1}^{g}(\lambda - \delta_j)$ belongs to $H^0(T)$. Thus, the trace $\mathrm{Tr}({}^{\tau}\psi\psi_t w)$ does not depend on λ. Computing the last function at $\lambda = \infty$ and $\lambda = 0$, we obtain the relation $(\psi_1)_t = ia$. Eliminating ψ_1, we get the equation $u_t + 2a_x = 0$. On the other hand, substituting $\lambda = 0$ in the formula for ψ_{xx}, we find that

$$\begin{aligned} u &= (\log\psi)_{xx} + ((\log\psi)_x)^2 \\ &= \tfrac{1}{2}(\log a)_{xx} + \tfrac{1}{4}((\log a)_x)^2 \\ &= -\frac{i}{2}\widetilde{\alpha}_{xx} - \frac{1}{4}(\widetilde{\alpha}_x)^2 \end{aligned}$$

for $\widetilde{\alpha} = i\log a$. Rewriting u and a via $\alpha(x,t) = -i\log c + \widetilde{\alpha}(x,(c^{-1}/4)t)$ (cf. §2 (2.25)), we have

$$\left(\frac{i}{2}\alpha_{xx} + \frac{1}{4}(\alpha_x)^2\right)_t = \frac{1}{2}(e^{-i\alpha})_x. \tag{4.10}$$

Using the fact that α is real, we sum (4.10) and its complex conjugate. Canceling α_x on both sides of the resulting equation, we arrive at the statement of the theorem. □

Exercise4.1. *Let $\mathcal{D}, \widehat{\mathcal{D}} \geqq 0$ be divisors of degree g which do not contain O_0, O_∞ on the curve Γ given by equation (4.1) (this time, without the reality condition on the coefficients). Under the assumption of non-speciality of $\mathcal{D}+Q-O_{0,\infty}$, construct Baker functions ψ, $\widehat{\psi}$ for $\mathcal{D}$ and $\widehat{\mathcal{D}} \sim \mathcal{D}+O_\infty - O_0$, as above. Replacing ψ^σ by ${}^\tau\widehat{\psi}$, prove equalities (4.7), (4.8). Check that functions α, $\widehat{\alpha}$ corresponding to $\mathcal{D}$, $\widehat{\mathcal{D}}$ (cf. (4.9)) are related by the relation $\alpha = -\widehat{\alpha}$ and each satisfy each (4.10). Summing (4.10) for α and the analogous equation for $\widehat{\alpha} = -\alpha$, show that α is a solution of the (complex) Sin-Gordon equation.* □

Exercise4.2. *Under the assumptions of Theorem4.2 (or Exercise4.1) show (cf. [Ch4]) that*

a)

$$a = (-1)^g \prod_{j=1}^{g} \gamma_j, \qquad {}^\tau\psi\psi = \prod_{j=1}^{g} \frac{\lambda - \gamma_j}{\lambda - \delta_j},$$

where $\gamma_j = \lambda(Q_j)$ and $\{Q_j\}$ are zeros of ψ (i.e., $(\psi) = \sum_{j=1}^g Q_j - \mathcal{D}$);

b)

$$a = (-1)^g \Delta \Big[\frac{\theta(w(\mathcal{D}) + v_q - e)\theta(w(\mathcal{D}))}{\theta(w(\mathcal{D}) - e)\theta(w(\mathcal{D}) + v_q)}\Big]^2,$$

where $P_0 = O_\infty$, $e = w(O_0)$, and the cycles $\{a_j\}$ are chosen τ-invariant (cf. §3.1). (Note that substitution of such a into (4.9) results in the formula in [KoK].) □

Lifting α to S^2-fields. We begin with the computation of two Wronskians. Define

$$\{f, g\}_x = f_x g - f g_x, \qquad \{f, g\}_t = f_t g - f g_t$$

for functions f, g of x, t. We keep all the previous conventions (though the reality condition is not used in the following proposition).

Proposition4.2. *a)* $\{\psi, {}^\tau\psi\}_x = 2i\mu \prod_{j=1}^g (\lambda - \delta_j)^{-1}$;
b) $\{\psi, {}^\tau\psi\}_t = 2i\mu\lambda^{-1} a \prod_{j=1}^g (\lambda - \delta_j)^{-1}$;

Proof. The principal part of $\{\psi, {}^\tau\psi\}_x$ at the point O_∞ is equal to $2ik_\infty^{-1}$. For generic x, t, the poles of $\psi_x\psi^{-1}$ at the point O_∞ coincide with the zeros of ψ. Therefore, the poles of $f = \psi_x\psi^{-1} - {}^\tau(\psi_x\psi^{-1})$ are contained in the set of zeros of ${}^\tau\psi\psi$. Since $\{O_j\}$ are fixed points of τ, all O_j, $1 \leqq j \leqq g$, must be contained in the set of zeros of f. There are no other zeros of f (we already know that the number of poles of f does not exceed $2g+1$). Therefore for general enough x, t the divisor $(\{\psi, {}^\tau\psi\}_x)$ of the function $\{\psi, {}^\tau\psi\}_x = {}^\tau\psi\psi f$ is equal to $\sum_{j=0}^{2g} O_j - \mathcal{D} - {}^\tau\mathcal{D} - O_\infty$. The condition that x, t are general enough can be omitted, since the last divisor is independent of x, t. Thus we get formula a). Analogously we obtain that

$(\{\psi, {}^{\tau}\psi\}_t) = \sum_{j=1}^{2g} O_j + O_\infty - \mathcal{D} - {}^{\tau}\mathcal{D} - O_0$ and at the point O_0 the principal part of $\{\psi, {}^{\tau}\psi\}_t$ is equal to $2ik_0^{-1}\, {}^{\tau}\psi\psi(O_0)$. Recalling the definition of k_0, we obtain b). □

Now for points $P \in \Gamma_{\mathbb{R}}$ ($\Leftrightarrow P^\sigma = P$), assuming that $v(P)\,{}^{\tau}v(P) > 0$ (cf. Corollary4.1), we put

$$\tfrac{1}{2}v(P)v(O_\infty)^{-1} = \beta_1^2, \qquad \tfrac{1}{2}v({}^{\tau}P)v(O_\infty)^{-1} = \beta_2^2,$$

where $\beta_1, \beta_2 \in \mathbb{R}$ and $\beta_1^2 + \beta_2^2 = 1$. This is possible since $v(P) + v({}^{\tau}P) = 2v(O_\infty)$ and $v(P)$, $v({}^{\tau}P)$, $v(O_\infty)$ must have the same sign. Set $\xi_i \stackrel{\text{def}}{=} \beta_1\psi(P)$, $\xi_2 \stackrel{\text{def}}{=} \beta_2\overline{\psi({}^{\tau}P)}$. Then $\overline{\xi}_1 = \beta_1\psi^\sigma(P)$, $\overline{\xi}_2 = \beta_2\psi({}^{\tau}P)$ and

$$\xi_1\overline{\xi}_1 + \xi\overline{\xi}_2 = 1 \tag{4.11}$$

by (4.7). Put

$$s_1 = \xi_1\xi_2 + \overline{\xi}_1\overline{\xi}_2, \quad s_2 = i(\xi_1\xi_2 - \overline{\xi}_1\overline{\xi}_2), \quad s_3 = \xi_1\overline{\xi}_1 - \xi_2\overline{\xi}_2,$$

$$\boldsymbol{s} = \begin{pmatrix} s_1 \\ s_2 \\ s_3 \end{pmatrix} \in \mathbb{R}^3.$$

Hereafter $(\,,\,)$ denotes the Euclidean scalar product, c, Δ the same as in Theorem4.2.

Theorem4.3. *a)* $(\boldsymbol{s}, \boldsymbol{s}) = 1$,
b) $(\boldsymbol{s}_x, \boldsymbol{s}_x) = 4\lambda(P)$,
c) $(\boldsymbol{s}_t, \boldsymbol{s}_t) = 4\lambda(P)^{-1}c^2$.

Proof. Statement a) is derived from (4.11) by direct calculation (which, by the way, establishes the local isomorphism of O_4 and $O_3 \times O_3$). Differentiating (4.11), we obtain the identities

$$(\boldsymbol{s}_x, \boldsymbol{s}_x) = 4\{\xi_1, \overline{\xi}_2\}_x\{\overline{\xi}_1, \xi_2\}_x, \tag{4.12a}$$

$$(\boldsymbol{s}_t, \boldsymbol{s}_t) = 4\{\xi_1, \overline{\xi}_2\}_t\{\overline{\xi}_1, \xi_2\}_t. \tag{4.12b}$$

For example, for the proof of (4.12), we use

$$((s_1)_x)^2 + ((s_2)_x)^2 = 4(\xi_1\xi_2)_x(\overline{\xi}_1\overline{\xi}_2)_x,$$
$$((s_3)_x)^2 = -4(\xi_1\overline{\xi}_1)x(\xi_2\overline{\xi}_2)_x.$$

Summing these identities and making simple transformations, we arrive at (4.12a). Replacing $\psi(P) = \beta_1^{-1}\xi_1$, ${}^{\tau}\psi(P) = \beta_2^{-1}\overline{\xi}_2$ in the formulae of Proposition4.2, we have

$$\{\xi_1, \overline{\xi}_2\}_x = 2i\beta_1\beta_2\mu(P)\prod_{j=1}^{g}(\lambda(P)-\delta_j)^{-1}, \tag{4.13a}$$

$$\{\xi_1, \overline{\xi}_2\}_t = \lambda(P)a\{\xi_1, \overline{\xi}_2\}_x. \tag{4.13b}$$

Substituting the above relations into (4.12), we obtain

(4.14a)

$$(\boldsymbol{s}_x, \boldsymbol{s}_x) = 16(\beta_1\beta_2)^2\mu(P)^2\prod_{j=1}^{g}(\lambda(P)-\delta_j)^{-1}(\lambda(P)-\overline{\delta}_j)^{-1},$$

(4.14b)

$$(\boldsymbol{s}_t, \boldsymbol{s}_t) = \lambda(P)^{-2}c^2(\boldsymbol{s}_x, \boldsymbol{s}_x).$$

For the derivation of (4.14b) we used the formulae (4.8) and (4.6). From (4.5) follows the relation

$$\begin{aligned}(\beta_1\beta_2)^2 &= \tfrac{1}{4}v(P)v({}^{\tau}P)v(\infty)^{-2} = \\ &= \prod_{j=1}^{g}(\lambda(P)-\delta_j)(\lambda(P)-\overline{\delta}_j)\mu(P)^{-2}\lambda(P),\end{aligned}$$

which gives statements b) and c) from (4.14). □

The vector function $\boldsymbol{s}$ satisfies the equations of the S^2-field (0.5) (cf. Introduction). For the S^2-field $\widetilde{\boldsymbol{s}}(x,t) = \boldsymbol{s}(x, (c^{-1}/4)t)$ the more symmetric relations

$$(\widetilde{\boldsymbol{s}}_x, \widetilde{\boldsymbol{s}}_x) = 4\lambda(P), \qquad (\widetilde{\boldsymbol{s}}_t, \widetilde{\boldsymbol{s}}_t) = (4\lambda(P))^{-1}$$

hold. By the coordinate change $x \mapsto (4\lambda(P))^{-1}x$, $t \mapsto (4\lambda(P))^{-1}t$ the field $\widetilde{\boldsymbol{s}}$ turns into an S^2-field in the normalized coordinates (cf. (0.6)). Thus, the function $\arccos(\widetilde{\boldsymbol{s}}_x, \widetilde{\boldsymbol{s}}_x)$ satisfies the Sin-Gordon equation. It is easy to check (cf. (4.12)) the formula

$$(\boldsymbol{s}_x, \boldsymbol{s}_t) = 2\{\xi_1, \overline{\xi}_2\}_x\{\overline{\xi}_1, \xi_2\}_t + 2\{\overline{\xi}_1, \xi_2\}_x\{\xi_1, \overline{\xi}_2\}_t.$$

Replacing $\{\xi_1, \overline{\xi}_2\}_t$ by means of (4.13b) and using Theorem4.3, we obtain the relation $(\boldsymbol{s}_x, \boldsymbol{s}_t) = 2(a+\overline{a})$, from which it follows that

$$\begin{aligned}(\widetilde{\boldsymbol{s}}_x, \widetilde{\boldsymbol{s}}_t) &= (c^{-1}/4)2(a+\overline{a})(x, (c^{-1}/4)t) \\ &= \cos\alpha.\end{aligned}$$

In this way, $\widetilde{\boldsymbol{s}}(x,t)$ *turns out to be a lifting of* $\alpha(x,t)$ *to an* S^2*-field*, and $\lambda(P) > 0$ is a parameter reflecting the non-uniqueness of this lifting. In Ch.II (§1.2) we construct a family of transformations of the PCF, which allow us, in particular, to calculate the S^2-field $\boldsymbol{s}$ for arbitrary (admissible) points P in terms of the field $\boldsymbol{s}$ constructed for one such $\lambda(P)$ by a certain general method which works for any S^{n-1}-fields.

Exercise4.3. *In the notation of Theorem4.3, for* $\lambda(P) < 0$ $(P \in \Gamma_{\mathbb{R}})$ *and for* $v(P)v(O_\infty) > 0$, $s_3 \overset{\text{def}}{=} \xi_1\overline{\xi}_1 + \xi_2\overline{\xi}_2$, $\frac{1}{2}v({}^\tau P)v(O_\infty)^{-1} \overset{\text{def}}{=} -\beta_2^2$ $(\beta_2 \in \mathbb{R})$, *show that*

$$(\boldsymbol{s},\boldsymbol{s})_* = 1,$$
$$(\boldsymbol{s}_x,\boldsymbol{s}_x)_* = 4\lambda(P), (\boldsymbol{s}_t,\boldsymbol{s}_t)_* = 4(\lambda(P))^{-1}c^2,$$

where $(\boldsymbol{s},\boldsymbol{s})_* \overset{\text{def}}{=} s_3^2 - s_2^2 - s_1^2$. □

Theorem4.3 and Exercise4.3 make it possible to construct algebraic-geometric Chebyshev nets (cf. Introduction and [H]) on the sphere S^2 and on the Lobachevskii plane. The Chebyshev nets for the latter arise as asymptotic lines for isometric regular embeddings of parts of the Lobachevskii plane into the Euclidean space E^3. The impossibility of such an embedding of all the Lobachevskii plane was proven by D. Hilbert (see [Tc]) as a corollary to the absence of solutions for all $x, t \in \mathbb{R}$ of the equation $\widetilde{\alpha}_{xt} = \sin\widetilde{\alpha}$ for the net angle $\widetilde{\alpha}$ with the natural restriction $0 < \widetilde{\alpha} < \pi$ (which reflects the regularity). The result of D. Hilbert (after the change $t \mapsto -t$) leads to the following claim: for an arbitrary divisor $\mathcal{D}$ of type (4.2), there are $\dot{x}, \dot{t} \in \mathbb{R}$ such that the divisor $\sum_{j=1}^{g} \dot{Q}_j$ (the zero divisor of $\psi(\dot{x},\dot{t})$) which is equivalent to $\mathcal{D} + \dot{Q}$ $(\dot{Q} \in \widetilde{q}(\dot{x},\dot{t}))$ satisfies the relation $(\prod_{j=1}^{g} \dot{\gamma}_j)^2 = c^2$, where $\dot{\gamma}_j = \lambda(\dot{Q}_j)$ (cf. Exercise4.2, a)). Apparently, this is not the only geometric fact possessing an interesting algebraic interpretation for solutions of equations (0.8) and (0.5) constructed in Theorem4.2, 4.3 (see also Exercise4.3).

4.2. VNS equation.

Let Γ be a curve with an anti-involution σ, λ a real function on Γ of degree n. We keep the notation of §§3.2,3. For simplicity we assume that $(\lambda)_\infty = \sum_{j=1}^{n} R_j$ is a simple fiber with real R_j. Pick out the point R_n and put $q = ik_n^{-1}x + ik_n^{-2}t$, where $k_n = \lambda^{-1}$ is the local parameter at R_n. Let $\mathcal{D} \geqq 0$ be a divisor of degree g on Γ not containing points of $\{R_j\}$ which satisfies the condition

$$\mathcal{D} + \mathcal{D}^\sigma \sim T - 2\sum_{j=1}^{n} R_j. \tag{4.15}$$

As in the previous section, (4.15) is consistent with the obvious relation $q^\sigma + q = 0$. Put $\mathcal{D} + \mathcal{D}^\sigma - T + 2\sum_{j=1}^n R_j = (w)$ for a rational function w normalized by the relation $w(R_n) = 1$. Define a real matrix $\Omega = \operatorname{diag}(\lambda^2 w(R_1), \ldots, \lambda^2 w(R_{n-1}))$ of order $n-1$. If $\mathcal{D} + Q - R_n$ is not special (as usual, $Q \in \widetilde{q}$), then $H^0(\mathcal{D}; q)$ is generated by the Baker function normalized by $\exp(-q)\psi(R_n) = 1$.

Theorem4.4. *a) Put* $r_j = \psi(R_j)$, $1 \leqq j \leqq n-1$, $\boldsymbol{r} = {}^t(r_j)$. *Then*

$$i\boldsymbol{r}_t = \boldsymbol{r}_{xx} + 2(\Omega \boldsymbol{r}, \overline{\boldsymbol{r}})\boldsymbol{r}, \tag{4.16}$$

where $(\,,)$ *is the Euclidean scalar product.*

b) There exists a divisor $\mathcal{D}$ *on* Γ *with condition* $\Omega > 0$, *if and only if* $\widetilde{\lambda}$ *does not have real branching points. The variety of classes of divisors* $\mathcal{D}$ *with* $\Omega > 0$ *is then connected and isomorphic to* $(S^1)^g$. *The functions* $\boldsymbol{r}(x,t)$ *corresponding to such* $\mathcal{D}$ *are regular and bounded for all* $x, t \in \mathbb{R}$.

Proof. As in the proof of Theorem4.2, we find that $\psi_{xx} - (u - k^{-2})\psi$ for $u = 2i(\psi_1)_x$ tends to zero in a neighbourhood of R_n once divided by $\exp(q)$, where $\exp(q)\psi \stackrel{\text{def}}{=} 1 + \psi_1 k_n + O(k_n^2)$. Consequently, $\psi_{xx} - u\psi - i\psi_t \in H^0(\mathcal{D} - R_n; q) = \{0\}$ and

$$i\psi_t = \psi_{xx} - u\psi. \tag{4.17}$$

Now we express u in terms of the coefficients of the expansion of the rational function $\psi\psi^\sigma$ in a neighbourhood of R_n. The differential form $w\psi_x\psi^\sigma d\lambda$ has a pole only at the point R_n. Hence $\operatorname{Res}_{R_n}(w\psi_x\psi^\sigma d\lambda) = 0$. Computing the residue, we get the relation

$$\operatorname{Res}_{R_n}(wi\lambda\psi\psi^\sigma d\lambda) - (\psi_1)_x = 0,$$

which implies $u = -2\operatorname{Res}_{R_n}(\lambda w\psi\psi^\sigma d\lambda)$. The form $\lambda w\psi\psi^\sigma d\lambda$ has poles only at $\{R_j\}$. Therefore, by the residue formula,

$$u = \sum_{j=1}^{n-1} \operatorname{Res}_{R_j}(\lambda w\psi\psi^\sigma d\lambda) = -2(\Omega\boldsymbol{r}, \overline{\boldsymbol{r}}).$$

Substituting the obtained expression for u and values of ψ at points $R_1, \ldots, R_{n-1}$ (instead of ψ) into (4.17), we have (4.16). Statement b) is directly derived from Proposition3.4 and Theorem3.5,6, because relation (4.15) for $\mathcal{D}$ is equivalent to condition (3.7) for $\mathcal{D} + \sum_{j=1}^{n-1} R_j$. □

It is easy to formulate and prove the theorem for divisors $\mathcal{D} \geqq 0$ containing R_j, $1 \leqq j \leqq n-1$ (but not R_n). Note that Theorem4.4 can be derived (for $(\lambda)_1$ instead of $(\lambda)_\infty$) from Theorem3.3 by the reduction procedure from the GHM to the VNS (cf. Introduction, Exercise2.1 in §2.1 and below §4.3). The direct proof mentioned becomes conceptually more transparent if we use the dual Jost functions (§2):

Exercise4.4. *a) For* Γ *with a marked point* P_0 *and a divisor* $\mathcal{D}$ *of degree* g *we put* $\widehat{\mathcal{D}} = \mathcal{K} - \mathcal{D} + 2P_0$, *where* $\mathcal{K}$ *belongs to the canonical class of* Γ. *Let* $q = k_0^{-1}x$ *(*k_0 *is a local parameter at* P_0*),* $\psi \in H^0(\mathcal{D}; q)$ *be normalized by* $\psi \exp(-q)(P_0) = 1$ *under the assumption of the non-speciality of* $\mathcal{D} + Q$. *Then* $\widehat{\mathcal{D}} - Q$ *is not special (Exercise3.1) and we can introduce* $\widehat{\psi} \in H^0(\widehat{\mathcal{D}}; -q)$ *by the condition* $\exp(1)\widehat{\psi}(P_0) = 1$. *For a differential form* ω *with the divisor* $(\omega) = \widehat{\mathcal{D}} + \mathcal{D} - 2P_0$ *the formal series* $-\widehat{\psi}\frac{\omega}{dk_0}$ *coincides with the function* ψ^* *defined in §§2.2, 2.5.* (Verify that

$$\operatorname{Res}_{P_0}\left(\frac{\partial^m \psi}{\partial x^m}\widehat{\psi}\omega\right) = 0$$

for any m — cf. [Ch5]).

b) If there is an anti-involution σ *on* Γ, $P_0^\sigma = P_0$ *and* $\mathcal{D} + \mathcal{D}^\sigma \sim \mathcal{K} + 2P_0$, *then we can take* $\mathcal{D}^\sigma$, ψ^σ *as* $\widehat{\mathcal{D}}$, $\widehat{\psi}$. □

Exercise4.5*. *In the notation of Theorem4.4, but for* $q = ik_n^{-1}x + ik_n^{-3}t$, *show that* $\boldsymbol{r}$ *satisfies the modified vector KdV equation*

$$r_t + r_{xxx} + 3(\Omega r, \overline{r})r_x + 3(\Omega r_x, \overline{r})r = 0.$$

(Follow the proof of Theorem4.4, or use Exercise2.5 (§2) in scalar variant, or the identity (2.19) — cf. [Ch5].) □

One advantage of using divisors $\mathcal{D}$ of degree g (not of $g+n-1$, as for the GHM) in the construction of solutions of the VNS is the possibility of obtaining the formula directly in terms of theta functions:

Exercise4.6. *a) In the notation of §3.1 for* $P_0 = R_n$, *show that*

$$r_j = \exp\left(\int_{R_n}^{R_j} \omega_q\right)\frac{\theta(\omega(R_j) - w(\mathcal{D}) - v_q)\theta(w(\mathcal{D}))}{\theta(\omega(R_j) - w(\mathcal{D}))\theta(w(\mathcal{D}) + v_q)}$$

(for $n = 2$ *this formula was obtained in [I]).*

b) Analyzing possible singularities of r_j *(cf. a)), give a direct proof of the regularity of any solutions of (4.16) in Theorem4.4 with* $\Omega > 0$. (If $\theta(w(\mathcal{D}) + v_q)$ has zero for x, then $u = -2\partial^2 \log \theta(w(\mathcal{D}) + v_q)/\partial x^2 +$ const. takes arbitrarily large positive value. But $u = -(\Omega r, \overline{r}) \leqq 0$. For $n = 2$ this argument is due to A. R. Its, the first proof of the regularity of r for $n = 2$ was given by V. P. Kotlyarov.) □

Case $n = 2$ (NS). It is convenient to give the curve Γ for the NS by the equation

$$\varepsilon^2 = \prod_{j=1}^{2g+2} (\lambda - \beta_j), \qquad \beta_j \neq \beta_l \in \mathbb{C} \text{ for } j \neq l, \tag{4.18}$$

where there are two points R_1, R_2 (with a local parameter λ^{-1}) over a point $\lambda = \infty$ on Γ, which are distinguished by the value of the function $\lambda^{-g-1}\varepsilon$. If the polynomial $\prod_{j=1}^{2g+2}(\lambda - \beta_j)$ has real coefficients, then an anti-involution σ can be introduced, putting $\lambda^\sigma = \lambda$, $\varepsilon^\sigma = \varepsilon$. We assume further that ${}^\tau\lambda = \lambda$, ${}^\tau\varepsilon = -\varepsilon$. Branching points of $\widetilde{\lambda}$ are points $E_j \in \Gamma$ for which $\lambda(E_j) = \beta_j$, $1 \leqq j \leqq 2g+2$. In this way, $\widetilde{\lambda}$ does not branch at infinity in contrast to the curve (4.1). Note that both points R_1, R_2 are real. The matrix Ω now reduces to one number $\nu \stackrel{\text{def}}{=} w\lambda^2(R_1) \in \mathbb{R}$, and equation (4.16) becomes the scalar NS:

$$ir_t = r_{xx} + 2\nu|r|^2 r.$$

ν can be positive only when there is no real number among $\{\beta_j, 1 \leqq j \leqq 2g+2\}$.

Let us suppose the last condition on $\{\beta_j\}$ is attained. Real points of second order of $\operatorname{Jac}\Gamma = \mathcal{J}^0$ are generated by divisors $E_j - E_j^\sigma$ for $1 \leqq j \leqq g+1$ and when $g = 2m+1$ is odd, by one more divisor $\sum_{j=1}^{m+1}(E_j + E_j^\sigma) - \sum_{j=m+2}^{2m+2}(E_j + E_j^\sigma)$, where $E_j^\sigma = E_{2g+2-j}$. The only relation among them is $\sum_{j=1}^{g+1}(E_j - E_j^\sigma) \sim 0$. By virtue of Proposition3.3 the number of connected components of the variety $\widehat{\mathcal{J}}_\sigma^g(R_2)$ of classes of divisors $\mathcal{D}$ with condition (4.15): $\mathcal{D} + \mathcal{D}^\sigma \sim \sum_{j=1}^{2g+2} E_j - 2R_1$ is equal to 1 for even $g = 2m$ and 2 for odd $g = 2m+1$. This is also verified in another way: tracing the signs of $\lambda^{-g-1}\varepsilon$ on each of two lcomponents of $\Gamma \setminus \{R_1, R_2\}$ when $\lambda \in \mathbb{R}$ changes from $-\infty$ to $+\infty$, and checking that for $g = 2m$, $2m+1$, the set $\Gamma_\mathbb{R}$ consists respectively of one and two ovals (cf. Exercise3.8). In the case $g = 2m+1$, as we showed above, $\nu > 0$ on one of the components of $\widehat{\mathcal{J}}_\sigma^g(R_2) \simeq \widehat{\mathcal{J}}_\sigma^{g+n-1}$; on the other component necessarily $\nu \leqq 0$ and there exists a non-regular solution r (Theorem3.6 and 4.4). If $g = 2m$, then always $\nu > 0$. Thus the case of imaginary $\{\beta_j\}$ is completely examined.

Now let us assume until the end of this section that at least one of β_j is real and, consequently $\nu \leqq 0$. Then there exists an isomorphism of the curves (4.1) and (4.18) which sends τ to τ and σ to σ. In addition, a pair of real points in $\{E_j\}$ corresponds to the points O_0 and O_∞ (if there exists one real β_j, then there exists a second one since their total number $2g+2$ is even). In particular, using the proof of Theorem4.1b) we can describe the group $\mathcal{E}$ of points of second order of $\widehat{\mathcal{J}}_\sigma^0 = \{x \in \mathcal{J}^0, x + x^\sigma = 0\}$, which coincides with the group of points of second order of $\mathcal{J}_\sigma^0 = \{x \in \mathcal{J}^0, x = x^\sigma\}$, in the following way:

First we arrange $\{\beta_j\}$: let $\beta_1 < \beta_2 < \ldots < \beta_{2p} \in \mathbb{R}$,

$$\beta_{2p+1}^\sigma = \beta_{2(p+1)}, \ldots, \beta_{2i+1}^\sigma = \beta_{2(i+1)}, \ldots, \beta_{2g+1}^\sigma = \beta_{2(g+1)}.$$

Then $\mathcal{E}$ is generated by $\mathbb{Z}_2$-linearly independent classes $\widetilde{e}_i$, $\widetilde{e}_i^0$ of the divisors

$$e_1 = E_2 - E_3, \ldots, e_i = E_{2i} - E_{2i+1}, \ldots, \quad e_{p-1} = E_{2(p-1)} - E_{2p-1},$$
$$e_1^0 = E_3 - E_4, \ldots, e_i^0 = E_{2i+1} - E_{2(i+1)}, \ldots, \quad e_g^0 = E_{2g+1} - E_{2(g+1)}.$$

Let us show that $\{\widetilde{e}_i^0, 1 \leqq i \leqq g\}$ generate the intersection $\mathcal{E}_0 = \mathcal{E} \cap (\widehat{\mathcal{J}}_\sigma^0)_0$, where $(\widehat{\mathcal{J}}_\sigma^0)_0$ is the connected component of zero of $\widehat{\mathcal{J}}_\sigma^0$.

Analyzing the sign of the right hand side of equation (4.18), we find that the ovals

$$\widehat{\Gamma}_{\mathbb{R}}^i \stackrel{\text{def}}{=} \{P \in \Gamma, \beta_{2i+1} \leqq \lambda(P) \leqq \beta_{2(i+1)}$$

for $0 \leqq i \leqq p-1$ exhaust all the set $\widehat{\Gamma}_{\mathbb{R}} = \{P \in \Gamma, P^\sigma = {}^\tau P\}$ of "imaginary" points of Γ. If $1 \leqq i \leqq p-1$, then the zero divisors 0 and e_i^0 are connected in $\mathcal{J}^0$ by a connected variety of divisors $P_1 - P_2$, $P_{1,2} \in \widehat{\Gamma}_{\mathbb{R}}^i$, the classes of which belong to $\widehat{\mathcal{J}}_\sigma^0$. For $g \geqq i \geqq p$ connect β_{2i+1} and $\beta_{2(i+1)} = \overline{\beta}_{2i+1}$ by an arc $\gamma \subset \mathbb{C}$ which is invariant under complex conjugation and intersects $\mathbb{R} \subset \mathbb{C}$ at the point β_1. Then $0, e_i^0 \in \widehat{\gamma} \stackrel{\text{def}}{=} \{P_z - P_z^\sigma, z \in \gamma\}$. The image of $\widehat{\gamma}$ in $\mathcal{J}^0$ is a connected curve in $\widehat{\mathcal{J}}_\sigma^0$. Therefore $\widetilde{e}_1^0, \ldots, \widetilde{e}_g^0 \in \mathcal{E}_0$.

As we found g independent elements $\widetilde{e}_1^0, \ldots, \widetilde{e}_g^0$ in $\mathcal{E}_0$, we obtain (cf. Proposition3.3) that *precisely one element of* $\widehat{\mathcal{E}} \stackrel{\text{def}}{=} \bigoplus_{i=1}^{p-1} \mathbb{Z}_2 \widetilde{e}_i$ *is contained in each component of* $\widehat{\mathcal{J}}_\sigma^0$. This corresponds to Exercise3.8, since p is equal to the number of real (or imaginary) ovals of Γ.

We will not use this statement. It is given to help the reader to understand the structure of $\widehat{\mathcal{J}}_\sigma^0$ clearly. We now turn to the description of the connected components of $\widehat{\mathcal{J}}_\sigma^g(R_2)$. It is more convenient to examine the variety $\widehat{\mathcal{J}}_\sigma^g$ of classes of divisors with the condition

$$\mathcal{D} + \mathcal{D}^\sigma \sim T - P_0 - P_0^\sigma, \qquad P_0 \in \widehat{\Gamma}_{\mathbb{R}}^0. \tag{4.19}$$

The map $\delta : \mathcal{D} \mapsto \mathcal{D} - R_2 + P_0$ establishes the isomorphism of $\widehat{\mathcal{J}}_\sigma^g(R_2)$ and $\widehat{\mathcal{J}}_\sigma^g$. Further we will assume that $\mathcal{D} = \sum_{i=1}^{p-1} \mathcal{D}_i + \mathcal{D}' \geqq 0$, $\operatorname{supp} \mathcal{D}_i \in \widehat{\Gamma}_{\mathbb{R}}^i$, $\operatorname{supp} \mathcal{D}' \cap (\widehat{\Gamma}_{\mathbb{R}} \setminus \widehat{\Gamma}_{\mathbb{R}}^0) = \emptyset$. Put

$$\Sigma(\mathcal{D}) = \{\sigma_i = \deg \mathcal{D}_i \mod 2, \quad 1 \leqq i \leqq p-1\}.$$

Proposition4.3. *If there exist real numbers among $\{\beta_j\}$, then the value of $\Sigma(\mathcal{D})$ depends only on the index of the connected component of $\widehat{\mathcal{J}}^g_\sigma$ containing the class of $\mathcal{D} \geqq 0$. The invariant Σ gives an isomorphism of the set of connected components of $\widehat{\mathcal{J}}^g_\sigma$ and $\mathbb{Z}_2^{p-1} = \{\sigma_i \in \mathbb{Z}_2, 1 \leqq i \leqq p-1\}$.*

Proof. (cf. [Fay]). Let us suppose that the classes $\mathcal{D}, \mathcal{D}' \geqq 0$ belong to the same connected component of $\widehat{\mathcal{J}}^g_\sigma$. Then for some distribution q' on Γ with the property $q' + (q')^\sigma = 0$ we can find the Baker function $\varphi \in H^0(\mathcal{D}; q')$ with the divisor $(\varphi) = \mathcal{D}' - \mathcal{D}$. We can take a σ-invariant subset of Γ as $\operatorname{supp} q'$. There exists a real rational function v on Γ with the divisor $(v) = \mathcal{D} + \mathcal{D}^\sigma + P_0 + P_0^\sigma - T$. Fix $P_0 \notin \operatorname{supp}\{\mathcal{D}, \mathcal{D}^\sigma\}$. The standard argument shows that the traces $\operatorname{Tr} v$ and $\operatorname{Tr}(\varphi\varphi^\sigma v)$ do not depend on λ. At points P_0, $P_0^\sigma = {}^\tau P_0$, however, the functions v and $\varphi\varphi^\sigma v$ are zero. Therefore at an arbitrary (generic enough) point $P \in \Gamma$ we have the relations

$$v(P)\varphi\varphi^\sigma(P) + v({}^\tau P)\varphi\varphi^\sigma({}^\tau P) = 0 = v(P) + v({}^\tau P).$$

Consequently, $\varphi\varphi^\sigma(P) = \varphi\varphi^\sigma({}^\tau P)$ and $\varphi\varphi^\sigma$ turns out to be a rational function of λ. For $z \leqq \beta_1$, $z \geqq \beta_2$, or for $\beta_{2i} \leqq z \leqq \beta_{2i+1}$, where $1 \leqq i \leqq p-1$, the points P with condition $\lambda(P) = z$ belong to $\Gamma_{\mathbb{R}}$. Therefore $\varphi\varphi^\sigma(z) = \varphi(P)\overline{\varphi(P)} \geqq 0$ for such values of z. But then $\varphi\varphi^\sigma(z)$ may have only an even number of zeros and poles (counted with multiplicities) in the interval $\beta_{2i+1} \leqq z \leqq \beta_{2(i+1)}$ for arbitrary indices $0 \leqq i \leqq p-1$. This means that $\Sigma(\mathcal{D}) = \Sigma(\mathcal{D}')$.

As a by-product, we have shown the existence of the representation $\mathcal{D} + \mathcal{D}^\sigma = {}^\tau\mathcal{D}^1 + \mathcal{D}^1$ for a certain divisor $\mathcal{D}^1 \geqq 0$ (using the fact that $\varphi\varphi^\sigma \in \mathbb{C}(\lambda)$). Hence, if $P \in \operatorname{supp}\mathcal{D}$, then either ${}^\tau P \in \operatorname{supp}\mathcal{D}$ (when $P = {}^\tau P$, the multiplicity of P in $\mathcal{D}$ should be even) or ${}^\tau P \in \operatorname{supp}\mathcal{D}^\sigma$ and ${}^\tau P^\sigma \in \operatorname{supp}\mathcal{D}$. This makes it possible to represent the divisor $\mathcal{D}$ in the form

$$\mathcal{D} = \widehat{\mathcal{D}} + \Delta_1 + {}^\tau\Delta_1 + \Delta_2 + {}^\tau\Delta_2^\sigma, \qquad \operatorname{supp}\widehat{\mathcal{D}} \in \widehat{\Gamma}_{\mathbb{R}}, \tag{4.20}$$

where $\mathcal{D}, \Delta_1, \Delta_2 \geqq 0$. Conversely, it is easy to see that any divisor $\mathcal{D} \geqq 0$ of degree g of the form (4.20) satisfies (4.19). In particular, as $g \geqq p-1$, any value of the invariant $\Sigma \in \mathbb{Z}_2^{p-1}$ is attained by some divisor $\mathcal{D} = \widehat{\mathcal{D}}$. We already know that the total number of components of $\widehat{\mathcal{J}}^g_\sigma$ is precisely 2^{p-1}. Therefore Σ is one-to-one on components of $\widehat{\mathcal{J}}^g_\sigma$. □

Corollary4.2. a) *If $p = g+1$ (i.e., all $\{\beta_j\}$ are real), then the support of any divisor $\mathcal{D} \geqq 0$ the class $\widetilde{\mathcal{D}}$ of which belongs to the component $(\widehat{\mathcal{J}}^g_\sigma)_1 \subset \widehat{\mathcal{J}}^g_\sigma$ where the invariant Σ takes the value $\{\sigma_i = 1, 1 \leqq i \leqq g\}$ is contained in the set $\cup_{j=1}^{g-1}\widehat{\Gamma}^j_{\mathbb{R}}$.*

On the other components for $p = g+1$ *and on any components of* $\widehat{\mathcal{J}}^g_\sigma$ *for* $p < g+1$ *there are classes* $\widetilde{\mathcal{D}}$ *with a representative* $\mathcal{D} \geqq 0$ *which contains the point* P_0.

b) Only in the case $p = g+1$ *and for a unique connected component of* $\widehat{\mathcal{J}}^g_\sigma(R_2)$, *namely for* $\delta^{-1}((\widehat{\mathcal{J}}^g_\sigma)_1)$, *is it true that arbitrary representatives* $\mathcal{D} \geqq 0$ *do not contain* R_2. *The solutions of the NS (with* $\nu < 0$*) corresponding to this component are defined and bounded for all* $x, t \in \mathbb{R}$.

Proof. For $p = g+1$ an arbitrary divisor $\mathcal{D}$ of the form (4.20) with $\Sigma(\mathcal{D}) = \{\sigma_i = 1\}$ must have at least one point on each oval $\widehat{\Gamma}^i_{\mathbb{R}}$ $(1 \leqq i \leqq g)$. On the other hand $\deg \mathcal{D} = g$ and $\mathcal{D} = \widehat{\mathcal{D}}$ in (4.20), as $\operatorname{supp}\mathcal{D} \subset \cup_{i=1}^g \widehat{\Gamma}^i_{\mathbb{R}}$. In all the other cases for arbitrary values of Σ we can find $\mathcal{D}$ of type (4.20), one point of which lies in $\widehat{\Gamma}^0_{\mathbb{R}}$. If $\operatorname{supp}\mathcal{D} \ni R_2$ for $\widetilde{\mathcal{D}} \in \widehat{\mathcal{J}}^g_\sigma(R_2)$, then the divisor $\delta(\mathcal{D}) = \mathcal{D} - R_2 + P_0$ is effective and contains P_0. The converse is also true. □

Exercise4.7. *Show that* Γ *is divided by* $\Gamma_{\mathbb{R}}$, *if and only if* $p = g+1$, *i.e.,* $\{\beta_j\} \subset \mathbb{R}$. (In the case $\beta_1 \notin \mathbb{R}$, any points $z_1, z_2 \in \mathbb{C} \setminus \mathbb{R}$ can be connected by a smooth path not intersecting with $\mathbb{P}^1_{\mathbb{R}} \setminus \widetilde{\lambda}(\Gamma_{\mathbb{R}})$. For $z_1 = z_2$ this path (cycle) can be chosen so that only the point β_1 among all $\{\beta_j\}$ is inside it.) □

4.3. Relations with the constructions in §3.

The use of divisors of degree g and a scalar Baker function ψ for constructing solutions of the VNS, in principle, does not supplement new features to the construction in §§3.2–3.3. Following the general recipe of the reduction from the GHM to the VNS and Proposition3.2, we can construct a vector Baker function $\boldsymbol{\psi} \overset{\text{def}}{=} \boldsymbol{\psi}^+ = (\Phi_{(+1)})^{-1}\varphi$ from q in §4.2 and $\mathcal{D}$ of type (3.7b). Then $\boldsymbol{\psi}_x$, $\boldsymbol{\psi}_t$ are expressed in terms of $\boldsymbol{\psi}$ and $(\Phi_{(+1)})^{-1}(\Phi_{(+1)})_x$. The latter matrix has the form $\sum_{j=1}^{n-1}(r_j \boldsymbol{I}_1^{j+1} - \overline{r}_j \boldsymbol{I}_{j+1}^1)$ for appropriate r_j (cf. Introduction). The vector $\boldsymbol{r} = {}^t(r_j)$ satisfies the VNS and is uniquely determined from the first component of $\boldsymbol{\psi}$ which is the function ψ also introduced in §4.2. Hence, essentially, Theorem4.4 is a specialized form of the theorems in §3. The above construction of solutions of the Sin-Gordon equation and the S^2-fields provides a more interesting example.

The formal mechanism which we actually made use of in the proof of Theorem4.2 is the interpretation of the Sin-Gordon equation as a higher KdV at a "finite" point $k = 0$ (§2.5). Let us recall that scalar Lax equations (in paticular, higher KdV's) can be written in the form (2.5) (cf. ibid.). Consequently, the Sin-Gordon equation is represented in the form (2.5a), which allows us to obtain the matrix zero curvature representation for it. The corresponding matrix formal Jost function $\Psi(k)$ is introduced by the formula

$$\Psi(k) = \begin{pmatrix} 1 & 1 \\ 1 & -1 \end{pmatrix}^{-1} \begin{pmatrix} \psi(k) & \psi(-k) \\ -ik\psi_x(k) & -ik\psi_x(-k) \end{pmatrix},$$

where $\psi(k)$ is the formal solution of the equation $-\psi_{xx}+u\psi = k^2\psi$ and u is related to α as in Theorem4.2.

Let $\psi(k)$ be the expansion of the Baker function $\psi(P)$ in §4.1 in a neighbourhood of O_∞ with respect to $k^{-1} \overset{\text{def}}{=} k_\infty = \lambda^{-1/2}$. Then $\Psi(k)$ can be expressed as an expansion of the matrix of the "values" of the vector Baker function,

$$\boldsymbol{\psi}(P) = \frac{1}{2}\begin{pmatrix} \psi(P) - i\lambda^{1/2}\psi_x(P) \\ \psi(P) + i\lambda^{1/2}\psi_x(P) \end{pmatrix},$$

in a neighbourhood of $\lambda = \infty$ according to the definitions in §3.2. However, $\boldsymbol{\psi}$ is defined not on Γ but on the two-fold covering Γ' of the curve Γ, for which $\mathbb{C}(\Gamma) \ni \lambda^{1/2},\ {}^{\tau}(\lambda^{1/2}) = \lambda^{1/2}$.

Next, we discuss in detail how to get $\boldsymbol{\psi}$ from ψ. Beginning with the construction of Γ', we establish the connection of the constructions in §3.3 and §4.1, and, in partcular, lift α of (4.9) to an SU_2-field. In order to study specific characters of the Sin-Gordon equation more flexibly, we make a slight change in the general procedure of §2.5.

Define a hyperelliptic curve Γ' by the equation

$$\varepsilon^2 = \prod_{j=1}^{2g}(\lambda' - \alpha'_j)(\lambda' + \alpha'_j), \tag{4.21}$$

where $\alpha'_j \overset{\text{def}}{=} \alpha_j^{1/2}$ for $1 \leqq j \leqq 2g$ and $\{\alpha_j\}$ are the same as in (4.1). In the notation of §4.2, the curve Γ is given by equation (4.18) for $1 \leqq j \leqq 2g'$, $g' = 2g-1$, $\{\beta_j\} = \{\pm\alpha_j\}$. Its genus is equal to g' and there is an involution τ (${}^{\tau}\lambda = \lambda$, ${}^{\tau}\varepsilon = -\varepsilon$) on it. If (cf. §4.1) Γ satisfies the conditions of Theorem4.1, Γ' with the anti-involution σ, $(\lambda')^\sigma = \lambda'$, $\varepsilon^\sigma = \varepsilon$, satisfies the conditions of Proposition3.4 for λ', $n = 2$. Indeed, arbitrary α'_j cannot be real, since $\alpha_j < 0$ if $\alpha_j \in \mathbb{R}$. There is a covering $\pi : \Gamma' \to \Gamma$ of degree 2, for which

$$(\lambda')^2 = \lambda, \qquad \mu = \varepsilon\lambda'.$$

This covering is extended to a mapping from $(\mathbb{P}^1)'$ with coordinate λ' to $\mathbb{P}^1$ with coordinate $\lambda = (\lambda')^2$; we also denote this by the letter π. The latter covering is ramified exactly over the points $\lambda = 0, \infty$, over which $\widetilde{\lambda} : \Gamma \to \mathbb{P}^1$ also ramifies. Therefore, $\pi : \Gamma' \to \Gamma$ is not ramified (this is also seen from the Hurwitz formula, $g' - 1 = 2(g-1)$. Let us bring together all the mappings in one diagram. On the right, we show what is occurring over the point $\infty \overset{\text{def}}{=} (\lambda = \infty)$, $\infty' \overset{\text{def}}{=} (\lambda' = \infty)$.

$$\begin{array}{ccccccc}
\Gamma' & \xrightarrow{\pi} & \Gamma & \qquad & R'_1, R'_2 & \xrightarrow{\pi} & O_\infty \\
{\scriptstyle\widetilde{\lambda}'}\downarrow & & \uparrow{\scriptstyle\widetilde{\lambda}} & & {\scriptstyle\widetilde{\lambda}'}\downarrow & & \uparrow{\scriptstyle\widetilde{\lambda}} \\
(\mathbb{P}^1)' & \xrightarrow[\pi]{} & \mathbb{P}^1 & & \infty' & \xrightarrow[\pi]{} & \infty
\end{array}$$

The covering π commutes with τ on Γ' and on Γ. All coverings commute with σ.

Let $\mathcal{D}$, q on Γ be the same as in §4.1. Put

$$\mathcal{D}' = \pi^{-1}\mathcal{D}, \; q' = ik_\infty^1 x + ik_\infty^2 t + ik_0^1 x + ik_0^2 t$$

(q' is a lift of q on Γ'). The parameters $k_\infty^{1,2}$ are obtained, by definition, from the local parameter k_∞ at the point O_∞ by lifting to the points R_1, R_2, respectively. If we choose a point R_1' by the condition that $k_\infty^1 = (\lambda')^{-1}$ in a neighbourhood of R_1', then $k_\infty^2 = -{}^\tau k_\infty^1 = -(\lambda')^{-1}$ in a neighbourhood of R_2' (${}^\tau k_\infty = -k_\infty$). Analogously the local parameters in a neighbourhood of points $(\widetilde{\lambda}')^{-1}(0') = \{R_1^{0'}, R_2^{0'}\}$, $k_0^{1,2}$ are defined by k_0, namely, $k_0^{1,2} = \lambda\varepsilon^{-1}$.

The degree of $\mathcal{D}'$ is equal to $2\deg\mathcal{D} = g'+1$. Because of (4.2),

$$\mathcal{D}' + (\mathcal{D}')^\sigma \sim T' = \sum_{j=1}^{2g'+2} E_j,$$

where $\widetilde{\lambda}'(E_j) = \beta_j \in \{\pm\alpha_j\}$, T' is the branching divisor of $\widetilde{\lambda}'$. Therefore, the divisor $\mathcal{D}'$ and the distribution q' satisfy (3.7) and we can apply Theorem3.4 to Γ', σ, $\mathcal{D}'$, q', $\lambda_{z'} = 1 - \frac{2z'}{\lambda'+z'}$ taken as the parameter λ ($z' \in \mathbb{R}$). The condition of non-speciality of the divisors $\mathcal{D} - O_0$, $\mathcal{D} - O_\infty$ is automatically satisfied. This condition is equivalent to non-speciality of the divisors $\mathcal{D}' = (R_1' + R_2') \sim \mathcal{D}' - (R_1^{0'} + R_2^{0'})$. The matrices U, V in Theorem3.4 corresponding to $\mathcal{D}'$ are diagonalized (cf. §3.2):

$$U \sim \frac{1}{2z'}\begin{pmatrix} i & 0 \\ 0 & -i \end{pmatrix}, \quad V \sim \frac{z'}{2c}\begin{pmatrix} i & 0 \\ 0 & -i \end{pmatrix}, \quad c^2 = \prod_{j=1}^{2g} \alpha_j. \tag{4.22}$$

This formula can be checked by rewriting (the principal part of) $k_\infty^{1,2}$ and $k_0^{1,2}$ via $1 - \lambda_{z'}$ and $1 + \lambda_{z'}$.

If the value of $\lambda_{z'}$ is equal to ∞, then $\lambda' = -z'$ and $\lambda = (z')^2 \stackrel{\text{def}}{=} z > 0$. Analogously, the point $\lambda_{z'} = 0 = \lambda' - z'$ is mapped by π to the same point $\lambda = z$ on $\mathbb{P}^1$ as $\lambda_{z'} = \infty$ does. Put $\widetilde{\lambda}^{-1}(z) = \{P_1, P_2\}$, where $\mu(P_1) > 0$. The points of the fibers of $\widetilde{\lambda}'$ over $\lambda' = -z$, $\lambda' = z'$ are denoted respectively by $\{R_1^{-z'}, R_2^{-z'}\}$, $\{R_1^{z'}, R_2^{z'}\}$ and ordered by the condition $\varepsilon(R_1^{\pm z'}) > 0$. The genus of Γ' is odd, hence the pairs $\{R_1^{\pm z'}\}$ and $\{R_2^{\pm z'}\}$ belong to different components of $\Gamma'_{\mathbb{R}}$ (there are exactly two of them). Note that $\pi(R_1^{z'}) = P_1$, $\pi(R_2^{z'}) = P_2$, but $\pi(R_1^{-z'}) = P_2$, $\pi(R_2^{-z'}) = P_1$.

We can take the function v lifted from Γ (§4.1) with the divisor $(v) = \mathcal{D} + \mathcal{D}^\sigma = \sum_{j=1}^{2g} O_j$ as the function w' on Γ' (cf. §3.3) with the divisor $(w') = \mathcal{D}' + (\mathcal{D}')^\sigma -$

$\sum_{j=1}^{2g} O_j$. Hereafter we will identify functions on Γ with their lifts onto Γ' with respect to π. Applying Corollary3.2, we obtain that

$$\Omega = \Omega_\infty = \operatorname{diag}(v(P_2), v(P_1)), \qquad \Omega_0 = \operatorname{diag}(v(P_1), v(P_2))$$

(see §3.3) are definite quadratic forms. As in §4.1, put

$$\beta_1^2 = \tfrac{1}{2} v(P_1) v(O_\infty)^{-1}, \beta_2^2 = \tfrac{1}{2} v(P_2) v(O_\infty)^{-1}.$$

To conclude the inspection, it remains to relate the vector Baker function $\varphi = {}^t(\varphi_1, \varphi_2)$ defined by the method in §3.3 by Γ', $\lambda_{z'}$ with the Baker function ψ in §4.1. If we set $\psi' = ({}^\tau\psi^\sigma) f$, where $f = {}^\tau v\varepsilon \prod_{i=1}^{g} (\lambda - \delta_i)^{-1}$, then ψ, ψ' form a basis in $H^0(\mathcal{D}'; q')$. Constructing the vector Baker function normalized at the fiber $(\lambda_{z'})_1$ or $(\lambda_{z'})_\infty$ with the aid of ψ, ψ', we easily obtain two types of zero curvature representations for α in (4.9). We do not do this here, but use ψ, ψ' to lift α to an SU_2-field, following Theorem3.3 and 3.4. As in the proof of Theorem4.3, put $\xi_1 = \beta_1 \psi(P_1)$, $\xi_2 = \beta_2 \overline{\psi(P_2)}$.

Theorem4.5. *The matrix function of $x, t \in \mathbb{R}$*

$$G = \begin{pmatrix} 2\xi_2\xi_1 & \xi_2\overline{\xi}_2 - \xi_1\overline{\xi}_1 \\ \xi_1\overline{\xi}_1 - \xi_2\overline{\xi}_2 & 2\overline{\xi}_1\overline{\xi}_2 \end{pmatrix}$$

is an SU_2-field (satisfies (0.1) with values in SU_2).

Proof. Put

$$\Psi_0 = \begin{pmatrix} \psi(P_1) & \psi(P_2) \\ ev(P_2)\overline{\psi(P_2)} & -ev(P_1)\overline{\psi(P_1)} \end{pmatrix}, \Psi_\infty = \begin{pmatrix} \psi(P_2) & \psi(P_1) \\ ev(P_1)\overline{\psi(P_1)} & -ev(P_2)\overline{\psi(P_2)} \end{pmatrix},$$

where $e = \prod_{i=1}^{g} (z - \delta_i)^{-1} \left(\prod_{j=1}^{2g} (z - \alpha_j) \right)^{1/2}$, and we take the positive sign of the root. The matrices Ψ_0, Ψ_∞ represent the matrices of "values" of ${}^t(\psi, \psi')$ at points $\lambda_{z'} = 0$, $\lambda_{z'} = \infty$ in the definition in §3.2. From Theorem3.3 it follows that $\Psi_\infty^{-1} \Psi_0$ satisfies (0.1). Putting $\Omega = B_\infty B_\infty^+$, $\Omega_0 = B_0 B_0^+$, where $B_\infty = \operatorname{diag}(\beta_2, \beta_1)$, $B_0 = \operatorname{diag}(\beta_1, \beta_2)$, we obtain by Theorem3.4 that $G = B_\infty^{-1} \Psi_\infty^{-1} \Psi_0 B_0$ is an SU_2-field. The rest of the proof is the direct calculation of G using the relations

$$\begin{aligned} \det \Psi_\infty &= -e(v(P_2)\psi(P_2)\overline{\psi(P_2)} + v(P_1)\overline{\psi(P_1)}\psi(P_1)) \\ &= -2ev(\infty) \end{aligned}$$

(Proposition4.1). □

Exercise4.8. *Relate the U_2-field $\widetilde{G} = G\begin{pmatrix} 0 & 1 \\ 1 & 0 \end{pmatrix}$ for G of Theorem4.5 with the property $\widetilde{G}^2 = I$ to the vector function s of Theorem4.3. Show that*

$$\widetilde{G} = s_1\begin{pmatrix} 0 & 1 \\ 1 & 0 \end{pmatrix} + s_2\begin{pmatrix} 0 & -i \\ i & 0 \end{pmatrix} + s_3\begin{pmatrix} -1 & 0 \\ 0 & 1 \end{pmatrix}.$$

Substituting $\widetilde{G}$ in (0.1), reduce Theorem4.3 to Theorem4.5. □

In the construction of Theorem4.3 we could use not only $\lambda(P) = z > 0$ but also $z < 0$. For such P the function s is a field on a hyperboloid (Exercise4.3). An analogous extension is also possible for Theorem4.5:

Exercise4.9. *Examine the anti-involution σ' on Γ' defined by (4.21): $(\lambda')^{\sigma'} = -\lambda'$, $\varepsilon^{\sigma'} = -\varepsilon$. Check that*

$$\{R \in \Gamma', R^{\sigma'} = R\} \Leftrightarrow \{\pi(R)^{\sigma} = \pi(R), \lambda(\pi(R)) < 0\}.$$

For $z < 0$ and σ' instead of σ, trace the construction of Theorem4.5 and show that the corresponding matrix Ω is not definite (cf. Corollary4.1). □

We will showed that the concrete methods of constructing algebraic-geometric solutions of §§4.1,2 can be deduced from the general method in §§3.2,3, based on vector Baker functions normalized at points of a certain real (σ-invariant) fiber. We will give one example for which another normalization is more convenient.

Let Γ be the curve in §4.1 with the real equation (4.1). Note that the negativity of $\alpha_j \in \mathbb{R}$ is not required below. We keep the same distribution $q = ik_\infty x + ik_0 t$ as that for the Sin-Gordon equation, but choose a divisor $\mathcal{D} \geqq 0$ of degree $g+1$ with the standard reality property (3.7a). Replacing it with an equivalent one, we may assume that $\mathcal{D}^\sigma = {}^\tau\mathcal{D} \geqq 0$ (cf. (4.20)). Put $\mathcal{D} + \mathcal{D}^\sigma - T = (w)$ for w with conditions $w^\sigma = w = -{}^\tau w$. We fix a real point P on Γ. We assume further that $\mathcal{D}$, P, x, t are in a generic position. The basis $\{\varphi_1, \varphi_2\} \subset H^0(\mathcal{D}; q)$ is normalized by the conditions

$$\exp(-q)\varphi_{1,2}(O_\infty) = 0, 1, \qquad \exp(-q)\varphi_{1,2}(O_0) = 1, 0.$$

Computing $\mathrm{Tr}(w\varphi_1\, {}^\tau\varphi_2)$, we obtain the relation

$$w(P)\varphi_1(P)\varphi_2({}^\tau P) + w({}^\tau P)\varphi_1({}^\tau P)\varphi_2(P) = 2w_\infty\varphi_1^\infty = -2w_0\varphi_2^0,$$

where $w_\infty = (k_\infty w)(O_\infty)$, $w_0 = (k_0 w)(O_0) \in \mathbb{R}$,

$$\varphi_1^\infty = \exp(-q)k_\infty^{-1}\varphi_1(O_\infty), \qquad \varphi_2^0 = \exp(-q)k_0^{-1}\varphi_2(O_0).$$

Taking the normalization of φ_1, φ_2 into account, we find that

$$(\varphi_1)_x = i\varphi_1^\infty \varphi_2, \qquad (\varphi_2)_t = i\varphi_2^0 \varphi_1.$$

Since $\mathcal{D}^\sigma = {}^\tau\mathcal{D}$, then $\varphi_1^\sigma = {}^\tau\varphi_1$, $\varphi_2^\sigma = {}^\tau\varphi_2$. Consequently,

$$i(\psi_1)_x = -\frac{w(P)}{2w_\infty} s\psi_2,$$
$$i(\psi_2)_t = -\frac{w(P)}{2w_0} s\psi_1,$$
$$s = \psi_1\overline{\psi}_2 + \overline{\psi}_1\psi_2,$$

where we set $i\psi_1 = \varphi_1(P)$, $i\psi_2 = \varphi_2(P)$. Removing the real factors

$$-(1/2)w(P)/w_\infty, \qquad -(1/2)w(P)/w_0$$

by a suitable substitution of x, t (they are easily calculated explicitly in terms of $\mathcal{D}$), we get the solution of the system of equations in [NP] with the symplectic group of chiral symmetry.

4.4. Application: the duality equation.

Let (x_j) be coordinates of $\mathbb{C}^4$, $\{A_j\}$ be functions on $\mathbb{C}^4$ with values in $\mathfrak{gl}_n$ ($j = 1, 2, 3, 4$). Put

$$\mathcal{D}_j = \frac{\partial}{\partial x_j} + A_j,$$
$$F_{jk} = [\mathcal{D}_j, \mathcal{D}_k] = (A_k)_{x_j} - (A_j)_{x_k} + [A_j, A_k].$$

The system of differential equations,

$$F_{12} = F_{34}, \qquad F_{13} = F_{42}, \qquad F_{14} = F_{23}, \tag{4.23}$$

on matrix elements of functions A_j is called the complexified Euclidean duality equation. It is convenient to introduce vector fields

$$\nabla_1 = \mathcal{D}_1 - i\mathcal{D}_2, \qquad \nabla_2 = \mathcal{D}_3 - i\mathcal{D}_4, \qquad \nabla_3 = -\mathcal{D}_1 - i\mathcal{D}_2, \qquad \nabla_4 = -\mathcal{D}_3 - i\mathcal{D}_4.$$

Following A. A. Belavin and V. E. Zakharov [BZ], we rewrite (4.23) in the form of the relation

$$[\nabla_1 + \lambda\nabla_2, \nabla_3 - \lambda^{-1}\nabla_4] = 0, \tag{4.24}$$

which is fulfilled for all values of the parameter λ (check the equivalence of (4.23) and (4.24)). Put further $\nabla_j = \partial/\partial z_j + B_j$ $(1 \leqq j \leqq 4)$ for the new variables

$$z_1 = \tfrac{1}{2}(x_1 + ix_2), \qquad z_2 = \tfrac{1}{2}(x_3 + ix_4),$$

$$z_3 = \tfrac{1}{2}(-x_1 + ix_2), \qquad z_4 = \tfrac{1}{2}(-x_3 + ix_4).$$

System (4.23) for real x_1, x_2, x_3, x_4 and anti-hermitian $\{A_j\}$ is called the *duality equation*. This equation is a reduction of more general Yang-Mills equations. If we extend the hermitian conjugation "+" to $\mathcal{D}_j$, ∇_j, setting $(\partial/\partial x_j)^+ = -\partial/\partial x_j$, the anti-hermitian condition of $\{A_j\}$ can be written in the form $\nabla_3 = \nabla_1^+$, $\nabla_2 = \nabla_4^+$ briefly. Correspondingly, A_j are expressed in terms of B_1, B_2 in this case by the formulae:

$$A_1 = (1/2)(B_1 - B_1^+),\ A_2 = (i/2)(B_1 + B_1^+),$$
$$A_3 = (1/2)(B_2 - B_2^+),\ A_4 = (i/2)(B_2 + B_2^+). \tag{4.25}$$

Below we sketch the construction of anti-hermitian solutions of system (4.23). Modification of this construction for obtaining complex solutions of (4.23) or A_j with values in other real Lie algebras (definite and indefinite – see §3.3,4) is not complicated and left to the reader as an exercise.

Let Γ be an algebraic curve with an anti-involution σ, for which $\lambda^\sigma = -\lambda^{-1}$ (in contrast to the situation of §3.3). This, in particular, means that there do not exist real points on Γ, since the values of λ at such points should satisfy the unsolvable equation $\overline{\lambda} = -\lambda^{-1}$. As in §3, we denote by T and Tr the branching divisor and the trace of the covering $\widetilde{\lambda} : \Gamma \to \mathbb{P}^1$ corresponding to λ. Let q_1, q_2 be two distributions with support in the set of zeros and poles of λ and $\mathcal{D}$ a divisor of degree $g + n - 1$ where g is the genus of Γ and n is the degree of λ. We suppose that the distribution $q_1 + \lambda q_2$ is regular at zero, and the distribution $\lambda^{-1}q_1 + q_2$ is regular at poles of λ. We also impose the condition of reality on $\mathcal{D}$: $\mathcal{D} + \mathcal{D}^\sigma - T = (w)$ for a real $(w^\sigma = w)$ rational function w on Γ.

Repeating the steps of the argument of §3.3, we consider the vector function $\varphi = {}^t(\varphi_1, \dots, \varphi_n)$, where $\{\varphi_1, \dots, \varphi_n\}$ is a certain basis of $H^0(\mathcal{D}; q)$ for the distribution

$$q = q_1 z_1 + q_2 z_2 - q_1^\sigma \overline{z}_1 - q_2^\sigma \overline{z}_2.$$

We assume hereafter that $\dim H^0(\mathcal{D}; q) = n$. This is true for (z_1, z_2) in an open dense subset of $\mathbb{C}^2$. Introduce the $n \times n$-matrix $U = (u_i^j)$, $u_i^j = \mathrm{Tr}(\varphi_i w \varphi_j^\sigma)$. Then U does not depend on λ and is hermitian and invertible (Theorem3.4). Let us suppose that U is definite, i.e., $VUV^+ = \pm I$ for a suitable matrix-valued function V of z_1, z_2.

Theorem4.6. *a) There exist unique* $\mathfrak{gl}_n$*-valued functions* $\widetilde{B}_1$, $\widetilde{B}_2$ *of* z_1, z_2 *such that*

$$(\widetilde{\nabla}_1 + \lambda\widetilde{\nabla}_2)\varphi = 0, \qquad \widetilde{\nabla}_{1,2} = \frac{\partial}{\partial z_{1,2}} + \widetilde{B}_{1,2}.$$

b) For functions

$$B_1 = V\widetilde{B}_1V^{-1} + \frac{\partial V}{\partial z_1}V^{-1}, \qquad B_2 = V\widetilde{B}_2V^{-1} + \frac{\partial V}{\partial z_2}V^{-1},$$

"gauge equivalent" to $\widetilde{B}_{1,2}$*, the corresponding* $\mathfrak{u}_n$*-valued* A_1, A_2, A_3, A_4 *of* x_1, x_2, x_3, x_4 *(cf. (4.25)) satisfy system (4.23).*

Proof. Statement a) is proved completely analogously to the proof of Theorem3.2 and based directly on the regularity of $q_1 + \lambda q_2$, $\lambda^{-1}q_1 + q_2$. Turning to b), we use the notation and the relations in Theorem3.4. For generic $c \in \mathbb{R}$, we have

$$\Phi_{(c)}\Omega_c\Phi^+_{(-c^{-1})} = U,$$

where $\Phi_{(c)}$, Ω_c are the matrices of values of φ, w at points of $(\lambda)_c$ (the fibers of $\widetilde{\lambda}$ over c and $-c^{-1}$ are ordered consistently with the action of σ). Consequently, together with the operator $\nabla_1 + c\nabla_2$ the operator $\nabla_3 - c^{-1}\nabla_4$ for $\nabla_3 = \nabla_1^+$, $\nabla_4 = \nabla_2^+$ must also be zero on $V\Phi_{(c)}$. From the invertibility of the latter matrix it follows that relation (4.24) and statement b) hold. □

In spite of the obvious resemblance between the present construction and the discussion in §§3,4, there are also some fundamental differences. The higher dimensionality of the duality equation becomes apparent in the fact that the distributions q_1, q_2 are not uniquely defined from the equation nor from its natural reduction. Let us recall that in the case of the general algebraic-geometric PCF (cf. Theorem3.3) corresponding distributions are recovered from the eigenvalues of functions U, V which are invariants of system (0.2). The reality condition ($\lambda^\sigma = -\lambda^{-1}$) apparently cannot be combined with any natural normalization of φ and functions $\{A_j\}$. The proof of regularity of the above-derived anti-hermitian solutions (cf. Theorem3.5) does not go through either (in contrast to the case of the PCF). If regularity (the absence of singularities with respect to $\{x_j\}$) holds for more particular classes of solutions, it should be under a more delicate algebraic-geometric mechanism. The problem of including the instanton solutions of the duality equation in the above framework is not solved (cf., for example, [DrM], [ADHM]). The last problem is closely related to the problem of constructing instantons by the method of "dressing" (cf. [B2]) or by the technique of Bäcklund-Darboux transformations (cf. [CGW], [Taka], [UN] and the next chapter).

Exercise4.10. *For the curve* $\varepsilon^2 = F(\lambda)$ *of the type (4.18), where* F *is a polynomial of degree* $2g+2$ *for odd* g *with the property* $\overline{F}(-\lambda^{-1}) = \lambda^{-2g-2}F(\lambda)$, *put* $\lambda^\sigma = -\lambda^{-1}$, $\varepsilon^\sigma = \varepsilon\lambda^{-g-1}$. *Then for a divisor* $\mathcal{D}$ *satisfying* $\mathcal{D}+\mathcal{D}^\sigma = T$, *and,* z_1, z_2 *close enough to zero, the* 2×2*-matrix* U *(cf. above) is definite (corresponding* $\{A_j\}$ *take values in* $\mathfrak{u}_2$*).* □

Exercise4.11. *(cf. Exercise3.3). The Lagrangian density*

$$\tfrac{1}{2}\sum_{j,k=1}^{4} \mathrm{Sp}(F_{jk}^2)$$

for self-dual Yang-Mills fields of Theorem4.6 is quasi-periodic function of x_1, x_2, x_3, x_4, *i.e., it can be written in terms of certain rational functions on the Jacobian variety of* Γ. □

4.5. Comments.

"Finite-zone" integration of the Sin-Gordon equation was done for the first time by V. O. Kozel and V. P. Kotlyarov [KoK] (cf. the formula of Exercise4.2 b) and (4.9)). They also considered the corresponding real solutions and showed their regularity. The method of [KoK] was simplified (in the complex version) by A. R. Its (cf. [Mat3]), and then generalized and refined in the works [Ch4], [Ch5] and [Ch2]. In [Ch5] on the basis of the general definition of the duality of Baker functions, the reality condition was interpreted algebraic-geometrically. In [Ch2] the number of connected components of the variety of real solutions of this equation was calculated (cf. also the author's paper in the Doklady of the Academy of Sciences, USSR, 1980, **252**-5, pp.1104–1108). Concerning the duality of Baker functions, some results can be found in [BC]. The lifting of algebraic-geometric real solutions of the Sin-Gordon equation to S^2-fields was done in [Ch5] using one idea of [NP]. Note that the construction of Kozel-Kotlyarov is connected with the problem of eigenfunctions of certain two-dimensional Schrödinger operators which was solved in [DuKN] (cf. also [DuNo], [N3]). Concrete formulae in theta functions for curves of small genus are given in [BBM], [BE], [DuNa], [ZJ] and others.

Algebraic-geometric solutions of the NS equation (cf. formula of Exercise4.6a) for $n = 2$) were constructed in [I], [IK]. This construction was made more invariant and generalized to the VNS in the work [Ch5]. After that in [Ch10], the description of the variety of such solutions of the VNS and NS equations was obtained with the aid of [Fay] for various Ω, ν and their regularity was studied in the development of the result by V. P. Kotlyarov (see Theorem4.4, Corollary4.2, and Exercise4.6). The correspondence of "finite-zone" solutions of the Sin-Gordon equation to suitable algebraic-geometric SU_2-fields (§4.3), induced by a certain unramified covering of the spectral curve was established (in the complex variant) by E. Date (cf. [Dat2]).

He also constructed "finite-zone" solutions of the so-called massive Thirring model (cf. [Dat1]) which is similar to the equations studied in this section, and found an explicit formula for invariant finite-zone SL_2-fields (e.g., for finite-zone solutions of the Pohlmeyer-Lund-Regge-Getmanov system). Note that statement b) of Corollary4.2 (in the case $p \leqq g$) can be derived from the result of [DuNo], where the reader can also find a discussion of reality properties of finite-zone solutions of the Sin-Gordon equation in the language of the period matrices and arguments of corresponding theta functions. The method of constructing algebraic-geometric Yang-Mills fields was proposed in [Ch7] (cf. also [Ch1]).

We mention before concluding this section that (complete or partial) trigonometric degenerations of solutions in theta functions and their relation with multi-soliton solutions are discussed in considerably many articles. One of the first works in this direction was done by A. R. Its, I. M. Krichever, V. B. Matveev (cf. [Mat3]). Among applications of non-degenerate and degenerate theta solutions of soliton equations, we note their use in the framework of the WKB method (Whitham method) in the works of V. P. Maslov, S. Yu. Dobrokhotov, H. Flashka, J. Forest, D. V. MacLaughlin, M. V. Karasyev and others. Theta solutions of the Sin-Gordon equation (Exercise4.2,b)) can be seen in Ch.11 of the book by G. Baker, "*Abel's Theorem and the allied theory including the theory of the Theta functions*", Cambridge, 1897 (cf. also the book [Mu]).

CHAPTER II

BÄCKLUND TRANSFORMS & INVERSE PROBLEM

The inverse scattering problem method occupies a central place in soliton theory. For the study of a soliton equation, it is not enough to know some of its exact solutions, rather we must describe classes of solutions with certain analytic properties. The inverse problem in principle allows us to do this for ***solitons***, i.e. for all solutions rapidly decreasing to zero (or stabilizing) as $x \to \pm\infty$. There are versions of the inverse problem method which allow us to study families of solutions of more general types, but here we do not discuss those methods. For instance, the results in §§3,4 of the previous chapter lead to an analogue of the inverse problem method for the class of almost-periodic solutions of the basic equations. See [ZMaNP], in which the KdV is discussed in detail, and [TF2].

The Bäcklund-Darboux transformations are closely related to the inverse problem method. By means of these transformations we can construct new solutions of a soliton equation from an arbitrary solution by solving an auxiliary system of linear differential equations. This system is in fact the associated linear problem. Therefore it is not surprising that Bäcklund-Darboux transformations are directly related to the inverse problem method. The so-called ***multi-soliton*** solutions turn out to be a result of the application of these transformations to the trivial (zero or constant) solutions. Generally speaking, they are superpositions of the simplest one-soliton solutions (of constant velocity and form) as $t \to \infty$ and can be written in terms of trigonometric functions of x and t. These solutions can also be obtained as the degeneration of the algebraic-geometric solutions of the previous chapter. The Bäcklund-Darboux transformations transform solitons into solitons and preserve multi-soliton solutions.

In §1 we examine the action of the Bäcklund-Darboux transformations on arbitrary solutions of the basic equations (PCF, GHM) and their reductions (S^{n-1}-fields, solutions of the Sin-Gordon and the NS) from various viewpoints. We demonstrate two different fundamental approaches to these transformations, extending the classical results of Bäcklund and Darboux and making them as explicit as possible. The group theoretic aspects of them are studied in [Ch11]. In the inverse problem the Bäcklund-Darboux transformations ("dressing transformations") control the discrete spectrum (§2, §3).

In §2 elementary facts about the inverse scattering problem are presented systematically up to the reduction to a suitable Riemann problem. The latter is based on the following. Solutions $\Phi_\pm$ of the equation

$$\frac{\partial \Phi}{\partial x} + Q\Phi = \alpha U_0 \Phi$$

which are analytic with respect to α in the upper and lower half planes respectively are constructed by the method of M. G. Krein (via the triangular factorization of Jost solutions). Here $Q(x)$ is a matrix function integrable over the x-axis and U_0 is a diagonal anti-hermitian matrix $U_0 = \mathrm{diag}(\mu_1, \dots, \mu_n)$ in which it is not necessary to suppose that the numbers $\{\mu_k\}$ are pairwise distinct. Then the ratio $S = (\Phi_-)^{-1}\Phi_+$ for $\alpha \in \mathbb{R}$ does not depend on x. Moreover, the map $Q(x) \mapsto S(\alpha)$ turns out to be one-to-one modulo "discrete spectral data" and is infinitesimally invertible. This theorem (together with the construction of $\Phi_\pm$) is fundamental in §2.

General theorems of §2 are applied to the basic equations and their reductions in §3. Here Q and $\Phi_\pm$ depend on t. We can find a simple transformation law of the entries of S and the discrete spectral data with respect to t, which then allows implementation of the corresponding integration procedure. For pairwise distinct $\{\mu_k\}$ the dependence on t results directly from the definition. When some of the μ_k coincide, we need a more detailed discussion. We consider S and the scattering data for O_n and S^{n-1}-fields in more detail. As an application, the trace formulae are derived which allows us to express the local integrals of motion of Ch.I in terms of the scattering data and then clarify and interpret some of the statements obtained in §1 (ibid). In particular, we solve the problem of their functional independence.

We note that the examples of the scattering data of concrete solutions of the NS equation and the so-called derivative NS equation (DNS) considered at the end of §3 are closely related to the numerical study of multi-soliton solutions of those equations as well as to the variational formulae established in §3. On one hand, these examples show how to get and study certain solutions with the scattering data expressed in terms of elementary functions. On the other hand, they illustrate how to use the inverse problem to describe nonlinear effects in soliton equations theoretically and numerically. The numerical (computer) simulation of these equations is very interesting, but we do not discuss this here.

The general goal of this chapter is to give a systematic algebraic exposition of the direct scattering problem oriented to applications. Our more concrete purpose is to connect the algebraic theory of the Bäcklund-Darboux transformations and the methods in Chapter I with the scattering theory.

§1. Bäcklund transformations

The first §1.1 contains the definition of the Bäcklund-Darboux transformations, "dressing" transformations of solutions of the basic equations (first without imposing reduction constraints). In §1.2 differential relations between the initial and transformed fields are calculated for transformations of the simplest type and specialized in the case of chiral U_n, O_n, S^{n-1}- fields. Then in §1.3 we demonstrate the connection of the transformations of §§1.1,2 with the classical Bäcklund transformation of solutions of the Sin-Gordon equation. §1.4 consists of applications and a study of local conservation laws. In §1.5 we show the relationship of our constructions to the classical Darboux transformations. As examples, the KdV and the NS equations are investigated.

The material is self-contained and does not use any special facts about differential algebras or the theory of differential equations. It is enough for the reader to know basic theorems on linear differential equations (cf., for example, [A2]).

1.1. Transformations of basic equations.

In this section $g(x,t)$ is a solution of the PCF equation (0.1). The pair $U(x,t)$ and $V(x,t)$ is a solution of system (0.2) corresponding to g. We also consider simultaneously the case when $U(x,t)$ satisfies the GHM equation (0.9) with conditions (0.10). Let us recall that in both cases the corresponding equations are the consistency conditions of the equations

$$\Phi_x = (1-\gamma)U\Phi = \frac{1}{1-\lambda}U\Phi, \tag{1.1}$$

$$\Phi_t = (1-\gamma^{-1})V\Phi = \frac{1}{1+\lambda}V\Phi, \tag{1.1A}$$

$$\Phi_t = \Big(\frac{c^2}{(1-\lambda)^2}U + \frac{1}{1-\lambda}[U, U_x]\Big)\Phi, \tag{1.1B}$$

where, as in Ch.I, the letters A,B correspond to the PCF and the GHM.

Using the conservation laws $\mathrm{Sp}(U^s)_t = 0 = \mathrm{Sp}(V^s)_x$ $(s = 1,2,\dots)$ for the PCF, we perform the following reduction. Fix the eigenvalues $(\mu_1(x),\dots,\mu_n(x))$, $(\nu_1(t),\dots,\nu_n(t))$ of the matrices U and V which do not depend on t and on x respectively. Let us assume that U and V are diagonalizable. For the GHM, the corresponding requirement is condition (0.10):

$$U \sim c_1 \sum_{i=1}^{p} I_i^i + c_2 \sum_{i=p+1}^{n} I_i^i,$$

where $c_{1,2} \in \mathbb{C}$, $c \overset{\mathrm{def}}{=} c_1 - c_2 \neq 0$, $1 \leqq p < n$. We do not specify the domains of U, V in this section. One may assume that the variables x, t belong to an appropriate open connected dense everywhere domain in $\mathbb{R}^2$.

A conjugation by a constant invertible matrix $F : U \mapsto F^{-1}UF$ and $V \mapsto F^{-1}VF$ is the simplest transformation of (0.1), (0.2) and (0.9) which preserves the eigenvalues of U and V in an obvious way. If U and V are $*$-anti-hermitian (cf. Introduction), then F should be chosen $*$-unitary (like g). This conjugation, called the ***chiral symmetry***, lifts to a transformation $g \mapsto F^{-1}g\mathcal{D}$ for an arbitrary invertible constant matrix $\mathcal{D}$. Our first goal is to construct a one-parameter family of transformations which is a generalization of the chiral symmetry and an analogue of the Pohlmeyer transformation [Poh] for S^{n-1}-fields.

We construct a solution $\Phi_0 = \Phi(x,t;\gamma_0)$ of system (1.1,2) for a fixed value of $\gamma = \gamma_0 \in \mathbb{C}^* = \mathbb{C}\setminus\{0\}$ which is uniquely determined by the value $\Phi(x_0,t_0;\gamma_0)$ at a certain point (x_0,t_0) in the domain of definition of U and V (or of only U for the GHM). In order to do this, we can first find $\Phi(x_0,t;\gamma_0)$ by (1.2A or B) and then (1.1) determines all Φ_0. It only remains to check that this Φ_0 satisfies (1.2) for all x, t. This follows from the uniqueness theorem of differential equations, since the difference of the left and right hand sides of (1.2) must be a solution of (1.1) with zero boundary value at $t = t_0$ (by the construction of Φ_0) via the zero curvature representation.

Proposition1.1. a) *Let $\Phi(x,t;\gamma)$ be an arbitrary matrix solution of system (1.1,2A). Then the function*

$$\widetilde{\Phi}(x,t;\gamma) = \Phi_0^{-1}\Phi(x,t;\gamma\gamma_0)$$

satisfies, in the case of the PCF, the same system for the pair of functions

$$\widetilde{U}(x,t) = \gamma_0\Phi_0^{-1}U\Phi_0, \qquad \widetilde{V}(x,t) = \gamma_0^{-1}\Phi_0^{-1}V\Phi_0,$$

corresponding to the principal chiral field $\widetilde{g} = \Phi_0^{-1}\Phi(x,t;-\gamma_0)$. The field $\widetilde{g}$ is related to $\widetilde{U}$ and $\widetilde{V}$ by

$$\widetilde{g}_x\widetilde{g}^{-1} = \widetilde{U}, \qquad \widetilde{g}_t\widetilde{g}^{-1} = \widetilde{V}.$$

The eigenvalues of $\widetilde{U}$ and $\widetilde{V}$ are obtained from $\{\mu_j\}$, $\{\nu_j\}$ by means of multiplication by $\gamma_0^{\pm 1}$ respectively.

b) For system (1.1,2B) the function $\widetilde{U}$ of a) satisfies

$$\gamma_0\widetilde{U}_t = [\widetilde{U},\widetilde{U}_{xx}] + 2\widetilde{c}^2(\gamma_0^{-1}-1)\widetilde{U}_x$$

and relation (1.10) with $\widetilde{c} = \gamma_0 c$.

Proof. We have

$$\begin{aligned}\widetilde{\Phi}_x\widetilde{\Phi}^{-1} &= \Phi_0^{-1}(1-\gamma\gamma_0)U\Phi_0 - \Phi_0^{-1}(\Phi_0)_x \\ &= (\gamma_0 - \gamma\gamma_0)\Phi_0^{-1}U\Phi_0.\end{aligned}$$

Analogously we can compute $\widetilde{\Phi}_t\widetilde{\Phi}^{-1}$ and check the relation for $\widetilde{g}$. □

Note that for real γ_0 (and real c^2 in case B) it is easy to get a solution of (0.2) and (0.9) with the same eigenvalues as those of U and V from $\widetilde{U}$ and $\widetilde{V}$ by means of a constant linear transformation of the variables x and t.

Exercise1.1. *Let us denote $\widetilde{g}$ of Proposition1.1 by $\widetilde{g} = g^{(\gamma)}$, and write $g \sim g'$ if g' is obtained from g by the chiral symmetry (see above). Then*

a) $g^{(\gamma_1\gamma_2)} \sim ((g^{(\gamma_1)})^{(\gamma_2)}$,

b) $g^{(1)} \sim g$,

c) $g^{(-1)} \sim g^{-1}$. □

Now we generalize the above construction, by multiplying Φ on the left by matrices depending on γ. Let K_1 and K_2 be two subspaces of $\mathbb{C}^n$ of complementary dimension, that is, $\dim K_1 + \dim K_2 = n$, and let $\lambda_1 \neq \lambda_2 \in \mathbb{C}$. For a given pair U and V fix two invertible solutions of system (1.1,2) at $\lambda = \lambda_1, \lambda_2$ respectively. We denote the projection onto the space $\Phi_1 K_1$ parallel to $\Phi_2 K_2$ by P. We assume further that $\lambda_{1,2} \neq \pm 1$ (PCF), $\neq 1$ (GHM) and $\Phi_1 K_1 \cap \Phi_2 K_2 = 0$. Put

$$B = I - \frac{\lambda_2 - \lambda_1}{\lambda_2 - \lambda} P.$$

Theorem1.1. a) *If Φ satisfies (1.1,2) in some neighbourhood of the point $\lambda' \in \mathbb{C}$, then $\widetilde{\Phi} = B\Phi$ is a solution of (1.1,2) in a neighbourhood of $\lambda' \neq \lambda_1$ and $\lambda_2, \pm 1$ for*

$$\widetilde{U} = U + P_x(\lambda_1 - \lambda_2), \qquad \widetilde{V} = V + P_t(\lambda_2 - \lambda_1). \tag{1.3}$$

b) Matrices $\widetilde{U}$ and $\widetilde{V}$ are equivalent to U and V respectively (by means of conjugation by matrices $B(\pm 1)$). The function

$$\widetilde{g} = \left(I - \frac{\lambda_2 - \lambda_1}{\lambda_2} P\right) g$$

is a solution of (0.1) corresponding to $\widetilde{U}$ and $\widetilde{V}$ in the case of the PCF.

Proof. The definition of $\widetilde{\Phi}$ implies the formulae

$$\widetilde{\Phi}_x\widetilde{\Phi}^{-1} = B_x B^{-1} + \frac{1}{1-\lambda} BUB, \tag{1.4}$$

(1.5A)

$$\widetilde{\Phi}_t\widetilde{\Phi}^{-1} = B_t B^{-1} + \frac{1}{1+\lambda} BVB,$$

(1.5B)

$$\widetilde{\Phi}_t\widetilde{\Phi}^{-1} = B_t B^{-1} + \frac{c^2}{(1-\lambda)^2} BUB^{-1} + \frac{1}{1-\lambda} B[U, U_x]B^{-1}.$$

We will show that the right hand side of (1.4) is rational with respect to λ and does not have poles at $\lambda = \lambda_1, \lambda_2$. In a neighbourhood of $\lambda = \lambda_1$ we can represent it in the form $\widehat{\Phi}_x\widehat{\Phi}^{-1}$ where $B^{-1}\widehat{\Phi}$ is a certain analytic continuation of Φ_1 in a neighbourhood of the point $\lambda = \lambda_1$. As $B^{-1} = \boldsymbol{I} - \frac{\lambda_1-\lambda_2}{\lambda_1-\lambda}P$,

$$\widehat{\Phi}^{-1}\mathcal{O}_1^n = K_1\mathcal{O}_1(\lambda-\lambda_1)^{-1} + \mathcal{O}_1^n \stackrel{\text{def}}{=} \mathcal{K}_1,$$

where $\mathcal{O}_1 = \mathcal{O}_{\lambda_1}$ is the ring of formal power series in $(\lambda - \lambda_1)$ with coefficients depending on x, $\mathcal{O}_1^n = \mathcal{O}_1 \otimes_{\mathbb{C}} \mathbb{C}^n$; $\widehat{\Phi}^{-1}\mathcal{O}_1^n$ is the image of $\mathcal{O}_1^n$ under multiplication of the matrix $\widehat{\Phi}^{-1}$ by all vector functions from $\mathcal{O}_1^n$. Differentiating $\widehat{\Phi}\mathcal{K}_1 = \mathcal{O}_1^n$ with respect to x (here $(\mathcal{O}_1^n)_x = \mathcal{O}_1^n$ and $(\mathcal{K}_1)_x = \mathcal{K}_1$ since K_1 does not depend on x), we obtain the relation

$$\widehat{\Phi}_x(\mathcal{K}_1) = \widehat{\Phi}_x\widehat{\Phi}^{-1}\mathcal{O}_1^n \subset \mathcal{O}_1^n.$$

It follows from the last inclusion that $\widehat{\Phi}_x\widehat{\Phi}^{-1}$ is analytic at the point $\lambda = \lambda_1$. An analogous argument holds for $\lambda = \lambda_2$ and for (1.5) instead of (1.4).

In this way we see that the right hand sides of (1.4) and (1.5) which depend rationally on λ do not have singularities at the points $\lambda = \lambda_1, \lambda_2$ (at the poles of the matrix elements of B, B^{-1}) and may have poles only at $\lambda = \pm 1$ of first order (or less) for (1.4,5A), and at $\lambda = 1$ of order $\leqq 2$ for (1.5B). Taking the equality, $B(\lambda = \infty) = \boldsymbol{I}$ into account we obtain the relations

$$\begin{aligned}
\widetilde{\Phi}_x\widetilde{\Phi}^{-1} &= \frac{1}{1-\lambda}\widetilde{U},\\
\widetilde{\Phi}_t\widetilde{\Phi}^{-1} &= \frac{1}{1+\lambda}\widetilde{V}(\text{ in the case of A }),\\
\widetilde{\Phi}_t\widetilde{\Phi}^{-1} &= \frac{c^2}{(1-\lambda)^2}\widetilde{U} + \frac{1}{1-\lambda}\widetilde{W}(\text{ case B})
\end{aligned}$$

for $\widetilde{U} = B(1)UB(1)^{-1}$ and $\widetilde{V} = B(-1)VB(-1)^{-1}$ and a suitable matrix function $\widetilde{W}$. In the case of (B) in a neighbourhood of $\lambda = 1$ we can express the right hand side of (1.5B) in the form $(B\check{\Phi})_t\check{\Phi}^{-1}B^{-1}$ where $\check{\Phi}$ is the formal solution of (1.1,2B) from Proposition1.2, Ch.I for U. Since $B\check{\Phi}$ is a solution of the same type for $\widetilde{U}$ and $\widetilde{W} = [\widetilde{U}, \widetilde{U}_x]$ (cf. §1.4, Ch.I). This argument proves all the claims of the theorem except for (1.3) since the relation of $\widetilde{g}$ with $\widetilde{U}$ and $\widetilde{V}$ is trivial, as $\widetilde{g} = B(0)g$. The remaining formulae are obtained from the expansion of (1.4) and (1.5A) with respect to λ^{-1} in a neighbourhood of $\lambda = \infty$. □

Exercise1.2. *If we determine $\widetilde{U}$ and $\widetilde{V}$ by U and V and get γ_0 by the construction of Proposition1.1, apply the transformation in Theorem1.1 to $\widetilde{U}$ and $\widetilde{V}$ for λ_1 and λ_2 corresponding to γ_1 and γ_2 and again let the transformation in Proposition1.1 for γ_0^{-1} act on the result, then we obtain matrix functions which come from U and V by the transformation of the type described in Theorem1.1 with $\gamma_1' = \gamma_1\gamma_0$ and $\gamma_2' = \gamma_2\gamma_0$. All transformations from the therorem for $\lambda_{1,2}' = (\gamma_{1,2}'+1)/(\gamma_{1,2}'-1)$ are given by the above construction for suitable initial K_1 and K_2 corresponding to λ_1 and λ_2.* □

Developing the proof of Theorem1.1 (cf. also [Ch11]), we arrive at the following proposition.

Proposition1.2. *Let Φ be a solution of system (1.1,2) which is defined and analytic in a neighbourhood of the points $\lambda_1, \dots, \lambda_N \neq \pm 1$ and $B(\lambda)$ a rational matrix function of λ for which*

a) $B(\infty) = I$,
b) *matrix elements of B and B^{-1} do not have poles at $\lambda \neq \lambda_1, \dots, \lambda_N$,*
c) $\Phi^{-1}B^{-1}\mathcal{O}_i^n \overset{\text{def}}{=} \mathcal{K}_i$ *do not depend on x, t, i.e., are generated as $\mathcal{O}_i$-modules by a vector-valued series in the parameters $\lambda - \lambda_i$ which are constant with respect to x and t.*

Here $\mathcal{O}_i = \mathcal{O}_{\lambda_i}$ is defined for the local parameter $\lambda - \lambda_i \leqq i \leqq N$ as for $\mathcal{O}_1$ (cf. above). Then $\widetilde{\Phi} = B\Phi$ is a solution of (1.1,2) which is analytic for λ in some domain and corresponds to certain solutions $\widetilde{U}$ and $\widetilde{V}$ of (0.2) or (0.9) with the same eigenvalues as U and V. □

In the next chapter we discuss the restrictions on $\{\mathcal{K}_i\}$ and Φ which ensure the existence of B. For the moment we turn to an example which illustrates Proposition1.2 and shows what happens with Theorem1.1 when the points λ_1 and λ_2 are equal.

Let $\Phi = \Phi_1 + (\lambda - \lambda_1)\Phi_1' + o(\lambda - \lambda_1)$ be the expansion of an invertible solution Φ of system (1.1,2) in a neighbourhood of the point $\lambda = \lambda_1 \neq \pm 1$. For $\sigma \in \mathbb{C}^*$ and a constant matrix Q with the property $Q^2 = 0$, put $F = \Phi_1 + \Phi_1' Q$ and denote a rational matrix function of λ (depending also on x, t) which transforms a vector of the form Fz, $z \in \mathbb{C}^n$ to $Fz - \sigma(\lambda - \lambda)^{-1}\Phi_1 Qz$ by B. Hereafter we suppose that the matrix F is invertible. Then

$$\begin{aligned} B^{-1}\mathcal{O}_1^n &= (F + \sigma(\lambda - \lambda_1)^{-1}\Phi_1 Q)\mathcal{O}_1^n \\ &= \Phi(I\sigma(\lambda - \lambda_1)^{-1}Q)\mathcal{O}_1^n. \end{aligned}$$

This makes it possible to apply Proposition1.2.

Corollary1.1. *For σ, λ_1, Q and F introduced above, the functions*

$$\widetilde{U} = U + \sigma(FQF^{-1})_x, \qquad \widetilde{V} = V - \sigma(FQF^{-1})_t$$

are solutions of (0.2) and (0.9) which are equivalent to U and V.

Proof. $B = \boldsymbol{I} - \sigma(\lambda - \lambda_1)^{-1}FQF^{-1}$. □

Exercise1.3. *Carry over Theorem1.1 and Proposition1.2 to equations (2.3) and (2.5) of Ch.I, constructing B from solutions at points $\lambda_1, \dots, \lambda_N$ of the proper analogues of system (1.1,2B) with the normalization $B(k = 0) = \boldsymbol{I}$ and $B(k = \infty) = \boldsymbol{I}$ for (2.3) and (2.5) respectively, where $k = (1 - \lambda)^{-1}$ (cf. §2, Ch.I).* □

Exercise1.4. *([BZ], [CGW], [Taka], [UN]). In the notation of §4.4, Ch.I, let $\{A_j\}$ be a solution of the complex duality equation (4.23) and Φ a solution of the system*

$$(\nabla_1 + \lambda\nabla_2)\Phi = 0, \qquad (\nabla_3 - \lambda^{-1}\nabla_4)\Phi = 0$$

for ∇_i constructed from $\{A_j\}$. If a rational function B of λ satisfies conditions b) and c) of Proposition1.2, then $\widetilde{\Phi} = B\Phi$ satisfies the same condition as Φ with an appropriate $\{A'_j\}$ which form a solution of (4.23). □

1.2. Bäcklund transformations of U_n, O_n, and S^{n-1}-fields.

In this section we describe a more traditional approach to Bäcklund transformations via a differential relation between initial and transformed solutions of the equations under consideration. Hereafter we denote the complex conjugate and the hermitian conjugate of a matrix A by $\overline{A}$ and A^+, respectively.

For the unitary PCF it is necessary in the construction of Theorem1.1 to choose conjugate parameters $\lambda_1 = \lambda_0$, $\lambda_2 = \overline{\lambda}_0$ for $\lambda_0 \neq \overline{\lambda}_0$, $\lambda_0 \neq \pm 1$, and then take one (constant) space $K \subset \mathbb{C}^n$ and construct the hermitian projection P_K onto $\Phi_1 K = \Phi(\lambda_0)K$. Then $P_K^+ = P_K \Rightarrow B(0)^+B(0) = \boldsymbol{I} \Rightarrow \widetilde{g}^+\widetilde{g} = \boldsymbol{I} \Rightarrow \widetilde{U}^+ + \widetilde{U} = 0 = \widetilde{V}^+ + \widetilde{V}$ in the notation of Theorem1.1 if $g^+g = \boldsymbol{I}$. However, the projection P_K admits another definition as a solution of a certain system of differential equations.

Proposition1.3. *The projection P_K is a solution of the system of equations*

$$PP_x = \frac{1}{1 - \overline{\lambda}_0} PU(P - \boldsymbol{I}), \tag{1.6a}$$

$$PP_t = \frac{1}{1 + \overline{\lambda}_0} PV(P - \boldsymbol{I}). \tag{1.6b}$$

Conversely, any hermitian projection P satisfying (1.6) for anti-hermitian solutions U and V of the PCF system (0.2) can be uniquely expressed in the form P_K for λ_0 and a fixed invertible solution $\Phi(\lambda_0)$ of system (1.1,2A).

Proof. Relations (1.6) together with their hermitian conjugates are equivalent to the absence of poles in the right hand sides of (1.4,5A) at $\lambda = \lambda_0, \overline{\lambda}_0$ and, consequently, they are satisfied by P_K. Using hermitian conjugation and the identity $P^2 = P$, we can rewrite (1.6) into an equivalent form

$$P_x = \frac{1}{1-\overline{\lambda}_0} PU(P-I) - (1-\lambda_0)(P-I)UP, \tag{1.7a}$$

$$P_t = \frac{1}{1+\overline{\lambda}_0} PV(P-I) - (1+\lambda_0)(P-I)VP. \tag{1.7b}$$

An arbitrary solution P of the above system is uniquely determined by the value $P(x_0, t_0)$ at an arbitrary point (x_0, t_0) in the domain of P. Therefore P is uniquely expressed in the form P_K. □

Theorem1.2. *Let $g(x,t)$ be a unitary solution of the PCF equation (0.1), $\lambda_0 \in \mathbb{C}$ with $\lambda_0 \neq \pm 1$. The system of equations*

$$\widetilde{g}_x\widetilde{g}^{-1} - g_x g^{-1} = \overline{\lambda}_0(\widetilde{g}g^{-1})_x, \tag{1.8a}$$

$$\widetilde{g}_t\widetilde{g}^{-1} - g_t g^{-1} = -\overline{\lambda}_0(\widetilde{g}g^{-1})_t, \tag{1.8b}$$

$$\overline{\lambda}_0\widetilde{g}g^{-1} + \lambda_0 g\widetilde{g}^{-1} = (\lambda_0 + \overline{\lambda}_0)I \tag{1.8c}$$

has a unique unitary solution ($\widetilde{g}\widetilde{g}^{+} = I$) in a neighbourhood of a point (x_0, t_0) for any initial value $\widetilde{g}(x_0, t_0)$ which is unitary and satisfies (1.8c). The function $\widetilde{g}$ satisfies (0.1). The eigenvalues of the matrices $U = g_x g^{-1}$ and $V = g_t g^{-1}$ and those of $\widetilde{U} = \widetilde{g}_x\widetilde{g}^{-1}$ and $\widetilde{V} = \widetilde{g}_t\widetilde{g}^{-1}$ coincide.

Proof. For a certain matrix function $P(x,t)$ we can put $\widetilde{g} = B_0 g$, where $B_0 = I - \frac{\overline{\lambda}_0 - \lambda_0}{\overline{\lambda}_0} P$. If $\widetilde{g}$ is a unitary solution of (1.8), then by (1.8c)

$$B_0^{-1} = g\widetilde{g}^{-1} = I - \frac{\lambda_0 - \overline{\lambda}_0}{\lambda_0} P.$$

Consequently, $P^2 = P$. Since $\widetilde{g}$ is unitary, B_0 is unitary and the projection P is hermitian. Substituting $\widetilde{g} = B_0 g$ into (1.8a,b), we get formulae (1.7) which are equivalent to (1.6). Hence $P = P_K$ for a certain subspace K (Proposition1.3). On the other hand, any unitary matrix $\widetilde{g}(x_0, t_0)$ satisfying (1.8c) can be extended to the function $\widetilde{g} = B_0 g$, as above, by choosing an appropriate projection $P = P_K$. Then $\widetilde{g}$ of Theorem1.1 satisfies all the requirements of the theorem. □

O_n-fields. Now let us assume that $g(x,t) \in O_n$, i.e., $\overline{g} = g$. The simplest example of the Bäcklund transformation of O_n-fields is constructed from the initial subspace $K \subset \mathbb{C}^n$ (cf. above) for which $(K, \overline{K}) = 0$, where $(,)$ is the standard hermitian form on $\mathbb{C}^n$ (which is anti-linear for the second argument). Fixing $\overline{\Phi}(\lambda_0)$, we can put $\Phi(\overline{\lambda}_0) = \overline{\Phi(\lambda_0)}$, since $\Phi(\lambda) \stackrel{\text{def}}{=} \overline{\Phi(\overline{\lambda})}$ together with $\Phi(\lambda)$ is a solution of system (1.1,2A) for real U and V.

We denote the hermitian projections onto $\Phi(\lambda_0)K$ and $\Phi(\overline{\lambda}_0)\overline{K}$ by P_K and $P_{\overline{K}}$. Then $P_{\overline{K}} = \overline{P}_K$. Now we take $\Phi(\lambda_0)$ with its values in the complex orthogonal group (i.e. $\Phi(\lambda_0)\,{}^t\Phi(\lambda_0) = \boldsymbol{I}$). This is possible, since ${}^tU + U = 0 = {}^tV + V$. In this case

$$\begin{aligned}(\Phi(\lambda_0)K, \Phi(\overline{\lambda}_0)\overline{K}) &= (\Phi(\overline{\lambda}_0)^+\Phi(\lambda_0)K, \overline{K}) \\ &= ({}^t\Phi(\lambda_0)\Phi(\lambda_0)K, \overline{K}) \\ &= (K, \overline{K}) = 0.\end{aligned}$$

Hence, $P_K P_{\overline{K}} = P_{\overline{K}} P_K = 0$.

Proposition1.4. *The function*

$$B(\lambda) = \boldsymbol{I} - \frac{\overline{\lambda}_0 - \lambda_0}{\overline{\lambda}_0 - \lambda} P - \frac{\lambda_0 - \overline{\lambda}_0}{\lambda_0 - \lambda} \overline{P}$$

for $P = P_K$ above satisfies the condition of Proposition1.2 and is orthogonal for $\lambda \in \mathbb{R}$. The orthogonal field $\widetilde{g}$ and the anti-hermitian currents corresponding to $B\Phi$ have the form

$$\widetilde{g}g^{-1} = \boldsymbol{I} - \frac{\overline{\lambda}_0 - \lambda_0}{\overline{\lambda}_0} P - \frac{\lambda_0 - \overline{\lambda}_0}{\lambda_0} \overline{P} = B(0), \tag{1.9a}$$

$$\widetilde{U} = U + (\lambda_0 - \overline{\lambda}_0)(P - \overline{P})_x, \qquad \widetilde{V} = V + (\overline{\lambda}_0 - \lambda_0)(P - \overline{P})_t. \tag{1.9b}$$

Proof. From the equation $P\overline{P} = 0$ follows the formula

$$B^{-1}(\lambda) = \boldsymbol{I} - \frac{\lambda_0 - \overline{\lambda}_0}{\lambda_0 - \lambda} P - \frac{\overline{\lambda}_0 - \lambda_0}{\overline{\lambda}_0 - \lambda} \overline{P}.$$

In particular, $B(\lambda)B(\overline{\lambda})^+ = \boldsymbol{I}$. Taking the trivial relation $\overline{B(\lambda)} = B(\overline{\lambda})$ into account, we see that $B(\lambda)$ is orthogonal for $\lambda \in \mathbb{R}$.

Properties b) and c) of Proposition1.2 should be checked for $\lambda = \lambda_0, \overline{\lambda}_0$. Since $\Phi(\overline{\lambda}_0) = \overline{\Phi(\lambda_0)}$ and $B(\overline{\lambda}_0) = \overline{B(\lambda_0)}$, it is enough to check it only for the point λ_0. We have:

$$B^{-1}\mathcal{O}_{\lambda_0} = \Phi(\lambda_0)((\lambda - \lambda_0)^{-1}K + \overline{K}^{\perp} + (\lambda - \lambda_0)\mathcal{O}^n_{\lambda_0}),$$

where $\overline{K}^{\perp}$ is the orthogonal complement of $\overline{K}$ in $\mathbb{C}^n$. □

Here we will not analyze the differential equations defining P (cf. Proposition1.3) and express the transformation $g \mapsto \widetilde{g}$ in the form of a differential relation (as in Theorem1.2) in the case of O_n-fields.

S^{n-1}-fields. We now impose one more reduction condition on an O_n-field g. Let $g^2 = I$. If we introduce the parameter $\gamma = \frac{\lambda+1}{\lambda-1}$ instead of λ, it follows that $g(x,t)\Phi(x,t;-\gamma)$ satisfies (1.1,2A) together with Φ, where $\Phi(\gamma) \stackrel{\text{def}}{=} \Phi(\lambda)$. The converse is also true up to the chiral symmetry of g. Put $\gamma_0 = (\lambda_0+1)/(\lambda_0-1)$. We keep all previous notation, in particular, $\Phi(\overline{\gamma}_0) = \overline{\Phi(\gamma_0)}$.

Proposition1.5. *Let us assume that $\gamma_0 \in i\mathbb{R}$ ($\Leftrightarrow |\lambda_0| = 1$) and $g\Phi(\gamma_0)K = \Phi(\overline{\gamma}_0)\overline{K}$ at some point (x_0, t_0) in a (connected) domain of definition of $\Phi(\gamma_0)$ for x, t. Then $\widetilde{g}^2 = 1$. Conversely, the last relation (the involutivity of $\widetilde{g}$) implies the above restriction for γ_0 and K, if the matrices U and V corresponding to g are not non-degenerate (do not preserve any proper subspace of $\mathbb{C}^n$ which is constant with respect to x and t).*

Proof. Let us suppose that $\widetilde{g}^2 = I$ and that U and V are non-degenerate. Choose one solution $\Phi^0(x;\gamma)$ of equation (1.1) normalized by the condition $\Phi^0(x_0;\gamma) = I$ (temporarily we omit t and do not consider equation (1.2A)). We may assume that this function Φ^0 is an analytic function of γ in a certain domain which is symmetric with respect to the reflection $\gamma \mapsto -\gamma$ and contains γ_0 and $\overline{\gamma}_0$ as a result of the well-known theorem regarding analytic dependence on parameters of solutions of differential equations. Then the function $\widetilde{\Phi}^0 = \widetilde{\Phi}(x;\gamma)B(x_0;\gamma)^{-1}$, constructed from the function $\widetilde{\Phi} = B\Phi^0$, is a solution of (1.1) for $\widetilde{U}$ with normalization $\widetilde{\Phi}^0(x_0;\gamma) = I$, which singles out $\widetilde{\Phi}^0$ uniquely among all solutions. The function $\widetilde{\Phi}^0$ is analytic and invertible in the same domain as Φ^0 except at the points γ_0 and $\overline{\gamma}_0$. Due to the uniqueness of solutions of differential equations,

$$\widetilde{\Phi}^0 = \widetilde{g}(x)\widetilde{\Phi}(x;-\gamma)\widetilde{\Phi}(x_0;-\gamma)^{-1}\widetilde{g}(x_0)^{-1}.$$

The above function is expressed in terms of $\Phi(x;-\gamma)$, $B(x;-\gamma)$ and hence is analytic and invertible everywhere except at $-\gamma_0$ and $-\overline{\gamma}_0$. Consequently, if $\overline{\gamma}_0 \neq \gamma_0$, $\widetilde{\Phi}^0$ must be invertible at γ_0, which contradicts the assumption that U and V are non-degenerate.

Moreover, if $\widetilde{g}^2 = I$, then

$$I - (\overline{\lambda}_0 - \lambda_0)(P/\overline{\lambda}_0 - \overline{P}/\lambda_0) = I - (\lambda_0 - \overline{\lambda}_0)(gPg/\lambda_0 - g\overline{P}g/\overline{\lambda}_0).$$

Therefore $gPg = \overline{P}$, since the pairs of projections P, $\overline{P}$ and gPg, $g\overline{P}g$ are orthogonal. This proves the necessity. Reversing the last discussion, we obtain that

$$g\Phi(\gamma_0)K = \Phi(\overline{\gamma}_0)\overline{K} \Leftrightarrow gPg = \overline{P} \Rightarrow \widetilde{g}^2 = I.$$

We can prove the remaining part, noting that $\Phi(\overline{\gamma}_0)^{-1}g\Phi(\gamma_0) = \Phi(-\gamma_0)^{-1}g\Phi(\gamma_0)$ does not depend on x, t, if $\gamma_0 \in i\mathbb{R}$. □

In this section we apply the last reduction, restricting ourselves to involutive g for which the corresponding projection $Q \stackrel{\text{def}}{=} (I-g)/2$ is an orthogonal projection onto a $\mathbb{R}^n$-valued function $q(x,t)$ with the condition that $(q,q) = 1$. As we checked in the Introduction, equation (0.1) then follows from the equation of S^{n-1}-fields (0.5). Let us recall that $U = 2q \wedge q_x$ and $V = 2q \wedge q_t$ for an S^{n-1}- field q. Also recall the notation:

$$(p \wedge q)z = (z,p)q - (z,q)p$$

for $p, q, z \in \mathbb{C}^n$.

Following Proposition1.4, we select the projection P onto the one-dimensional subspace $\mathbb{C}\varphi$ for the solution $\varphi(x,t) \in \mathbb{C}^n$ of (1.1,2A) with the condition $(\varphi, \overline{\varphi}) = 0$. Here the one-dimensionality of P is connected to the one-dimensionality of Q. In addition, γ_0 must be taken from $i\mathbb{R}$, and the function $r \stackrel{\text{def}}{=} \sqrt{2}\varphi(\varphi,\varphi)^{-1/2}$ must have the form $r = p + iq$, where $(p,p) = 1$, $(p,q) = 0$, and q is the initial S^{n-1}-field (Proposition1.5). Let us comment on the latter conditions.

Regarding the last restriction, we see that since g permutes P and $\overline{P}$, $g\varphi = u\varphi$, where $|u| = 1$. Generally speaking, u may be some scalar function of x and t. But, as $g\varphi$ and $\overline{\varphi}$ satisfy the same system of differential equations (1.1,2A), u must be constant. Put $u^{-1} = v^2$ for $|v| = 1$. Then

$$g(v\varphi) = uv^2(\overline{v\varphi}) = (\overline{v\varphi}).$$

Hence, multiplying φ by a suitable constant, we achieve the relation $gr = \overline{r}$, which is equivalent to the following two conditions on the real and imaginary parts of r:

$$g\,\mathrm{Re}\,r = \mathrm{Re}\,r \Leftrightarrow (q, \mathrm{Re}\,r) = 0, \qquad g\,\mathrm{Im}\,r = -\,\mathrm{Im}\,r \Leftrightarrow \mathrm{Im}\,r = \pm q.$$

Here we use the identities $(\mathrm{Re}\,r, \mathrm{Re}\,r) = 1 = (\mathrm{Im}\,r, \mathrm{Im}\,r)$ and $(\mathrm{Re}\,r, \mathrm{Im}\,r) = 0$, which are equivalent to the constraints, $(r,r) = 2$ and $(r,\overline{r}) = 0$.

Thus we have shown that we can find the solution φ for which the function r has the required form. The construction from Proposition1.5 of the projection P corresponding to φ gives an O_n-field $\widetilde{g}$ with the condition $\widetilde{g}^2 = I$ which is connected with a certain S^{n-1}-field $\widetilde{q}$. Our goal here is to give an alternative description of the function p analogous to that given in Proposition1.3 and to calculate $\widetilde{q}$ in terms of p. Put $i\gamma_0 \stackrel{\text{def}}{=} \kappa \in \mathbb{R}$. We apply the relation $(q, q_x) = 0 = (q, q_t)$ derived from the equation $(q, q) = 1$.

Differentiating $r = \sqrt{2}\varphi(\varphi, \varphi)^{-1/2}$, we have the following equations:

$$r_x = \frac{1}{2}(1 + i\kappa)Ur - \frac{i\kappa}{4}(Ur, r)r, \tag{1.10a}$$

$$r_t = \frac{1}{2}(1 - i\kappa^{-1})Vr + \frac{i\kappa^{-1}}{4}(Vr, r)r, \tag{1.10b}$$

derived from the formula

$$(\varphi, \varphi)_x = \frac{1}{2}(1 + i\kappa)(U\varphi, \varphi) + \frac{1}{2}(1 - i\kappa)(\varphi, U\varphi) = i\kappa(U\varphi, \varphi)$$

and the analogous formula for $(\varphi, \varphi)_t$. Substituting $r = p + iq$ into (1.10), we get the equations

$$(p + \kappa q)_x = (p, q_x)(\kappa p - q), \tag{1.11a}$$

$$(\kappa p - q)_t = -(p, q_t)(p + \kappa q), \tag{1.11b}$$

$$(p, p) = 1, \qquad (p, q) = 0. \tag{1.11c}$$

Let us derive, for instance, (1.11a) from (1.10a). We have

$$\begin{aligned} 2p_x &= Up - \kappa Uq + \kappa(Uq, p)p \\ &= 2(q \wedge q_x)(p - \kappa q) + \kappa(2(q \wedge q_x)q, p)p \\ &= -2(q_x, p)q - 2\kappa q_x + 2\kappa(q_x, p)p \\ &= 2\kappa q_x + 2(p, q_x)(\kappa p - q). \end{aligned}$$

Exercise1.5. *Show that any solution p of system (1.11) can be obtained by the above procedure from the solution $\varphi(x, t) \in \mathbb{C}^n$ of system (1.1,2A) with the conditions $(\varphi, \overline{\varphi}) = 0$, and $g(x_0, t_0)\varphi(x_0, t_0) = \overline{\varphi}(x_0, t_0)$ at a certain point (x_0, t_0) in the domain where φ is defined.* □

Now we calculate $\widetilde{g}$ from formula (1.9a). Put, for the sake of clarity, $(1 - \kappa^2)/(1 + \kappa^2) = \cos\beta$ and $2\kappa/(1 + \kappa^2) = \sin\beta$ for $0 \leqq \beta < 2\pi$. Then

$$\begin{aligned} \lambda_0 &= \frac{\gamma_0 + 1}{\gamma_0 - 1} = \frac{i\kappa - 1}{i\kappa + 1} = -\cos\beta + i\sin\beta, \\ \widetilde{g}r &= \left(\frac{i\kappa + 1}{i\kappa - 1}\right)^2 \overline{r}, \\ \widetilde{g}\overline{r} &= \left(\frac{i\kappa - 1}{i\kappa + 1}\right)^2 r. \end{aligned}$$

Consequently,

$$\widetilde{g}r' = \overline{r}, \qquad \widetilde{g}\overline{r}' = r'$$

for $r' = (i\kappa - 1)(i\kappa + 1)^{-1}r = (i\sin\beta - \cos\beta)r$. Since the image of the projection $\widetilde{Q} = (I - \widetilde{q})/2$ is included in the two dimensional space spanned by the vectors r and $\overline{r}$ (or p and q), $\widetilde{Q}$ is the projection onto $\widetilde{q} \stackrel{\text{def}}{=} \operatorname{Im} r' = p\sin\beta - q\cos\beta$. By Proposition1.5 the function $\widetilde{q}$ must satisfy (0.5). The points $\widetilde{U} = 2\widetilde{q} \wedge \widetilde{q}_x$ and $\widetilde{V} = 2\widetilde{q} \wedge \widetilde{q}_t$ corresponding to $\widetilde{q}$ can be easily expressed in terms of p, using (1.9b):

$$\widetilde{U} = U + 2\sin\beta(q \wedge p)_x, \qquad \widetilde{V} = V - 2\sin\beta(q \wedge p)_t.$$

Theorem1.3. *a) If q is a solution, the system of equations*

$$(\widetilde{q} + q)_x = \frac{\kappa^2 + 1}{2}(\widetilde{q}, q_x)(\widetilde{q} - q), \tag{1.12a}$$

$$(\widetilde{q} - q)_t = -\frac{\kappa^{-2} + 1}{2}(\widetilde{q}, q_t)(\widetilde{q} + q), \tag{1.12b}$$

$$(\widetilde{q}, \widetilde{q}) = 1, \qquad (\widetilde{q}, q) = \frac{\kappa^2 - 1}{\kappa^2 + 1} \tag{1.12c}$$

in $\widetilde{q}(x,t) \in \mathbb{R}^n$ has a unique solution for $\kappa \in \mathbb{R}$, $\kappa \neq 0$ and an arbitrary initial value $\widetilde{q}(x_0, t_0)$ satisfying (1.12c) at a certain point in a (connected) domain where q is defined.

b) Each solution $\widetilde{q}$ of system (1.12) satisfies the equation of the S^{n-1}-field. Moreover,

$$\tfrac{1}{2}\widetilde{U} = \widetilde{q} \wedge \widetilde{q}_x = q \wedge q_x + (q \wedge \widetilde{q})_x, (\widetilde{q}_x, \widetilde{q}_x) = (q_x, q_x),$$

$$\tfrac{1}{2}\widetilde{V} = \widetilde{q} \wedge \widetilde{q}_t = q \wedge q_t + (\widetilde{q} \wedge q)_t, \quad (\widetilde{q}_t, \widetilde{q}_t) = (q_t, q_t),$$

Proof. The existence of solutions of (1.12) for an arbitrary initial condition of type (1.12c) is proven above. The uniqueness reduces to standard properties of differential equations. Therefore it follows from Proposition1.5 that any solution $\widetilde{q}$ of system (1.12) satisfies (0.5). The formulae for $\widetilde{U}$ and $\widetilde{V}$ were obtained before and the norms of derivatives of q and $\widetilde{q}$ coincide since, by Proposition1.2, $\operatorname{Sp} U^2 = -2(q_x, q_x)$ and $\operatorname{Sp} V^2 = -2(q_t, q_t)$. □

Exercise1.6. *Check directly that solutions of system (1.12) satisfy (0.5),*

$$(\widetilde{q}_x, \widetilde{q}_x) = (q_x, q_x) \qquad (\widetilde{q}_t, \widetilde{q}_t) = (q_t, q_t).$$

(Using $\wedge$, multiply (1.12a) by q_x and $\widetilde{q}_x$, sum these equations and then compare the result with (0.2). Multiplying (1.12a) scalarly by q_x and $\widetilde{q}_x$, derive the statement on the norms. The proof is analogous for t.) □

Exercise1.7. *Modify Theorem1.2 and 1.3 for the case of the $*$-unitary fields, i.e., for the case of indefinite hermitian forms.*

Exercise1.8. *Show that, in contrast to the constructions in §1.1, in the present section transformations do not diminish the defining domain of the initial field for x and t due to the hermitian condition.* (Singularities of the transformed solutions $\widetilde{U}$ and $\widetilde{V}$ of Theorem1.1 might appear only if $\Phi_1 K_1 \cap \Phi_2 K_2 \neq \{0\}$ for some x, t. In the hermitian case, $\Phi_1 K_1$ and $\Phi_2 K_2$ are always orthogonal and, consequently, do not intersect.) □

In conclusion, we describe the action of the transformation of Proposition1.1 on an S^{n-1}-field. It is easy to show that for $\gamma_0 \in \mathbb{R}$ (in the notation of that proposition) these transformations are compatible with the restriction to an arbitrary real subgroup of the group GL_n of complex matrices. More exactly, if g is a G-field and $\Phi(x, t; \pm\gamma_0)$ has its values in G, then $\widetilde{g}$ is also a G-field.

Proposition1.6. *Let $\kappa \in \mathbb{R}$, $\kappa \neq 0$ and let $\Phi_{(\kappa)} \overset{\text{def}}{=} \Phi(x, t; \kappa) \in O_n$ be an arbitrary solution of system (1.1,2A) for currents $U = 2q \wedge q_x$ and $V = 2q \wedge q_t$ corresponding to an S^{n-1}-field q. Then $q^{(\kappa)} = \Phi_{(\kappa)}^{-1} q$ is an S^{n-1}-field; currents $U^{(\kappa)}$ and $V^{(\kappa)}$ constructed from $q^{(\kappa)}$ are obtained from U and V by the transformation of Proposition1.1 for $\gamma_0 = \kappa$,*

$$q_x^{(\kappa)} = \kappa \Phi_{(\kappa)}^{-1} q_x, \qquad q_t^{(\kappa)} = \kappa^{-1} \Phi_{(\kappa)}^{-1} q_t.$$

Proof. First, we check the last formulae:

$$\begin{aligned} q_x^{(\kappa)} - \Phi_{(\kappa)}^{-1} q_x &= -\Phi_{(\kappa)}^{-1} (\Phi_{(\kappa)})_x \Phi_{(\kappa)}^{-1} q \\ &= \tfrac{1}{2}(\kappa - 1) \Phi_{(\kappa)}^{-1} U q = \tfrac{1}{2}(\kappa - 1) \Phi_{(\kappa)}^{-1} (2q_x). \end{aligned}$$

This implies $q_x^{(\kappa)} = \kappa \Phi_{(\kappa)}^{-1} q_x$. An analogous calculation proves the formula for t. Using the orthogonality of Φ and the above formulae for $q_x^{(\kappa)}$ and $q_t^{(\kappa)}$, we can compute $U^{(\kappa)}$, $V^{(\kappa)}$ and $(q^{(\kappa)}, q^{(\kappa)})$. □

Exercise1.9. *Denote the function $\widetilde{q}$ for $\gamma = -i$ by q^+. Derive from Exercise1.2 that $((q^{(\kappa)})^+)^{(1/\kappa)}$ satisfies system (1.12), where the transformation $q \mapsto q^{(\kappa)}$ is as defined in Proposition1.6. Any solution of (1.12) can be thus obtained for a suitable initial value of q^+.* □

Exercise1.10*. *In the notation of Theorem4.3, Ch.I, take two algebraic-geometric S^2-fields constructed for two points $P = P_1$ and $P = P_2$ (for the same curve Γ, the function λ and corresponding divisors). Then the second solution is obtained by applying the transformation $q \mapsto q^{(\kappa)}$ for $\kappa = (\lambda(P_2)\lambda(P_1)^{-1})^{1/2}$ to the first solution.* □

1.3. Sin-Gordon equation.

We specialize Theorem1.3 to the case of S^2-fields in coordinates x, t normalized by the conditions $(q_x, q_x) = 1 = (q_t, q_t)$ (cf. Introduction). Then $(\widetilde{q}_x, \widetilde{q}_x) = 1 = (\widetilde{q}_t, \widetilde{q}_t)$ (Theorem1.3, b)). Putting $(q_x, q_t) = \cos\alpha$, $(\widetilde{q}_x, \widetilde{q}_t) = \cos\widetilde{\alpha}$ we have that α and $\widetilde{\alpha}$ satisfy the Sin-Gordon equation (0.8). Our goal here is to derive differential relations on α and $\widetilde{\alpha}$. This can be done by direct calculation using (1.12) but we take another approach which is not direct but more "geometric."

Put $q^{(\kappa)} = \widehat{q}$ and $\widetilde{q}^{(\kappa)} = \widehat{p}$. Then $\widehat{q}$ and $\widehat{p}$ satisfy the following system derived from (1.12) or (1.11) (see Exercise 1.9):

$$(\widehat{p} + \widehat{q})_x = (\widehat{p}, \widehat{q}_x)(\widehat{p} - \widehat{q}), \tag{1.13a}$$

$$(\widehat{p} - \widehat{q})_t = (\widehat{p}, \widehat{q}_t)(\widehat{p} + \widehat{q}), \tag{1.13b}$$

$$(\widehat{p}, \widehat{p}) = 1, \qquad (\widehat{p}, \widehat{q}) = 0. \tag{1.13c}$$

By Proposition1.6,

$$(\widehat{q}_x, \widehat{q}_x) = \kappa^2 = (\widehat{p}_x, \widehat{p}_x), \qquad (\widehat{q}_x, \widehat{q}_t) = \cos\alpha = (q_x, q_t),$$
$$(\widehat{q}_t, \widehat{q}_t) = \kappa^{-2} = (\widehat{p}_t, \widehat{p}_t), \qquad (\widehat{p}_x, \widehat{p}_t) = \cos\widetilde{\alpha} = (\widetilde{q}_x, \widetilde{q}_t).$$

As $(\widehat{q}_x, \widehat{q}) = (\widehat{q}_t, \widehat{q}) = 0 = (\widehat{p}_x, \widehat{p}) = (\widehat{p}_t, p)$, the vector $\widehat{q}$ belongs to the plane $\widehat{p}^{\perp}$ spanned by $\widehat{p}_x$ and $\widehat{p}_t$ by definition, and $\widehat{p} \in \widehat{q}^{\perp}$ is expressed as a linear combination of $\widehat{q}_x$ and $\widehat{q}_t$. Here we use the condition $n = 3$ and the equation $(\widehat{p}, \widehat{q}) = 0$ (which, by the way, characterize the cases $\kappa = \pm 1$ in Theorem1.3 and is the main reason why we turn to $\widehat{q}$ and $\widehat{p}$ from q and $\widetilde{q}$).

Let us denote by $\{\boldsymbol{a}, \boldsymbol{b}\}$ the angle from the vector $\boldsymbol{a}$ to the vector $\boldsymbol{b}$ with respect to a certain orientation on the plane spanned by $\boldsymbol{a}, \boldsymbol{b} \in \mathbb{R}^3$. We choose an orientation in $\mathbb{R}^3$ and then define the orientation of frames $(\widehat{q}, \widehat{q}_x, \widehat{q}_t)$ and $(\widehat{p}, \widehat{p}_x, \widehat{p}_t)$ consistently with this orientation, and finally define the orientation of planes $\widehat{p}^{\perp}$ and $\widehat{q}^{\perp}$ accordingly. Changing signs of α and β, if necessary (these angles are defined up to signs by their cosines), we can put, mod 2π,

$$\alpha = \{\widehat{p}, \widehat{q}_t\} - \{\widehat{p}, \widehat{q}_x\}, \tag{1.14a}$$

$$\widetilde{\alpha} = \{\widehat{q}, \widehat{p}_t\} - \{\widehat{q}, \widehat{p}_x\}. \tag{1.14b}$$

Differentiating the equation $(\widehat{p}, \widehat{q}) = 0$, we obtain the relations

$$\cos\{\widehat{p}, \widehat{q}_t\} = -\cos\{\widehat{q}, \widehat{p}_t\},$$
$$\cos\{\widehat{p}, \widehat{q}_x\} = -\cos\{\widehat{q}, \widehat{p}_x\}.$$

If $\{\widehat{q}, \widehat{p}_t\} - \{\widehat{p}, \widehat{q}_t\} = \pi$, then inevitably $\{\widehat{p}, \widehat{q}_x\} + \{\widehat{q}, \widehat{p}_x\} = \pi$. The converse case (mod 2π) is also possible. We suggest that the reader draw the figures for this and check this geometric statement. The last equation makes it possible to determine the sum and the difference of formulae (1.14a) and (1.14b) resulting in the formulae

$$\alpha + \widetilde{\alpha} = 2\{\widehat{p}, \widehat{q}_t\}, \qquad \alpha - \widetilde{\alpha} = 2\{\widehat{p}, \widehat{q}_x\}.$$

Recalling that $(\widehat{q}_x, \widehat{q}_x) = \kappa^2$ and $(\widehat{q}_t, \widehat{q}_t) = \kappa^{-2}$, we finally get the relations

$$\begin{aligned} \kappa \cos\Big(\frac{\alpha - \widetilde{\alpha}}{2}\Big) &= (\widehat{p}, \widehat{q}_x), \\ \kappa^{-1} \cos\Big(\frac{\alpha + \widetilde{\alpha}}{2}\Big) &= (\widehat{p}, \widehat{q}_t). \end{aligned} \tag{1.15}$$

In the last formulae, in accordance with the sign of κ and the choice of the signs of α, $\widetilde{\alpha}$ and $\alpha \pm \widetilde{\alpha}$ may be interchanged and κ may be multiplied by -1.

We use (1.15) and the fact that $\widehat{q}_{xt}$ and $\widehat{p}_{xt}$ are parallel to $\widehat{q}$ and $\widehat{p}$ respectively (equation (0.5)):

$$\begin{aligned} (\widehat{p}_x, \widehat{q}_t) = (\widehat{p}, \widehat{q}_t)_x &= -\kappa^{-1} \sin\Big(\frac{\alpha + \widetilde{\alpha}}{2}\Big)\Big(\frac{\alpha + \widetilde{\alpha}}{2}\Big)_x, \\ (\widehat{p}_t, \widehat{q}_x) = (\widehat{p}, \widehat{q}_x)_t &= -\kappa \sin\Big(\frac{\alpha - \widetilde{\alpha}}{2}\Big)\Big(\frac{\alpha - \widetilde{\alpha}}{2}\Big)_t. \end{aligned}$$

We substitute the above relations into the scalar product of formula (1.13a) and $\widehat{q}_t$, and into the scalar product of formula (1.13b) and $\widehat{q}_x$. As a result we obtain the classical *Bäcklund transformation*

$$\Big(\frac{\widetilde{\alpha} + \alpha}{2}\Big)_x = \kappa \sin\Big(\frac{\widetilde{\alpha} - \alpha}{2}\Big), \tag{1.16a}$$

$$\Big(\frac{\widetilde{\alpha} - \alpha}{2}\Big)_t = -\kappa^{-1} \sin\Big(\frac{\widetilde{\alpha} + \alpha}{2}\Big). \tag{1.16b}$$

Strictly speaking, we deduce (1.16) from (1.12) for a certain choice of the angles α and $\widetilde{\alpha}$ and for a suitable location of the frames $(\widehat{q}, \widehat{q}_x, \widehat{q}_t)$ and $(\widehat{p}, \widehat{p}_x, \widehat{p}_t)$ (cf. (1.15)). Again we suggest that the reader check that in other cases $\widetilde{\alpha} + \alpha$ and $\widetilde{\alpha} - \alpha$ may be interchanged and the sign of κ may also be different.

Exercise1.11. *By generalizing the previous observation or by direct analysis, show that for any solution α of the Sin-Gordon equation (0.8) with arbitrary initial value $\widetilde{\alpha}(x_0, t_0)$ there exists a solution $\widetilde{\alpha}$ of system (1.16) and that it satisfies equation (0.8).* □

Relation with the Lax formalism. Above, we obtained the Bäcklund transformation (1.16) as a corollary to the existence of the zero curvature representation for the PCF equation, performing five (or rather six) auxiliary reductions. It is natural to expect that (1.16) can be obtained directly from the analogous representation of the Sin-Gordon equation itself. Now we turn to the description of such representations (exactly speaking, to the description of one of them). Put

$$H = \begin{pmatrix} 1 & 0 \\ 0 & -1 \end{pmatrix} \frac{\partial}{\partial x} + \frac{\alpha x}{2} \begin{pmatrix} 0 & 1 \\ 1 & 0 \end{pmatrix},$$
$$W = \frac{1}{2} \begin{pmatrix} \cos\alpha & \sin\alpha \\ \sin\alpha & -\cos\alpha \end{pmatrix}.$$

The function $\alpha(x,t) \in \mathbb{C}$ satisfies equation (0.8) if and only if the differential operator $[\partial/\partial t + \kappa^{-1}W, H]$ is divisible by the operator $H - \frac{1}{2}\kappa I$ from the right. In fact, this commutator is equal to

$$\kappa^{-1}\sin\alpha \left(\begin{pmatrix} 0 & -1 \\ 1 & 0 \end{pmatrix} \frac{\partial}{\partial x} + \frac{1}{2} \begin{pmatrix} \alpha_x & -\kappa \\ -\kappa & \alpha_x \end{pmatrix} \right) + \frac{1}{2}(\alpha_{xt} + \sin\alpha) \begin{pmatrix} 0 & 1 \\ 1 & 0 \end{pmatrix},$$

where the operator in the parentheses coincides with $\begin{pmatrix} 0 & 1 \\ 1 & 0 \end{pmatrix}(H - \frac{i}{2}\kappa I)$. As a corollary of this statement we get the *zero curvature representation* of the Sin-Gordon equation:

$$\left[\frac{\partial}{\partial x} + \frac{1}{2} \begin{pmatrix} -\kappa & \alpha_x \\ -\alpha_x & \kappa \end{pmatrix}, \frac{\partial}{\partial t} + \frac{\kappa^{-1}}{2} \begin{pmatrix} \cos\alpha & \sin\alpha \\ \sin\alpha & -\cos\alpha \end{pmatrix} \right] = 0.$$

Equation (0.8) is also interpreted through the above representation as the compatibility condition of the following system of equations:

(1.17a) $$\psi_x = M\psi \stackrel{\text{def}}{=} \frac{1}{2} \begin{pmatrix} \kappa & -\alpha_x \\ \alpha_x & -\kappa \end{pmatrix},$$

(1.17b) $$\psi_t = -\kappa^{-1} W \psi.$$

Using these calculations, we can rewrite the Sin-Gordon equation in the Lax form (cf. §2, Ch.I):

$$(0.8) \Leftrightarrow (H^2)_x = -\tfrac{1}{2}[WH^{-1}, H^2].$$

The operator WH^{-1} here is an integro-differential operator. Therefore it is more convenient to use (and check) the last relation multiplied by the "denominator":

$$(H^2)_x H = [W, H^2].$$

Another way to construct the Lax pair for the Sin-Gordon equation uses only differential operators, but in 4×4 matrices. Note that the problem of searching for the Lax pair is somewhat abstract, since for any concrete purposes system (1.17) or the zero curvature representation of the Sin-Gordon equation are enough.

Thus, let α satisfy (0.8) and let $\psi = {}^t(\psi_1, \psi_2)$ be a solution of system (1.17) for $\kappa \in \mathbb{R}$, $\kappa \neq 0$, where $\psi_{1,2}(x,t) \in \mathbb{R}$. Below we may regard α and κ not only as real numbers, but also as complex numbers. Then ψ is taken from $\mathbb{C}^2$. Put $\varphi = \psi_1 \psi_2^{-1}$.

Theorem1.4. *a) The function φ satisfies the system*

$$\varphi_x + \frac{\alpha_x}{2}(1+\varphi^2) = \kappa\varphi, \tag{1.18a}$$

$$\varphi_t + \frac{\kappa}{2}(1-\varphi^2) = -\kappa^{-1}\varphi\cos\alpha. \tag{1.18b}$$

b) The function $\tilde{\alpha} = \alpha + 2\beta$, where $\varphi = \tan(\beta/2)$, satisfies system (1.16) as well as α, and, in particular, is a solution of equation (0.8).

Proof. By (1.17a) we have

$$(\psi_1)_x + \frac{\alpha_x}{2}\psi_2 = \frac{\kappa}{2}\psi_1,$$

$$(\psi_2)_x + \frac{\alpha_x}{2}\psi_1 = \frac{\kappa}{2}\psi_2.$$

Multiplying these equations crosswise and subtracting the result multiplied by $2\kappa^{-1}$ from the first equation of the system multiplied by $2\psi_2$, we obtain the equality

$$\psi_2(\psi_1)_x - \psi_1(\psi_2)_x + \frac{\alpha_x}{2}(\psi_1^2 + \psi_2^2) = \kappa\psi_1\psi_2.$$

Dividing this by ψ_2^2, we get (1.18a).

Rewrite equation (1.17a) in the following way:

$$-\frac{\kappa^{-1}}{2}\cos\alpha\psi_1 - \frac{\kappa^{-1}}{2}\sin\alpha\psi_2 = (\psi_1)_t,$$

$$-\frac{\kappa^{-1}}{2}\sin\alpha\psi_1 + \frac{\kappa^{-1}}{2}\cos\alpha\psi_2 = (\psi_2)_t.$$

Multiplying the first equation by ψ_2 and subtracting the second multiplied by ψ_1, we obtain the formula:

$$(\psi_1)_t\psi_2 - (\psi_2)_t\psi_1 + \frac{\kappa^{-1}}{2}\sin\alpha(\psi_2^2 - \psi_1^2) = -\kappa^{-1}\cos\alpha\psi_1\psi_2.$$

As above, we divide this by ψ_2^2 and get (1.18b).

After substituting $\varphi = \tan(\beta/2)$ and $u = \alpha + \beta$, system (1.18) has the following form:

$$u_x = \kappa \sin\beta,\qquad \beta_t = -\kappa^{-1}\sin u. \tag{1.19}$$

It is easy to see that the transformations $\widetilde{u} = u$, $\widetilde{\beta} = -\beta$ and $\widetilde{\kappa} = -\kappa$ bring (1.19) to itself. Lifting conversely from (1.19) to (1.18), we obtain that $\widetilde{\alpha} \stackrel{\text{def}}{=} \widetilde{u} - \widetilde{\beta} = \alpha + 2\beta$, $\widetilde{\varphi} = -\varphi$ and $\widetilde{\kappa} = -\kappa$ satisfy system (1.18). Consequently, $\widetilde{\alpha} = \alpha + 2\beta$ is a solution of (0.8). In order to find the relation connecting α and $\widetilde{\alpha}$, we substitute $\varphi = \tan((\widetilde{\alpha} - \alpha)/4)$ into (1.18) and come exactly to (1.16). □

To systematize the construction in Theorem1.14, we find $\widetilde{\alpha}$ by the method of §1.1 (using the B-factors), starting from system (1.17) (cf. Exercise1.3). Put $\boldsymbol{\psi}^\iota = {}^t(\psi_2, \psi_1)$ and $\Psi^\iota = \begin{pmatrix} 0 & 1 \\ -1 & 0 \end{pmatrix} \Psi \begin{pmatrix} 0 & -1 \\ 1 & 0 \end{pmatrix}$ for vectors $\boldsymbol{\psi}$ and matrices Ψ of rank 2. Then both matrix functions $\Psi(\kappa)$ and $\Psi^\iota(-\kappa)$ are solutions of system (1.17) (check this). We follow the construction in §1.2, but with the involution defined above, rather than complex conjugation. All functions and parameters may be considered to be real, or, if necessary, complex numbers.

For $\kappa \neq -\kappa_0$, Ψ_0 is an invertible solution of (1.17) for $\kappa = \kappa_0$, $\boldsymbol{\psi}_0 = {}^t(\psi_1, \psi_2)$ is the first column of Ψ_0, and P is the projection onto $\boldsymbol{\psi}_0$ parallel to $\boldsymbol{\psi}_0^\iota$. Explicitly,

$$\begin{aligned} P &= \frac{1}{\psi_1^2 + \psi_2^2} \begin{pmatrix} \psi_1^2 & \psi_1\psi_2 \\ \psi_2\psi_1 & \psi_2^2 \end{pmatrix} \\ &= \frac{1}{\psi_1/\psi_2 + \psi_2/\psi_1} \begin{pmatrix} \psi_1/\psi_2 & 1 \\ 1 & \psi_2/\psi_1 \end{pmatrix}. \end{aligned}$$

Note that $P + P^\iota = \boldsymbol{I}$. Put (cf. Theorem1.1)

$$B(\kappa) = \boldsymbol{I} - \frac{2\kappa_0}{\kappa_0 + \kappa} P.$$

Then B satisfies the conditions of Proposition1.2 for system (1.17) and $\kappa_1 = \kappa_0$ and $\kappa_2 = -\kappa_0$ (instead of (1.1,2), and the values λ_1 and λ_2 of the parameter λ). Consequently, for $\widetilde{\Psi}(\kappa) \stackrel{\text{def}}{=} B(\kappa)\Psi(\kappa)$ functions $\widetilde{\Psi}_x\widetilde{\Psi}^{-1}$ and $\widetilde{\Psi}_t\widetilde{\Psi}^{-1}$ are analytically continued to rational functions of κ with poles (of first order) only at the points $\kappa = 0$, $\kappa = \infty$, if Ψ is an invertible solution of (1.17) on some domain for κ.

As $B(\infty) = I$, the principal part of $\widetilde{\Psi}_x\widetilde{\Psi}^{-1}$ at $x = \infty$ has the same form as M and $\widetilde{\Psi}_t\widetilde{\Psi}^{-1} = -\kappa^{-1}\widetilde{W}\widetilde{\Psi}$, where $\widetilde{W}$ is a matrix which does not depend on κ. It remains to check the consistency of B with the involution:

$$\begin{aligned} B(-\kappa)^\iota &= I - \frac{2\kappa_0}{\kappa_0 - \kappa} P^\iota \\ &= I - \frac{2\kappa_0}{\kappa_0 - \kappa}(I - P) = \frac{\kappa + \kappa_0}{\kappa - \kappa_0} B(\kappa). \end{aligned}$$

The appearance of the scalar factor $(\kappa + \kappa_0)/(\kappa - \kappa_0)$ for B does not affect the formulae of the type (1.4,5). Hence, $\widetilde{W}^\iota = -\widetilde{W}$ and $\widetilde{M}(-\kappa)^\iota = \widetilde{M}(\kappa)$ for a matrix $\widetilde{M}$ which is M for the function $\widetilde{\Psi}$ instead of ψ in the right hand side of (1.17a). Let us compute $\widetilde{M}$ and $\widetilde{W}$.

Proposition1.7. *The matrix functions $\widetilde{M}$ and $\widetilde{W}$ defined by the construction in §1.1 from the solution α of equation (0.8) (and from $\varphi = \psi_1/\psi_2$ for the solution ψ of system (1.17) at the point κ_0) have the same form as M and W for $\widetilde{\alpha} = \alpha + 2\beta$, where $\varphi = \tan(\beta/2)$ as in Theorem1.4.*

Proof. We check the statements of the proposition only for $\widetilde{W}$ since the case of $\widetilde{M}$ is analogous. By definition $\widetilde{W}$ coincides with $B(0)WB(0)^{-1}$ (cf. proof of Theorem1.1). We have:

$$\begin{aligned} B(0) &= 1 - 2P = B(0)^{-1} \\ &= \frac{1}{\varphi + \varphi^{-1}} \begin{pmatrix} \varphi^{-1} - \varphi & -2 \\ -2 & \varphi - \varphi^{-1} \end{pmatrix} = \begin{pmatrix} \cos\beta & -\sin\beta \\ -\sin\beta & -\cos\beta \end{pmatrix}. \end{aligned}$$

The rest of the proof is a direct calculation using the fact that $B(0)$ and W are matrices of orthogonal reflection in in the lines of $\mathbb{R}^2$. □

1.4. Application: local conservation laws for the Sin-Gordon equation and S^{n-1}-fields.

Let $\alpha(x,t)$ be a solution of the Sin-Gordon equation (0.8). An equation $\zeta_t = \eta_x$ is called a *local conservation law* for (0.8), where ζ and η are, generally speaking, certain "local" functions of α (for instance, elementary functions of α and derivatives of α with respect to x). We call the term ζ the local density of the conservation law. Note the asymmetric role of x and t in this definition. If the function ζ is absolutely integrable in x from $-\infty$ to ∞ and η has limits $\eta(\pm\infty, t)$ as $x \to \pm\infty$ and $\eta(-\infty, t) = \eta(\infty, t)$, then

$$\begin{aligned} \frac{d}{dt}\int_{-\infty}^{\infty} \zeta(x,t)\,dx &= \int_{-\infty}^{\infty} \eta_x(x,t)\,dx \\ &= \eta(+\infty, t) - \eta(-\infty, t) = 0 \end{aligned}$$

(cf. Ch.I). Consequently, $\int_{-\infty}^{\infty} \zeta\, dx$ is an integral of equation (0.8). It is always so for an exact differential $\frac{d}{dx}\zeta$ of an arbitrary function ζ of the above type. Therefore densities of local conservation laws are interesting only modulo exact derivatives of absolutely integrable functions of x.

The procedure of constructing local conservation laws from the Bäcklund transformation of the Sin-Gordon equation reduces to the following two steps. First, we construct a solution $\widetilde{\alpha}$ of equation (1.16) in the form of the formal power series

$$\widetilde{\alpha} = \widetilde{\alpha}_0 + \widetilde{\alpha}_1\kappa^{-1} + \dots + \widetilde{\alpha}_s\kappa^{-1} + \cdots .$$

It is easy to show by expanding (1.16a) in the series of κ^{-1} and comparing its coefficients that the coefficients $\widetilde{\alpha}_s$ are uniquely determined from this equation. Moreover, the $\widetilde{\alpha}_s$ are expressed as differential polynomials in α (i.e., polynomials in α and derivatives of α with respect to x). Particularly, $\widetilde{\alpha}_0 = \alpha$ and $\widetilde{\alpha}_1 = 2\alpha_x$. Secondly, we substitute the above series $\widetilde{\alpha}$ into the conservation law

$$\kappa \cos\Big(\frac{\alpha - \widetilde{\alpha}}{2}\Big)_t + \kappa^{-1} \cos\Big(\frac{\alpha + \widetilde{\alpha}}{2}\Big)_x = 0, \tag{1.18}$$

which is easily derived from (1.16). Then we equate the coefficients of the expansion of (1.18) with respect to κ^{-1}. As a result we obtain infinitely many series of local conservation laws. We list the formulae for the first three of them:

$$(\alpha_x^2/2)_t = (\cos\alpha)_x, \tag{1.18a}$$

$$(\alpha_x\alpha_{xx})_t + (\alpha_x \sin\alpha)_x = 0, \tag{1.18b}$$

$$(\alpha_x^4/8 + \alpha_{xx}^2/2 + \alpha_x\alpha_{xx})_t + ((\alpha_x^2/2)\cos\alpha + \alpha_{xx}\sin\alpha)_x = 0. \tag{1.18c}$$

Law (1.18b) is trivial, since $\alpha_x\alpha_{xx} = (\alpha_x^2)_x/2$ is an exact derivative. The densities of all these laws are differential polynomials (without constant terms) not only in α but also in α_x. Therefore, if we assume that the values of $\alpha \bmod 2\pi$ for $x = \pm\infty$ coincide, then we can construct integrals of equation (0.8) from conservation laws (1.18) under the assumption of the absolute integrability of $\alpha_x, \alpha_{xx}, \ldots$.

Exercise1.12. a) *Using the uniqueness theorem for differential equations, show that the formal power series for $\widetilde{\alpha}$ in terms of κ^{-1} defined by (1.16) satisfies (1.16b).*

b) *Check that the density ζ_k for the coefficient of κ^{-k} ($k = 1, 2, \ldots$) in conservation law (1.18) is written as a differential polynomial in α_x, and that η_k is a sum of (two) differential polynomials in α_x multiplied by $\sin\alpha$ or by $\cos\alpha$.*

c)* *Show that the densities ζ_k for even k are exact derivatives, as for (1.18b).* □

Further we show that the construction above and the statement of Exercise1.12 reduce to results in Ch.I. Now we discuss other methods of enumerating local conservation laws of equation (0.8). For example, we can make use of the zero curvature representation (1.17) and directly apply the methods of §§1,2 of Ch.I for equation (1.17a) or for the operator H. For the computation of the densities ζ it is more convenient to use the interpretation of the Sin-Gordon equation as a "higher KdV" equation (cf. §2.5, Ch.I (2.25)).

Let us recall that (0.8) is equivalent to the pair of equations

$$(1.19) \qquad L_t = [A, L], \qquad L_{\bar{t}}^c = [A^c, L^c],$$

where $L = -\frac{\partial^2}{\partial x} + u$, $u = \alpha_x^2/4 + i\alpha_{xx}/2$ and $A = -\left(\frac{\partial}{\partial x} + \frac{i}{2}\alpha_x\right)^{-1} \frac{e^{-i\alpha}}{4}$ is an integro-differential operator and L^c and A^c are formal complex conjugation ($i^c = -i$ and $\alpha^c = \alpha$). Independent of the form of the operator A (for the KdV it is a third order differential operator), the densities h of conservation laws (2.21a) of Ch.I are given by the same formulae as for the KdV. We can also take the formal complex conjugate of (2.21a). Thus we obtain the following recurrence algorithm for computing the densities h_k of local conservation laws for the Sin-Gordon equation:

$$h_1 = u,$$

$$h_{m+1} = -(h_m)_x + \sum_{k=1}^{m-1} h_k h_{m-k},$$

where $u = -u_2$ is expressed via α_x by the above formula.

S^{n-1}-fields. Regarding (0.8) as a reduction of the equation of an S^{n-1}-field q (0.5), we construct an analogue of conservation law (1.18). In the notation of §1.2, by differentiating the function $\log(\varphi, \varphi)$ with respect to x and t (cf. formulae (1.10)) we get the relation

$$\kappa(Ur, r)_t + \kappa^{-1}(Vr, r)_x = 0.$$

Rewriting it in terms of p and q ($r = \sqrt{2}\varphi(\varphi, \varphi)^{-1/2} = p + iq$), we derive a conservation law

$$(1.20) \qquad \kappa(p, q_x)_t + \kappa^{-1}(p, q_t)_x = 0,$$

where p satisfies system (1.11). We are going to show that conservation law (1.20) corresponds to (1.18) after the reduction. In the notation of §1.3 the functions $\widehat{p}$ and $\widehat{q}$ satisfy (1.20) for $\kappa = 1$. On the other hand, (cf. (1.15))

$$(\widehat{p}, \widehat{q}_x) = \kappa \cos\Big(\frac{\alpha - \widetilde{\alpha}}{2}\Big), \qquad (\widehat{p}, \widehat{q}_t) = \kappa^{-1} \cos\Big(\frac{\alpha + \widetilde{\alpha}}{2}\Big).$$

Hence we indeed obtain (1.18).

Thus the construction for the Sin-Gordon equation described above reduces to Theorem1.4, §1, Ch.I (§1.5). We called the corresponding conservation laws local Pohlmeyer laws. Moreover, (1.20) is a direct consequence of formula (1.20) of Ch.I (surprising coincidence of the numbers!). In particular, Corollary1.2 and 1.3 of Ch.I generalize the statements of Exercise1.12 (we will discuss it in detail later). Let us examine the procedure for computing local conservation laws for S^{n-1}-fields which generalizes the computations for the Sin-Gordon equation.

Proposition1.8. *For an S^{n-1}-field q there exists the unique solution p of the equation (1.11a) of the form $p = \sum_{k=0}^{\infty} p_k \kappa^{-k}$, where $p_k(x,t) \in \mathbb{R}^n$. Such a p satisfies equations (1.11b) and (1.11c).*

Proof. Not using the results of Ch.I we will prove this and give an algorithm for constructing $\{p_k\}$.

Rewrite equation (1.11a) in the form

$$\kappa^{-1}(p, q_x) + \kappa^{-1} p_x + q_x = (p, q_x)p. \tag{1.21}$$

Then it is easy to see that $p_0 = q_x/|q_x|$, where we set hereafter $|z| = (z, z)^{1/2} \geqq 0$ for $z \in \mathbb{R}^n$. Substituting $q_x = p_0|q_x|$ into (1.21) and collecting the terms for κ^{-k}, we obtain the following relation:

(1.22)

$|q_x|((p_k, p_0)p_0 + p_k) = \{$ expression of q, $p_0, \dots, p_{k-1}$ and their derivatives by x $\}$.

Assuming that $p_1, \dots, p_{k-1}$ are already found, we can define p_k by this relation.

Taking the scalar product of both sides of (1.22) with p_0, we get the equation

$$|q_x|(p_k, p_0) = \{ \text{ expression of } q,\ p_0, \dots, p_{k-1} \text{ and their derivatives by } x \ \}.$$

Express (p_k, p_0) in terms of q and the previous $\{p_j\}$ then substitute it into (1.22). Then equation (1.22) transformed in such a way can be solved with respect to p_k. It is easy to check that the obtained p_k also satisfies the original equation (1.22) (i.e. before the substitution of (p_k, p_0)). Thus we proved the uniqueness and found an algorithm for computing $\{p_k\}$. Let us denote the series generated by $\{p_k\}$ by p.

Let us check that $(p, q) = 0$. Taking the scalar product of (1.21) with q, we get the formula

$$\kappa^{-1}(p, q)_x = (p, q)(p, q_x).$$

As the constant term of (p, q_x) is equal to $|q_x| \neq 0$, the series (p, q) must be a zero series in κ^{-1}. Analogously, multiplying (1.21) by p and using the orthogonality of p and q just proven, we have the identity

$$\kappa^{-1}|p|_x^2 = 2(p, q_x)(|p|^2 - 1),$$

from which it follows that $|p|^2 = 1$. By a similar method we can check that this p satisfies (1.11b). Note that the first equation of system (1.11) is enough to define p as a formal power series (cf. Theorem1.3). □

Substituting the above p into (1.20) and expanding this relation into power series, we obtain an infinite series of local conservation laws for (0.5). Here are the first two nontrivial laws (cf. [Poh]):

$$\left\{\frac{1}{2|q_x|}\left|\frac{\partial}{\partial x}\frac{q_x}{|q_x|}\right|^2\right\}_t = \left\{\frac{(q_x, q_t)}{|q_x|}\right\}_x,$$

$$\left\{\frac{1}{2|q_x|}\left|\frac{\partial}{\partial x}\left(\frac{1}{|q_x|}\frac{\partial}{\partial x}\frac{q_x}{|q_x|}\right)\right|^2 - \frac{5}{8|q_x|^3}\left|\frac{\partial}{\partial x}\frac{q_x}{|q_x|}\right|^4\right\}_t + \left\{\frac{(q_x, q_t)}{2|q_x|^3}\left|\frac{\partial}{\partial x}\frac{q_x}{|q_x|}\right|^2\right\}_x = 0.$$

In the coordinates x, t normalized by the conditions $|q_x|^2 = 1 = |q_t|^2$, they are written in the particularly simple form:

$$\{|q_{xx}|^2/2\}_t = \{(q_x, q_t)\}_x, \tag{1.23a}$$

$$\{|q_{xxx}|^2/2 - 5|q_{xx}|^4/8\}_t + \{(q_x, q_t)|q_{xx}|^2/2\}_x = 0. \tag{1.23b}$$

Exercise1.13. *a) Following the proof of Proposition1.8, write out the recurrence relations for* p_k *(cf. §1, Ch.I, or [OPSW]).*

b) Show that in the normalized coordinates the densities ζ *of the local conservation laws constructed above are polynomials of scalar products of* q_{xx}, q_{xxx} *and so on (cf. Exercise1.6, Ch.I), and that the* η *are also this kind of polynomial multiplied by* (q_x, q_t). (Use the relations $(q_x, q_x) = 1 = (q_t, q_t)$). □

Note that the densities ζ (in the normalized coordinates) satisfy a kind of homogeneity. Namely, the differential form $\zeta(x)\,dx$ does not change under a coordinate change $dx \mapsto f(x)\,dx$. Otherwise the integrals $\int_{-\infty}^{\infty} \zeta\,dx$ would be useless.

Before ending this section, we briefly explain how to extract the local conservation laws of the Sin-Gordon equation from those of S^2-fields. Let q be an S^2-field in the normalized coordinates. Take an orthogonal basis of $\mathbb{R}^3$, q, q_x and $q + q_{xx}$ ($(q, q_{xx}) = -|q_x|^2 = -1$). Starting from (0.7) (Introduction), we see that

$$|q_{xx}|^2 = \alpha_x^2 + 1, \qquad |q + q_{xx}|^2 = \alpha_x^2.$$

Differentiating the first relation, we get the formula $(q_{xxx}, q_{xx}) = \alpha_x\alpha_{xx}$. Further: $(q_{xxx}, q_x) = -(q_{xx}, q_{xx}) = -\alpha_x^2 - 1$, $(q_{xxx}, q) = -(q_{xx}, q_x) = 0$ ($|q_x|^2 = 1 \Rightarrow (q_{xx}, q_x) = 0$). Consequently,

$$\begin{aligned} |q_{xxx}|^2 &= (q_{xxx}, q)^2 + (q_{xxx}, q_x)^2 + \alpha_x^{-2}(q_{xxx}, q + q_{xx})^2 \\ &= \alpha_{xx}^2 + \alpha_x^4 + 2\alpha_x^2. \end{aligned}$$

As $(q_x, q_t) = \cos\alpha$, (1.23) implies relation (1.18a): $(\cos\alpha)_x = (\alpha_x^2/2)_t$, and we get a conservation law which is a linear combination of (1.18a), (1.18c) and the derivative of (1.18c) with respect to x.

Exercise1.14. *Show that the orthogonality of the basis q, q_x and $(q+q_{xx})\alpha_x^{-1}$ in $\mathbb{R}^3$ allows us to express any scalar products of derivatives of q with respect to x in terms of differential polynomials of α_x.* □

1.5. Darboux transformation; Nonlinear Schrödinger equation.

In the notation of §2.5, Ch.I, let $\psi = (1+\sum_{s=1}^{\infty} k^{-s}\psi_s)e^{kx}$ be an abstract formal power series in the parameter k^{-1} with indeterminate scalar coefficients $\psi_j(x)$. In §2.5, we constructed (differential) fractional powers of L_r from ψ, for which $L_r\psi = k^r\psi + O(k^{-1})e^{kx}$. Here

$$L_r = \left(\frac{\partial}{\partial x}\right)^r + \sum_{i=2}^{r} u_{i,r}\left(\frac{\partial}{\partial x}\right)^{r-i},$$

for $r \geqq 1$, and the $u_{i,r}$ are expressed as differential polynomials of $\psi_1, \psi_2, \ldots$. With the help of L_r we can introduce pairwise commutative differentiations of the ring of differential polynomials of $\{\psi_s\}$ by (2.17) of Ch.I:

$$\frac{\partial\psi}{\partial t_r} = L_r\psi - k^r\psi. \tag{1.24}$$

These differentiations commute, by definition, with $\partial/\partial x$; and $\partial e^{kx}/\partial t_r = 0$. Note that the differentiation $\partial/\partial t_1$ acts on $\{\psi_s\}$ as $\partial/\partial x$.

Let $\psi^1, \psi^2, \ldots, \psi^m$ be a set of functions of $x, t_1, t_2, \ldots$ that are solutions of (1.24) at $k = k_1, k_2, \ldots, k_m \in \mathbb{C}$, where (1.24) is regarded as an (infinite) system of equations with functions $u_{i,r}$ as coefficients. For concreteness we could restrict ourselves to a finite number of indices r. There may be repeats among k_i, but $\{\psi^i\}$ should be in generic position (see below). Put

$$\check{\psi}(k) = \psi(k)\exp\left(\sum_{r=1}^{\infty} k^r t_r\right)$$

and analogously define $\check{\psi}^i$ (here, the exponential factor plays its role only in the differentiations with respect to t_r).

We call the function

$$\check{\psi}'(k) = k^{-m-1}\frac{W_{m+1}(\check{\psi}^1, \ldots, \check{\psi}^m, \check{\psi}(k))}{W_m(\check{\psi}^1, \ldots, \check{\psi}^m)} \tag{1.25}$$

the *Darboux transformation* of the function $\check{\psi}(k)$ for $m \geqq 1$. Here

$$W_l(f_1, \ldots, f_l) \overset{\text{def}}{=} \det\left(\left(\frac{\partial}{\partial x}\right)^{i-1} f_j\right)$$

is the Wronskian of the set of functions $\{f_i\}$, $1 \leqq i, j \leqq l$. Just like ψe^{-kx}, we have $\psi' e^{-kx} = \check{\psi}' \exp(-kx - \sum_{r=1}^{\infty} k^r t_r)$ is a formal power series of k^{-1} with constant term equal to 1, where ψ' is defined in terms of ψ by the same formula (1.25).

Theorem1.5. *Define the fractional powers L'_r and the differentiations $\partial/\partial t'_r$ using ψ'. Then for $r \geqq 1$*

$$\frac{\partial \psi'}{\partial t_r} = \frac{\partial \psi'}{\partial t'_r},$$

that is, the differentiations $\partial/\partial t_r$ and $\partial/\partial t'_r$ coincide on the ring of differential polynomials in the coefficients ψ'_s of the series $\psi' e^{-kx}$. Here we assume that $W_m(\psi^1, \ldots, \psi^m) \not\equiv 0$.

Proof. Set $f = W_{m+1}(\check{\psi}^1, \ldots, \check{\psi}^m, \check{\psi})$. By the commutativity of $\partial/\partial x$ and $\partial/\partial t_r$ we have

$$\begin{aligned}\frac{\partial f}{\partial t_r} &= \sum_{i=1}^{m} W_{m+1}\left(\ldots, \frac{\partial \check{\psi}^i}{\partial t_r}, \ldots, \check{\psi}\right) + W_{m+1}\left(\check{\psi}^1, \ldots, \check{\psi}^m, \frac{\partial \check{\psi}}{\partial t_r}\right) \\ &= \sum_{i=1}^{m} W_{m+1}\left(\check{\psi}^1, \ldots, L_r\check{\psi}^i, \ldots, \check{\psi}^m, \check{\psi}\right) + W_{m+1}\left(\check{\psi}^1, \ldots, \check{\psi}^m, L_r\check{\psi}\right).\end{aligned}$$

Let us fomulate one simple property of Wronskians. Let φ, $a_1, \ldots, a_m$ be functions of x, $W_m(a_1, \ldots, a_m) \not\equiv 0$. Then each of the derivatives $\partial^s\varphi/\partial x^s$ are expressed linearly in terms of $1, \varphi, \partial\varphi/\partial x, \ldots, \partial^{m-1}\varphi/\partial x^{m-1}$, $W_{m+1}(a_1, \ldots, a_m, \varphi)$ and its derivatives with respect to x. The coefficients are differential polynomials in $a_1, \ldots, a_m$ divided by powers of $W_m(a_1, \ldots, a_m)$. We can check this claim by induction: write $\partial^m\varphi/\partial x^m$ in terms of W_{m+1} and $\partial^s\varphi/\partial x^s$, $0 \leqq s < m$, and then compute $\partial^{m+1}\varphi/\partial x^{m+1}$ in terms of $\partial W_{m+1}/\partial x$, and so on.

By the above statement,

$$\frac{\partial f}{\partial t_r} = Lf + \sum_{i=1}^{m} \frac{\partial^{i-1}\check{\psi}}{\partial x^{i-1}},$$

where L is a differential operator of order r in x, and the coefficients of L and the functions $\{c_i\}$ are differential polynomials in $\check{\psi}^1, \ldots, \check{\psi}^m$ divided by powers of

$W_m(\check{\psi}^1, \dots, \check{\psi}^m)$. Let us show that $c_i = 0$. First, substitute $\check{\psi} = \check{\psi}^1, \dots, \check{\psi}^m$ one by one in the above formula. Then $f = 0$ and consequently

$$\sum_{i=1}^{m} c_i \frac{\partial^{i-1}\check{\psi}^j}{\partial x^{i-1}} \equiv 0$$

for $1 \leqq j \leqq m$. As $W_m(\check{\psi}^1, \dots, \check{\psi}^m) \not\equiv 0$, it is necessary that $c_i \equiv 0$ for $1 \leqq i \leqq m$. Thus

$$\frac{\partial f}{\partial t_r} = Lf, \qquad \frac{\partial \psi'}{\partial t_r} = L'\psi' - k^r\psi',$$

where L' is a differential operator of order r which does not depend on k (with the highest coefficient equal to unity). Since $\partial\psi'/\partial t_r = O(k^{-1})e^{kx}$, it follows from the above relation for L' that this operator must coincide with the operator L'_r constructed from ψ'. □

Corollary1.2. *Let us assume that functions $u_{i,s}$ $(s = p, r)$ satisfy the Zakharov-Shabat equation which is the consistency condition of the pair of equations*

$$\frac{\partial \check{\psi}}{\partial t} = L_p\check{\psi}, \qquad \frac{\partial \check{\psi}}{\partial y} = L_r\check{\psi}$$

for some p, r, or that they satisfy the Lax equation which is the consistency condition of equations

$$\frac{\partial \psi}{\partial t} = L_p\psi - k^p\psi, \qquad L_r\psi = k^r\psi. \tag{1.26}$$

Then the same holds for the coefficients $u'_{i,s}$ corresponding to the function $\check{\psi}'$ constructed above. □

KdV equation. As is well-known (cf. §2.5, Ch.I), up to normalization this equation appears as the consistency condition of equations (1.26) for $p = 3$, $r = 2$. Let

$$L_2 = \left(\frac{\partial}{\partial x}\right)^2 + u.$$

Then

$$L_3 = \left(\frac{\partial}{\partial x}\right)^3 + \frac{3}{2}u\frac{\partial}{\partial x} + \frac{3}{4}u_x.$$

It follows directly from formula (1.25) that the Darboux transformation of the function u for the function ψ^1 which satisfies equation (1.26) at a point $k = k_1$ is the function

$$u' = u + 2v_x,$$

where $v = (\log \psi^1)_x$. Indeed,

$$\begin{aligned} \psi' &= \psi_x - \psi v, \\ \psi'_x &= -\psi_x v + (v^2 + k^2 - k_1^2)\psi, \\ \psi'_{xx} + (u - k^2)\psi' &= (2v^2 - 2(k_1^2 - u))\psi'. \end{aligned}$$

Here we use the formulae $\psi_{xx} = (k^2 - u)\psi$, $v_x = (k_1^2 - u) - v^2$.

Actually we are now repeating the classical argument by Darboux (and partly reproving Corollary1.2). Note that for arbitrary m the transformed function u is equal to

$$u' = u + 2(\log W_m(\psi^1, \dots, \psi^m))_{xx},$$

which is a result of Crum (cf., for example, [DT]).

Now we connect the Darboux transformation with the Bäcklund transformation in §1.1. For this purpose, following the method in §2.5, Ch.I, it is necessary to turn to the first order system from the equation $\psi_{xx} = (k^2 - u)\psi$ and the corresponding relation for ψ_t. We have:

$$(1.27) \qquad \Psi_x \Psi^{-1} = \begin{pmatrix} 0 & 1 \\ k^2 & 0 \end{pmatrix} - \begin{pmatrix} 0 & 0 \\ u & 0 \end{pmatrix}, \qquad \Psi \stackrel{\text{def}}{=} \begin{pmatrix} \psi(k) & \psi^*(k) \\ \psi_x(k) & \psi^*_x(k) \end{pmatrix},$$

where $\psi(k)$, $\psi^*(k)$ are two non-proportional solutions of system (1.26) for $p = 3$, $r = 2$ at the point k. For $\lambda = k^2$, $\lambda_1 = k_1^2$ let

$$B = \begin{pmatrix} -v & 1 \\ (\lambda - \lambda_1) + v^2 & -v \end{pmatrix}, \qquad \Psi' \stackrel{\text{def}}{=} B\Psi.$$

We check that B satisfies the conditions of Proposition1.2, i.e., $\Psi^{-1}B^{-1}\mathcal{O}_l^2$ does not depend on x, t in a neighbourhood of any point $\lambda = l \in \mathbb{C} \cup \infty$. Indeed,

$$B^{-1} = \frac{1}{\lambda - \lambda_1} \begin{pmatrix} v & 1 \\ (\lambda - \lambda_1) + v^2 & v \end{pmatrix},$$

$$\begin{aligned} B^{-1}\mathcal{O}_1^2 &= \mathbb{C}(\lambda - \lambda_1)^{-1} \begin{pmatrix} 1 \\ v \end{pmatrix} + \mathcal{O}_1^2 \\ &= \Psi(k_1)\left(\begin{pmatrix} 1 \\ 0 \end{pmatrix} \mathbb{C} + \mathcal{O}_1^2\right), \end{aligned}$$

$\mathcal{O}_1 = \mathcal{O}_{\lambda_1}$ in a neighbourhood of the point $\lambda = \lambda_1$, where the module in the parentheses does not depend on x and t. It is necessary to examine the point $\lambda = \infty$ since B has a pole there. We have

$$B^{-1}\mathcal{O}^2_\infty = \begin{pmatrix} 0 \\ 1 \end{pmatrix} \mathbb{C} + \lambda^{-1}\mathcal{O}^2_\infty = \Psi(\lambda^{-1}\mathcal{O}^2_\infty),$$

because the expansion of $\Psi \operatorname{diag}(e^{-kx}, e^{kx})$ into a series in a neighbourhood of $k = \infty$ is easily computed and begins with $\begin{pmatrix} 1 & 1 \\ k & -k \end{pmatrix}$. Strictly speaking, in Proposition1.2 "critical" points of B (points $\lambda_1, \dots, \lambda_N$) must not coincide with singular points of Ψ (points $\lambda = \pm 1$), as is the case here. However the method of the proof of the proposition does not change in the given situation (check this).

Thus, by Proposition1.2 $\Psi'_x(\Psi')^{-1}$ is a rational function of λ with a unique pole of first order at the point $\lambda = \infty$. (The estimate of the order of the pole at ∞ follows from the independence of $\Psi^{-1}B^{-1}\mathcal{O}^2_\infty$ of x.) Moreover it is easy to prove that $\Psi'_x(\Psi')^{-1}$ coincides with the right hand side of (1.27) for $u' = u + 2v_x$ instead of u. An analogous computation works also for $\Psi'_t(\Psi')^{-1}$. Of course, the last statement can be checked by a direct method. We presented the above reasononing in order to illustrate *the coincidence of the transformations given in* §*1.1 and the Darboux transformations in the case of Lax equations.*

The following example is more interesting. Apply the result of Corollary1.1 for Ψ instead of Φ and $Q = \begin{pmatrix} 0 & 1 \\ 0 & 0 \end{pmatrix}$ with $\sigma = 1$. Set as before $\lambda = k^2$, $\lambda_1 = k_1^2$. Then

$$F = \begin{pmatrix} \psi(k_1) & \psi^*(k_1) + \psi_\lambda(k_1) \\ \psi_x(k_1) & \psi^*_x(k_1) + \psi_{\lambda x}(k_1) \end{pmatrix},$$
$$FQF^{-1} = \begin{pmatrix} -\psi(k_1)\psi_x(k_1) & \psi^2(k_1) \\ -(\psi_x(k_1))^2 & \psi(k_1)\psi_x(k_1) \end{pmatrix} d^{-1},$$

where $\psi_\lambda \overset{\text{def}}{=} \partial\psi/\partial\lambda$, and $d = \det F = W_2(\psi(k_1), \psi^*(k_1)) + W_2(\psi(k_1), \psi_\lambda(k_1))$. Note that $W_2(\psi(k_1), \psi^*(k_1))$ does not depend on x.

It follows from Corollary1.1 that

$$\widetilde{\Psi}_x\widetilde{\Psi}^{-1} = B_xB^{-1} + B(\Psi_x\Psi^{-1})B^{-1}$$

(for $B = \boldsymbol{I} - (\lambda - \lambda_1)FQF^{-1}$, $\widetilde{\Psi} = B\Psi$) is a rational function of λ with a unique pole at $\lambda = \infty$. In addition, the principal part of the above expression is equal to $\begin{pmatrix} 0 & 0 \\ \lambda & 0 \end{pmatrix}$. The constant term of $-\widetilde{\Psi}_x\widetilde{\Psi}^{-1}$ can be easily calculated. It is equal to

$$\begin{pmatrix} 0 & 1 \\ u & 0 \end{pmatrix} + \left[FQF^{-1}, \begin{pmatrix} 0 & 0 \\ 1 & 0 \end{pmatrix}\right] = \begin{pmatrix} 0 & 1 \\ u & 0 \end{pmatrix} + \frac{1}{d}\begin{pmatrix} \psi(k_1)^2 & 0 \\ 2\psi\psi^*_x(k_1) & -\psi(k_1) \end{pmatrix}.$$

Let us denote the first column of the matrix $\widetilde{\Psi}$ by ${}^t(\varphi_1, varphi_2)$. Then

$$\begin{aligned}(\varphi_1)_x &= \varphi_2 - \psi(k_1)^2\varphi_1 d^{-1},\\ (\varphi_2)_x &= (\lambda - u)\varphi_1 + \psi(k_1)^2\varphi_2 d^{-1} - 2(\psi(k_1))_x\psi(k_1)\varphi_1 d^{-1},\\ (\varphi_1)_{xx} &= (\lambda - u)\varphi_1 - 2(d_x/d)_x\varphi_1.\end{aligned}$$

In the derivation of the last formula we use the relation $d_x = \psi(k_1)^2$ which follows from the equation $(\psi_\lambda)_{xx} = \psi + (\lambda - u)\psi_\lambda$.

In this way, starting with the transformation of §1.1 with "confluent zeros" $\lambda_1 = \lambda_2$, we construct from a pair of functions ψ and u connected by the relation $\psi_{xx} = (\lambda - u)\psi$ the functions φ_1 and $u' \stackrel{\text{def}}{=} u + 2(\log d)_{xx}$ satisfying the same relation, where $d = W_2(\psi(k_1), \psi(k_2))$. The previous formula is obtained, as is easily seen, by degenerating the last one as $k_2 \to k_1$. Analogously we can generalize Crum's formulae with arbitrary numbers $k_1, \ldots, k_m$ via suitable matrix Bäcklund transformations (but only for Lax type equations).

NS equation. As a conclusion of this section, we deduce the Darboux type formulae for the NS equation. We omit the complete derivation in general cases, referring the reader to appropriate literature. Continuing in the line of Proposition1.7, we apply the reduction of Theorem1.1 to this equation.

The NS equation

$$ir_t = r_{xx} + 2\omega r|r|^2$$

($\omega = \pm 1$ correspond to the "attractive" and "repulsive" cases) turns out to be (cf. [ZMaNP], [TF2]) the consistency condition of the system

$$\frac{\partial \Psi}{\partial x} = M\Psi, \qquad \frac{\partial \Psi}{\partial t} = N\Psi \tag{1.28}$$

for

$$M = \begin{pmatrix} -i\lambda & -\omega\overline{r} \\ r & i\lambda \end{pmatrix}, \quad N = \begin{pmatrix} -2i\lambda^2 + i\omega r\overline{r} & -2\omega\lambda\overline{r} - i\omega\overline{r}_x \\ 2\lambda r - i r_x & 2i\lambda^2 - i\omega r\overline{r} \end{pmatrix},$$

which is called the *zero curvature representation* of the NS. Together with $\psi(\lambda) = {}^t(\psi_1, \psi_2)$, the function $\psi(\lambda)^* \stackrel{\text{def}}{=} {}^t(\overline{\psi_x(\lambda)}, -\omega\overline{\psi_1(\lambda)})$ is also a solution of system (1.28) with the parameter $\overline{\lambda}$ instead of λ. Let P be the projection onto the direction of $\psi(\lambda_0) \stackrel{\text{def}}{=} {}^t(\psi_1^0, \psi_2^0)$ parallel to $\psi(\lambda_0)^*$. Such a projection P satisfies the conditions of Theorem1.1. We can write an explicit formula for it:

$$P = \frac{1}{\psi_1^0\overline{\psi}_1^0 + \omega\psi_2^0\overline{\psi}_2^0} \begin{pmatrix} \psi_1^0\overline{\psi}_1^0 & \omega\psi_1^0\overline{\psi}_2^0 \\ \psi_2^0\overline{\psi}_1^0 & \omega\psi_2^0\overline{\psi}_2^0 \end{pmatrix}.$$

Following Theorem1.1, let us introduce $B = I - \frac{\overline{\lambda}_0-\lambda_0}{\overline{\lambda}_0-\lambda}P$ and $\Psi' = B\Psi$. Analysis of the relation

$$\Psi'_x(\Psi')^{-1} = B_x B^{-1} + BMB^{-1}$$

(cf. (1.4,5)) in a neighbourhood of the point $\lambda = \infty$ shows that $M' = \Psi'_x(\Psi')^{-1}$ has the same form as M:

$$\begin{pmatrix} 0 & i\omega(\overline{r}' - r) \\ -i(r' - r) & 0 \end{pmatrix} = (\overline{\lambda}_0 - \lambda_0)\left[P, \begin{pmatrix} -1 & 0 \\ 0 & 1 \end{pmatrix}\right]$$
$$= \frac{2(\overline{\lambda}_0 - \lambda_0)}{\psi_1^0\overline{\psi}_1^0 + \omega\psi_2^0\overline{\psi}_2^0}\begin{pmatrix} 0 & \omega\psi_1^0\overline{\psi}_2^0 \\ -\psi_2^0\overline{\psi}_1^0 & 0 \end{pmatrix}.$$

As a result we construct a new solution of the NS equation from r:

$$r' = r - \frac{4\varphi \operatorname{Im}\lambda_0}{1 + \omega\varphi\overline{\varphi}}, \tag{1.29}$$

where $\varphi = \psi_2^0/\psi_1^0$.

It is easy to write the differential equation for φ (the Riccati equation) and to prove the analogue of Theorem1.4. Note that in the case of $\omega = +1$ the function r' is defined on the same domain of x and t, as r, since $1 + \varphi\overline{\varphi} \neq 0$ (cf. Exercise1.8).

We can also apply Proposition1.2 (instead of Theorem1.1) with appropriate generalization for a set of points $\lambda_1, \overline{\lambda}_1, \dots, \lambda_N, \overline{\lambda}_N$, and vector functions $\boldsymbol{\psi}(\lambda_i)$, $\boldsymbol{\psi}(\lambda_i)^*$ as the images of principal parts of matrices B, B^{-1} at the points $\overline{\lambda}_1, \dots, \overline{\lambda}_N$ and $\lambda_1, \dots, \lambda_N$ respectively. We search for B of the form

$$B = I - \sum_{i=1}^{N} \frac{Q_i}{\overline{\lambda}_i - \lambda}.$$

Calculating B^{-1}, we obtain a system of linear equations determining matrices Q_i via $\{\boldsymbol{\psi}(\lambda_i)\}$. After that we get an N-fold iteration of formula (1.29). In the following exercise, we formulate the final result, leaving the proof to the reader (cf. Crum's formula (1.25)).

Exercise1.15*. *Let $\boldsymbol{\psi}(\lambda_i) \stackrel{\text{def}}{=} {}^t(\psi_1^i, \psi_2^i)$ be solutions of system (1.28) for a solution r of the NS at pairwise distinct points $\lambda = \lambda_1, \dots, \lambda_N$ ($\overline{\lambda}_i \neq \lambda_j$). Put*

$$\varphi_i = \psi_2^i/\psi_1^i,$$
$$\widetilde{\varphi}_i = \overline{\varphi}_i, \qquad \widetilde{\varphi}_{N+i} = -\omega\varphi_i^{-1},$$
$$\widetilde{\lambda}_i = \overline{\lambda}_i, \qquad \widetilde{\lambda}_{N+i} = \lambda_i,$$

for $1 \leqq i \leqq N$. Consider a $2N \times 2N$-matrix, the k-th row of which has the form

$$(1, \widetilde{\varphi}_k, \widetilde{\lambda}_k, \widetilde{\lambda}_k\widetilde{\varphi}_k, \widetilde{\lambda}_k^2, \widetilde{\lambda}_k^2\widetilde{\varphi}_k, \dots, \widetilde{\lambda}_k^{N-1}, \widetilde{\lambda}_k^N),$$

and a matrix of the same size with k-th row

$$(1, \widetilde{\varphi}_k, \widetilde{\lambda}_k, \widetilde{\lambda}_k\widetilde{\varphi}_k, \widetilde{\lambda}_k^2, \widetilde{\lambda}_k^2\widetilde{\varphi}_k, \dots, \widetilde{\lambda}_k^{N-1}, \widetilde{\lambda}_k^{N-1}\widetilde{\varphi}_k),$$

Let us denote the determinant of the first by Δ_0 and of the second by Δ_1. Then $r = r_0 - 2i\omega\Delta_0/\Delta_1$ is a solution of the NS equation. □

Exercise1.16. *Following Proposition1.7, show that formula (1.29) can be obtained as a corollary of Theorem1.1 specialized for the Heisenberg magnet*

$$\boldsymbol{s}_t = \boldsymbol{s} \times \boldsymbol{s}_{xx}, \qquad (\boldsymbol{s}, \boldsymbol{s}) = 1$$

(cf. Introduction §0.3). □

1.6. Comments.

Bäcklund and Darboux transformations of integrable equations are a brilliant part of the soliton theory. They are closely connected with local conservation laws, the inverse problem method and the algebraic-geometric technique. In more algebraic versions of the soliton technique, the infinite dimensional groups of Bäcklund-Darboux transformations (actually, they are loop groups -see [Ch11]) appear even before introducing soliton equations. It is remarkable that only after the appearance of soliton theory did the equivalence (in principle) of classical Bäcklund transformations [Bä] for the Sin-Gordon equation (1.16) and Darboux transformations (see [Dar]) for the equation $\psi_{xx} = (k^2 - u)\psi$ (§1.5) became completely clear, though they were invented more than one hundred years ago.

The work by K. Pohlmeyer [Poh] plays an important role in the matrix generalization of Bäcklund transformations. He considers transformations of S^{n-1}-fields in the form of an overdetermined system of differential relations on initial and transformed fields. In the article [Ch6] this result was generalized to orthogonal fields and moreover was attached to the solutions Φ of the corresponding linear system (1.4,5). We mostly follow this paper and [Ch2] in this section. Compared to [Ch2,6] or [Poh], however, we make the constructions more transparent, because today we better understand the invariant meaning of Bäcklund transformations. Theorem1.1 and Proposition1.2 are taken from [Ch3] (results close to Theorem1.1 can also be found in works by other authors, for instance, in one of the papers by Chudonovsky). Theorem1.2 is contained in [OPSW], in which results of [Ch6] and [Poh] are transfered to U_n-fields, though the approach is different from ours (it is less constructive and more in the style of Pohlmeyer). In the same paper the statement of Theorem1.2 is formulated (which is in fact a complete version of the results by K. Pohlmeyer). Note also the work [U] which is devoted to Bäcklund transformations for principal chiral fields (in this article, particularly, the relation of infinitesimal Bäcklund transformations to the Lie algebra of formal currents is established). In the works [CGW], [UN] (and others) the duality equation is examined by an analogous technique (cf. also [Taka]).

It is important to note the close relation of Bäcklund transformations to the "dressing" transformations of Zakharov-Shabat [ZS1,2] which correspond to adding discrete scattering data of the associated linear problem (see the next section). This relation was not immediately clear, although it was rather well-known that the construction of multi-soliton solutions by the inverse problem method coincides with the application of Bäcklund transformations to trivial solutions of integrable equations. In articles from [Mi] (and other papers) the examples that the operation of adding "zero" to the coefficient a of the corresponding linear problem induces a certain Bäcklund transformation were considered. Now all these important pioneering observations are included in the general scheme and the connection is obvious.

The constructive form of the classical Bäcklund transformation for the Sin-Gordon equation (Theorem1.4) is from [Ch2]. (I used the analysis of this transformation performed by K. Pohlmeyer). Similar arguments are contained in [AKNS2], [Lu] and in the paper by H. Flashka, D. McLaughlin [Mi]. It might be that such an interpretation is contained in some classical works on the Sin-Gordon equation. The result of Theorem1.4 was reproven by V. B. Matveev and M. A. Sall who generalized the Darboux transformation (cf. Proposition1.7). As to the zero curvature representation of the Sin-Gordon equation, we refer to [Takh], [AKNS1] and [ZTF], [TF1]. For local conservation laws of the Sin-Gordon equation, see [DB], [ZMaNP], [TF2] and [Poh], [Ch2].

The connection of Bäcklund transformations and the Darboux-Crum transformations was established (in principle) by F. Estabrook and H. Wahlquist, in particular, in the paper by H. Wahlquist in [Mi]. Note also Hirota's formula (cf. [Mi]) which reflects this connection, and a series of other close observations (cf. also [DT]). B. V. Matveev extended the technique of Darboux transformations to the Zakharov-Shabat equations (Corollary1.2) and several similar soliton equations. Theorem1.5 is a generalization of the result by B. V. Matveev (see, for example, [Mat1,2]). At the present time there are many papers devoted to Bäcklund-Darboux transformations of concrete equations, which we do not review here (there are also works on the "invariant" theory of such transformations -see [Ch11]). We also mention a series of results by M. A. Sall [Sa], results by A. R. Its, D. Levi, A. I. Bobenko on the Bäcklund transformation. The statement of Exercise1.5 is from the work by M. A. Sall [Sa] (cf. also the paper by H. Wahlquist in [Mi]). It seems that the

Bäcklund transformation for the NS equation was considered for the first time by G. Lamb.

We do not touch upon the conceptually important papers by R. Hirota, works by M. Sato and the group of Japanese mathematicians, in which Bäcklund-Darboux transformations appear as a result of the action of a central extension of the group GL_∞ on the τ-functions (cf. for example, [DKM] or [Ch11]). We also do not mention the beautiful differential-geometric interpretation of Bäcklund transformations and their applications to soliton surfaces (started with the classical results by E. Beltrami, L. Bianki and others). See, for instance, [LR] and [Bä], [Ei], [H], [Tc]. We also refer the reader to more recent works by R. Sasaki and A. Sym (cf. [Sy]) on this subject. The results by F. Griffiths generalizes this direction. Works by H. Wahlquist, F. Estabrook and others on pseudo-potentials and the "theory of continuation" for soliton equations have direct relation to Bäcklund transformations (cf. [EsW], [C]). Certain geometric aspects of these transformations are connected with the work by J.-L. Verdier on the classification and description of instanton chiral fields on spheres and with a series of results by A. A. Belavin, A. M. Polyakov, V. L. Golo, A. M. Perelomov, H. Eichenherr and others on instantons in chiral models. Ending this far-from-complete list of references, we mention works by R. Anderson, I. H. Ibragimov, G. Neugebauer, S. Orfanadis, M. A. Semenov-Tian-Shansky and A. B. Shabat.

Recently, transformations of solutions of integrable equations connected with the action of the Virasoro algebra on compact and non-compact Riemann surfaces have been studied. In contrast to classical "pointwise" Bäcklund transformations it is necessary here to use the Φ in a neighbourhood or on certain contours in the spectral parameter.

§2. Introduction to the scattering theory

In §§2.1 and 2.2 we introduce the monodromy matrix T for equation (1.5) of Ch.I in the case that the diagonal matrix U_0 has purely imaginary entries and that the matrix function Q has absolutely integrable entries. (Note that in the literature, the matrix T is also called the scattering matrix, though "monodromy" dominates.) Herein we construct solutions of (1.5) Ch.I which are analytic on the upper and lower half-planes of the spectral parameter and study analytic continuations of the principal minors of T. In §2.3 we briefly discuss variations of the afforementioned construction in the case when some of eigenvalues of U_0 coincide and in the case of a general (not necessarily purely imaginary) complex matrix U_0. §2.4 contains the "discrete scattering data," a formulation of the inverse scattering problem for (1.5) (a Riemann problem on the real axis), and certain results on its solvability. Then in §2.5 we calculate the variational derivatives of the entries of T, allowing us to prove the "infinitesimal" solvability of the inverse problem and to compute the Poisson brackets of the entries of T.

The reader does not need any preliminary knowledge of scattering theory. We use only the theory of linear differential equations (cf., for example, [A1]) and fundamental facts from the theory of complex functions of one variable (cf., for example, [Ca]).

2.1. Monodromy matrix.

In this section, $Q = (q_r^s)$ is a matrix function of $x \in \mathbb{R}$ with values in the set of $n \times n$-matrices whose entries $q_r^s = q_r^s(x)$ (located at the crossing of the r-th row and the s-th column) are absolutely integrable functions of x from $-\infty$ to ∞. Put

$$|q(x)| = \max_{r,s} |q_r^s(x)|, \qquad c_Q = \int_{-\infty}^{\infty} |q(x)|\, dx.$$

Throughout this section except at the end of §2.3 we assume that the constant diagonal matrix $U_0 = \operatorname{diag}(\mu_1, \dots, \mu_n)$ (cf. Ch.I) has purely imaginary entries μ_j which are ordered in the following way:

$$\mu_j \stackrel{\text{def}}{=} i a_j, \quad a_j \in \mathbb{R}, \quad j > s \quad \Rightarrow \quad a_j \geqq a_s. \tag{2.1}$$

Note that there can be repeats among the μ_j. To simplify formulae we denote the matrix-valued function $\exp(\alpha U_0 x)$ by $R_0 = R_0(x;\alpha)$. Finally, following the notation in Ch.I, we put $Q' = (q'^s_r)$, where $q'^s_r = q_r^s$ if $\mu_s \neq \mu_r$, and $q'^s_r = 0$ otherwise, i.e., $q'^s_r = q_r^s(1 - \delta_{\mu_r}^{\mu_s})$, or in other words,

$$Q'(x) \in [U_0, \mathfrak{gl}_n], \qquad [Q - Q', U_0] = 0.$$

Proposition2.1. *There exist unique invertible matrix-valued solutions $E_+(x;\alpha)$ and $E_-(x;\alpha)$ (where $x, \alpha \in \mathbb{R}$) of the equation*

$$E_x + QE = \alpha U_0 E, \tag{2.2}$$

which satisfy the normalization conditions

$$E_\pm \to R_0 \overset{\text{def}}{\Leftrightarrow} E_\pm(x;\alpha) - R_0(x;\alpha) = o(1)$$

as $x \to \infty$ (for E_+) and as $x \to -\infty$ (for E_-). These are called the Jost functions.

Proof. Arbitrary solutions of (2.2) can be written in the form EC via an invertible solution of this equation, where C is a matrix which is constant with respect to x. The existence of $E_\pm$ is checked constructively in a simple way, by explicitly writing the function in the form of "ordered exponentials".

Put (formally) for $p \geqq 0$

$$\begin{aligned} K_\pm^{(p+1)}(x;\alpha) &= \int_x^{\pm\infty}\int_{x_1}^{\pm\infty}\cdots\int_{x_p}^{\pm\infty}\prod_{s=0}^{p} R_0^{-1}QR_0(x_{s+1};\alpha)\,dx_{p+1}\ldots dx_1, \\ K_\pm^{(0)} &= I, \\ E_\pm(x;\alpha) &= R_0(x;\alpha)\sum_{p=0}^{\infty} K_\pm^{(p)}(x;\alpha). \end{aligned} \tag{2.3}$$

As usual, by the definition of integrals, x_1 runs in the interval $[x, \pm\infty)$ from x to $\pm\infty$; $x_2 \in [x_1, \pm\infty)$ and so on; $\int_x^{-\infty} f(y)\,dy \overset{\text{def}}{=} -\int_{-\infty}^{x} f(y)\,dy$. The existence of each of the integrals in the formulae for $K_\pm^{(p+1)}$ follows immediately from the absolute integrability of the entries of Q and the fact that R_0 (for $|\exp(\alpha\mu_j x)| = 1$) is unitary. Moreover, it is easy to prove the following estimate for the entries $(k_\pm^{(p+1)})_j^r$ of the matrix $K_\pm^{(p+1)}$:

$$|k_\pm^{(p+1)}|_j^r \leqq \frac{n^p c_Q^{p+1}}{(p+1)!}, \quad 1 \leqq j, r \leqq n, \tag{2.4}$$

where we abbreviated $|(k_\pm^{(p+1)})_j^r|$ to $|k_\pm^{(p+1)}|_j^r$.

Indeed, $|k_\pm^{(p+1)}|_j^r$ does not exceed

$$I \overset{\text{def}}{=} n^p \left| \int_x^{\pm\infty}\int_{x_1}^{\pm\infty}\cdots\int_{x_p}^{\pm\infty}\prod_{s=0}^{p} |q(x_{s+1})|\,dx_{p+1}\ldots dx_1 \right|.$$

The factor n^p appears here due to the matrix multiplication; it is equal to the number of products of matrix entries of $R_0QR_0^{-1}$ in the sum for $(k_\pm^{(p+1)})_j^r$. The integrand of I is invariant under an arbitrary permutation of the arguments $x_1, \dots, x_{p+1}$. Consequently we can change the integration over the domain $\{x_{s+1} \in [x_s, \pm\infty), 0 \leqq s \leqq p\} \subset \mathbb{R}^{p+1}$, where we put $x_0 = x$ for consistency, to be integration over all the space $\mathbb{R}^{p+1}$, and

$$I = \frac{n^p}{(p+1)!} \int_{-\infty}^{\infty} \cdots \int_{-\infty}^{\infty} \prod_{s=0}^{p} |q(x_{s+1})| \, dx_{p+1} \dots dx_1 = \frac{n^p c_Q^{p+1}}{(p+1)!}.$$

Using (2.4), we obtain that series (2.3) for $E_\pm$ converge absolutely and uniformly for $x, \alpha \in \mathbb{R}$. Due to the uniform convergence, we can differentiate these series term by term with respect to x. Thus we get

$$(K_\pm^{(p+1)})_x = -R_0^{-1} Q R_0 K_\pm^{(p)}, \qquad (R_0)_x = \alpha U_0 R_0,$$

since $R_0 = \exp(\alpha U_0 x)$. Therefore functions $E_\pm$ satisfy equation (2.2).

Integrating with respect to x_1 in the formulae for $K_\pm^{(p+1)}$, it is easy to obtain the following variant of estimate (2.4):

$$\text{(2.4a)} \qquad |k_\pm^{(p+1)}(x)|_j^r \leqq n^p c_Q^\pm(x) \frac{c_Q^p}{p!}.$$

Hereafter we denote the absolute value of the integral $\int_x^{\pm\infty} |q(x)|\, dx$ by $c_Q^\pm(x)$. Obviously $c_Q^\pm(x) \leqq c_Q$ and $c_Q^\pm(x) \to 0$ when $x \to \pm\infty$. Consequently, (2.4a) implies

$$E_\pm(x; \alpha) \to R_0(x; \alpha) \qquad \text{as } x \to \pm\infty.$$

To complete the proof we must check the invertibility of $E_\pm$. Considering the exterior power $\bigwedge^n \mathbb{C}^n$ instead of $\mathbb{C}^n$, we get a differential equation for the determinant, $\det E$, where E is a solution of equation (2.2):

$$\text{(2.5)} \qquad (\det E)_x + \operatorname{Sp} Q \det E = \alpha \operatorname{Sp} U_0 \det E.$$

Since $\det E_\pm \to \det R_0 \neq 0$ as $x \to \pm\infty$, it follows from (2.5) that $\det E_\pm$ do not vanish for any $x, t \in \mathbb{R}$. □

Definition2.1. *The invertible matrix-valued function $T(\alpha)$ for $\alpha \in \mathbb{R}$, determined uniquely by the relation $E_+T = E_-$ and expressed in terms of $K_\pm$ of (2.3) by*

$$T = R_0^{-1}E_-(x = +\infty) = \sum_{p=0}^{\infty} K_-^{(p)}(x = +\infty)$$

is called the monodromy matrix of equation (2.2).

Set $Q^0 = Q - Q'$ and denote solutions of equation (2.2) by $\widehat{E}_\pm^\infty$ normalized as in Proposition2.1 for Q^0 at $\alpha = 0$.

Lemma2.1. a) *Let $\widehat{E}_\pm \overset{\text{def}}{=} E_\pm R_0^{-1}$. The entries of $\widehat{E}_\pm - \widehat{E}_\pm^\infty$ and $T - \widehat{E}_-^\infty(x = +\infty)$ as functions of α are Fourier transformations of certain absolutely integrable functions on the real axis.*

b) The functions $\widehat{E}_\pm$ and T are continuous and invertible on $\mathbb{R} \cup \infty$ and

$$\widehat{E}_\pm(\alpha \to \infty) = \widehat{E}_\pm^\infty = \widehat{E}_\pm(\alpha \to -\infty),$$
$$T(\alpha \to \infty) = \widehat{E}_-^\infty(x = +\infty) = T(\alpha \to -\infty).$$

In particular, when $Q = Q'$ we have $\widehat{E}_\pm^\infty = I$ and $\widehat{E}_\pm, T \to I$ as $|\alpha| \to \infty$.

Proof. We set

$$\widehat{K}_\pm^{(p+1)}(x;\alpha) = R_0 K_\pm^{(p+1)} R_0^{-1}(x;\alpha)$$

for $p \geqq 0$ and, use (2.3), to write an explicit formula for the entry $(\widehat{k}_\pm^{(p+1)})_i^j$ of $\widehat{K}_\pm^{(p+1)}$. For notational ease let $i = i_0$ and $j = i_{p+1}$, $x = x_0$. Then the entry above is equal to the sum of the following terms over all sets $1 \leqq i_1, i_2, \ldots, i_p \leqq n$:

$$(2.6)\quad (\pm 1)^{p+1} \int_{-\infty}^{\infty} \int_{-\infty}^{\infty} \exp\Big(i\alpha \sum_{m=0}^{p} (a_{i_m} - a_{i_{p+1}})(x_m - x_{m+1})\Big) \cdot$$
$$\cdot \prod_{m=0}^{p} q_{i_m}^{i_{m+1}}(x_{m+1}) \prod_{m=0}^{p} \theta(\pm(x_{m+1} - x_m))\, dx_{p+1} \ldots dx_1,$$

where $\theta(x) = 1$ for $x \geqq 0$ and $\theta(x) = 0$ for $x < 0$.

The integrand in (2.6) is an absolutely integrable function on $\mathbb{R}^{p+1}$ (see estimate (2.4)). If $a_{i_{m'}} \neq a_{i_{m'+1}}$ for some m', then we can make the non-degenerate coordinate transformation

$$\widetilde{x}_m = x_m, \qquad \text{for } m \neq m',$$
$$\widetilde{x}_{m'} = \sum_{m=0}^{p} (a_{i_m} - a_{i_{p+1}})(x_m - x_{m+1}).$$

Integrating with respect to all $\widetilde{x}_m$, $m > 0$, without $\widetilde{x}_{m'}$, we can write term (2.6) in the form

$$\int_{-\infty}^{\infty} e^{i\alpha\widetilde{x}} f(\widetilde{x})\, d\widetilde{x}$$

for a certain absolutely integrable function f of $\widetilde{x} = \widetilde{x}_{m'}$ which also depends on the indices $\{i_m\}$, p and $x_0 = x$. We can estimate $\int_{-\infty}^{\infty} |f(\widetilde{x})|\, d\widetilde{x}$ as in (2.4). As a result, we see that this integral does not exceed $c_Q^{p+1}/(p+1)!$.

Thus $(\widehat{k}_{\pm}^{(p+1)})_i^j$ is represented as the sum of the Fourier transformation

$$\int_{-\infty}^{\infty} \exp(i\alpha\widetilde{x}) g(\widetilde{x})\, d\widetilde{x}$$

of a function g, for which

$$\int_{-\infty}^{\infty} |g(\widetilde{x})|\, d\widetilde{x} \leqq \frac{n^p c_Q^{p+1}}{(p+1)!},$$

and the "anomalous" terms of (2.6) such that $a_{i_m} = a_{i_0}$ for all $0 \leqq m \leqq p+1$ (recall that $(\widehat{k}_{\pm}^{(p+1)})_i^j$ is a sum of n^p terms of the form (2.6)). Returning to the summation over p ($\widehat{E}_{\pm} = \boldsymbol{I} + \sum_{p=1}^{\infty} \widehat{K}_{\pm}^{(p+1)}$), we obtain the first statements of the lemma. Indeed, the sum of the "anomalous" terms over all sets $\{i_m\}$ and indices p is precisely equal to $\widehat{E}_{\pm}^{\infty}$. The invertibility of $\widehat{E}_{\pm}^{\infty}$ follows immediately from equation (2.5) for det $\widehat{E}_{\pm}^{\infty}$ ($\widehat{E}_{\pm}$ are invertible for $\alpha \in \mathbb{R}$ by Proposition2.1).

If $Q' = Q$, then $\widehat{E}_{\pm}^{\infty} = \mathbf{I}$. This is obvious, since $a_{i_m} = a_{i_{m+1}} \Rightarrow q_{i_m}^{i_{m+1}} = 0 \Rightarrow$ (2.6) $= 0$. In order to finish the proof, we use a well-known property of Fourier transformations of absolutely integrable functions. □

Before ending this section we construct and study analytic continuations of the boundary columns of $E_{\pm}$ (corresponding to eigenvalues μ_1 and μ_n) by direct analysis of formula (3.3). In the next section we complete these results using purely algebraic methods.

Set $\widehat{E}_{\pm} = E_{\pm} R_0^{-1}$ as before. Let us denote the k-th column of $E_{\pm}$ by $\mathbf{e}_{\pm}^k$. Define analogously

$$\widehat{e}_{\pm}^k \stackrel{\text{def}}{=} \exp(-i\alpha a_k x) \mathbf{e}_{\pm}^k.$$

Set $T^{-1} = (t_{-i}^j)$ and, for consistency of notation, $T = (t_{+i}^j)$. We will assume that $\mu_s = \mu_n$ and $\mu_r = \mu_n$ for the column $\widehat{e}_+^s$ and the entry t_{-r}^s, respectively, $\mu_s = \mu_1 = \mu_r$ when $\widehat{e}_-^s$, t_{+r}^s are considered. We call such indices r and s *boundary indices.* Let $\widehat{e}_{\pm}^{\infty s}$ denote the s-th column of the matrix $\widehat{E}_{\pm}^{\infty}$ of Lemma2.1 and $\widehat{e}_{\pm r}^{\infty s}$ denote the corresponding entry of $\widehat{E}_{\pm}^{\infty}$. We remind that

$$\widehat{e}_{\pm}^{\infty s} = 1^s, \qquad \widehat{e}_{\pm r}^{\infty s} = \delta_r^s,$$

when $Q' = Q$, where

$$1^1 = \begin{pmatrix} 1 \\ 0 \\ \vdots \\ 0 \end{pmatrix}, \qquad 1^2 = \begin{pmatrix} 0 \\ 1 \\ \vdots \\ 0 \end{pmatrix}, \qquad \dots, 1^n = \begin{pmatrix} 0 \\ 0 \\ \vdots \\ 1 \end{pmatrix}.$$

Lemma2.2. *a) The vectors $\widehat{e}_{+}^{s}$, $\widehat{e}_{-}^{s}$, and t_{-r}^{s}, t_{+r}^{s} for boundary indices r, s may be continued to analytic functions of α on the upper half-plane $\operatorname{Im}\alpha > 0$.*

b) For boundary indices r and s, and for $\operatorname{Im}\alpha \geqq 0$ we have:

$$|\alpha| \to \infty \Rightarrow \widehat{e}_{\pm}^{s} \to \widehat{e}_{\pm}^{\infty s}, t_{\pm r}^{s} \to \widehat{e}_{-r}^{\infty s},$$

where we denote the analytic continuations by the same letters.

c) For boundary indices r and s, and for $\operatorname{Im}\alpha > 0$,

$$x \to +\infty \Rightarrow \widehat{e}_{+}^{s} \to 1^s, \widehat{e}_{-r}^{s} \to t_{+r}^{s},$$
$$x \to -\infty \Rightarrow \widehat{e}_{-}^{s} \to 1^s, \widehat{e}_{+r}^{s} \to t_{-r}^{s}.$$

If $\mu_j \neq \mu_1$, then $\lim_{x\to+\infty} \widehat{e}_{-j}^{s} = 0$; if $\mu_j \neq \mu_n$ then $\lim_{x\to-\infty} \widehat{e}_{+j}^{s} = 0$, where $\operatorname{Im}\alpha > 0$.

d) Statements a), b), c) hold if we replace the condition that $\operatorname{Im}\alpha \geqq 0$ ($\operatorname{Im}\alpha > 0$ for $\mu_j \neq \mu_1$ in c)) by $\operatorname{Im}\alpha \leqq 0$ (respectively $\operatorname{Im}\alpha < 0$) and exchange the places of μ_1 and μ_n in the definition of boundary indices (i.e., regard that $\mu_r = \mu_1 = \mu_s$ for $\widehat{e}_{+}^{s}$, t_{-r}^{s}; and $\mu_r = \mu_n = \mu_s$ for $\widehat{e}_{-}^{s}$, t_{+r}^{s}).

Proof. For simplicity, let $s = n = r$ and sign $= +$. Any other combinations of the sign $\pm$, boundary indices r, s, and the sign of $\operatorname{Im}\alpha$ are considered in the same way. We follow the proof of Proposition2.1 and Lemma2.1.

First of all, the absolute value of the exponential factor $\exp\Big(i\alpha \sum_{m=0}^{p}(a_{i_m} - a_{i_n})(x_m - x_{m+1})\Big)$ in formula (2.6) for $j = i_{p+1} = n$ does not exceed 1 for $\operatorname{Im}\alpha \geqq 0$ (recall that $a_i \leqq a_n$, $x_{m+1} \geqq x_m$ for the index sign $+$). Consequently, estimate (2.4) is valid and the n-th column of the series $I + \sum_{p=1}^{\infty} \widehat{K}_{\pm}^{(p)}$ (see (2.3)) converges uniformly for x and α to a continuation of $\widehat{e}^n_+$ in the upper half-plane $\operatorname{Im}\alpha \geq 0$. Moreover, this vector function is represented in the form

$$\widehat{e}^{\infty n}_+ + \int_{-\infty}^{\infty} e^{i\alpha\widetilde{x}} f(\widetilde{x})\, d\widetilde{x}$$

for $\operatorname{Im}\alpha \geqq 0$ (see the proof of Lemma2.1) for a suitable vector function f which is absolutely integrable. Here, $f(\widetilde{x}) = 0$ if $\widetilde{x} < 0$, since $\sum_{m=0}^{p}(a_{i_m} - a_{i_n})(x_m - x_{m+1}) \geqq 0$ for an arbitrary set $\{i_1, \dots, i_p\}$.

If $\operatorname{Re}\alpha \overset{\text{def}}{=} a$, $\operatorname{Im}\alpha \geqq 0$, then

$$\widehat{e}^n_+ = \widehat{e}^{\infty n}_+ + \int_{-\infty}^{\infty} e^{ia\widetilde{x}} e^{-i\operatorname{Im}\alpha\widetilde{x}} f(\widetilde{x})\, d\widetilde{x}.$$

Thus $\widehat{e}^n_+$ is analytic for $\operatorname{Im}\alpha > 0$ and $\widehat{e}^n_+ \to \widehat{e}^{\infty n}_+$ as $|\alpha| \to \infty$, $\operatorname{Im}\alpha \geqq 0$ (statements a) and b) for $\widehat{e}^n_+$).

Now let us use estimate (2.4a) for $\operatorname{Im}\alpha \geqq 0$ and $r = n$ and an analogous estimate for $x < 0$ taking the exponential factor into account:

$$|k_{\pm}^{(p+1)}|_j^n \leqq \exp(-\operatorname{Im}\alpha(a_n - a_j)\frac{|x|}{2})\frac{n^p c_Q^{p+1}}{(p+1)!} + \frac{n^p c_Q^-(x/2) c_Q^p}{p!}, \qquad (\operatorname{Im}\alpha \geqq 0, x < 0) \tag{2.4b}$$

Its derivation does not differ at all from those of (2.4), (2.4a) for $\alpha \in \mathbb{R}$. We find that $\widehat{e}^n_+$ converges uniformly to 1^n as $x \to +\infty$ and that for $h > 0$ and $\operatorname{Im}\alpha \geqq h$, the function $\widehat{e}^n_{+j}$ converges uniformly to zero if $\mu_j \neq \mu_n$ ($\Leftrightarrow a_j \neq a_n$). This proves statement c) for $\widehat{e}^n_+$.

Analogous estimates also hold for the entry t^n_{-n} of the matrix T^{-1}, but we do not need this. By the definition of T for $\alpha \in \mathbb{R}$ the entry $\widehat{e}^n_{+n}$ converges uniformly to t^n_{-n} when $x \to -\infty$. This is also true for $\alpha = \infty$ (Lemma2.1). As the function $\widehat{e}^n_{+n}$ is analytic on the upper half-plane and continuous (including at the point $\alpha = \infty$), the limit of $\widehat{e}^n_{+n}$ exists as $x \to -\infty$ and for $\operatorname{Im}\alpha \geqq 0$ by a standard theorem of complex analysis. Moreover, this limit is the analytic continuation of t^n_{-n}, for which statement b) follows from the property of $\widehat{e}^n_+$ proven above. □

Lemma2.3. *For boundary indices s, the vector functions $\widehat{e}^s_+$ and $\widehat{e}^s_-$ have no zeros (componentwise) and are linearly independent for arbitrary $x \in \mathbb{R}$ and α in the upper half-plane, $\operatorname{Im}\alpha \geqq 0$ including the point $\alpha = \infty$.*

Proof. Let us suppose that for a certain sign, $\widehat{e}^s_\pm(x;\alpha_0) = 0 = e^s_\pm(x_0;\alpha_0)$ where $\alpha_0 \geqq 0$, $x_0 \in \mathbb{R}$. The vector function $e(x) \overset{\text{def}}{=} e^s_\pm(x;\alpha_0)$ defined for all $x \in \mathbb{R}$

satisfies equation (2.2) for $\alpha = \alpha_0$. Hence by the uniqueness theorem for differential equations, $e(x_0) = 0$ implies $e(x) \equiv 0$. Therefore α_0 is a zero of $e^s_\pm$ and $\widehat{e}^s_\pm$ for all x (not only for $x = x_0$). But this is impossible since $\widehat{e}^s_\pm(x;\alpha) \to 1^s$ as $x \to \pm\infty$ and for any α by Lemma2.2. The linear independence of $\{e^s_+\}$ or $\{e^s_-\}$ is established analogously. By Lemma 2.1, $\widehat{e}_\pm$ are linearly independent also at $\alpha = \infty$. □

2.2. Analytic continuations.

For ease of notation we assume that $Q' = Q$, in particular $\widehat{E}^\infty_\mp = I$ (see §2.1). We keep all previous assumptions and notation.

For $\alpha \in \mathbb{R}$ construct from $T(\alpha) = (t^j_i)$ the ***principal "upper" minors***

$$t_0 = 1,\ t_1 = t^1_1,$$
$$t_2 = t^1_1 t^2_2 - t^2_1 t^1_2,\ \ldots\ t_n = \det T.$$

Here and futher, abusing the standard terminology, by minors we mean the determinants of the minors. Let us show that $t_n \equiv 1$. Really, it follows from differential equation (2.5) that $\det E = c \det R_0$ for a suitable $c \in \mathbb{C}$, where E is a certain solution of (2.2). Here we use the equation $\operatorname{Sp} Q = 0$ which is derived from the condition $Q' = Q$. Consequently, $\det E_- = \det R_0$ by the normalization of E_-. Hence,

$$\det T = \det(R_0^{-1} E_-(x = +\infty)) = 1.$$

Moreover, $\prod_{i=1}^{n-1} t_i(\alpha) \not\equiv 0$ for the remaining minors which results from Lemma2.1: $t_i(\alpha) \to 1$ as $|\alpha| \to \infty$.

Define the principal "lower" minors of T, changing the order of rows and columns:

$$\widetilde{t}_n = 1,\ \widetilde{t}_{n-1} = t^n_n,$$
$$\widetilde{t}_{n-2} = t^n_n t^{n-1}_{n-1} - t^{n-1}_n t^n_{n-1},\ \ldots\ \widetilde{t}_0 = \det T = 1$$

Similarly,

$$t_{-1} = t^1_{-1},\ \ldots\ t_{-n} = \det T^{-1} = 1,$$
$$\widetilde{t}_{-n} = t_0 = 1,\ \widetilde{t}_{-n+1} = t^n_{-n},\ \ldots\ \widetilde{t}_0 = \det T^{-1} = 1$$

for $T^{-1} = (t^j_{-i})$. The relations among minors are given by the following formulae from linear algebra:

$$\widetilde{t}_{-m} \det T = t_m, \qquad \widetilde{t}_m \det T^{-1} = t_{-m}, \tag{2.7}$$

where $0 \leqq m \leqq n$ and the determinant of T may be any arbitrary invertible element (not necessarily the unity). It is probably helpful to give a sketch of the proof of these beautiful formulae.

First note that

$$(\det T)\mathcal{P}_n = (T^{\otimes n})\mathcal{P}_n,$$

where $T^{\otimes n} = T\otimes\cdots\otimes T$ is the n-th tensor power of T and $\mathcal{P}_n$ is the antisymmetrizer (the canonical projection of $(\mathbb{C}^n)^{\otimes n}$ onto the exterior power $\bigwedge^n \mathbb{C}^n$). Therefore

$$T^{-1}\det T = \mathrm{Sp}_{n-1}((\mathbf{I}\otimes T^{\otimes(n-1)})\mathcal{P}_n),$$

where Sp_{n-1} is the trace over the last $n-1$ matrix indices of the tensor products. Here we apply the relation $\mathrm{Sp}_{n-1}\mathcal{P}_n = 1$. Analogously, using the identity $\mathrm{Sp}_{n-m}\mathcal{P}_n = \mathcal{P}_m$, we can check the formula

$$(T^{-1})^{\otimes m}\det T\mathcal{P}_m = \mathrm{Sp}_{n-m}((\boldsymbol{I}^{\otimes m}\otimes T^{\otimes(n-m)})\mathcal{P}_n),$$

where Sp_{n-m} is taken over the last $(n-m)$ matrix indices. By calculating the trace of the last formula (for all m matrix indices), we get relation (2.7).

We call the decomposition $T = T^0T_0$ the *triangular factorization* of T, where T^0 is a lower triangular matrix and T_0 is an upper triangular matrix. In order for this decomposition to be unique, we need to fix the diagonal of T^0 or T_0. Let $\operatorname{diag} T^0 = \operatorname{diag}(t_1,\dots,t_n)$. Then it follows that $\operatorname{diag} T_0 = \operatorname{diag}(1, t_1^{-1},\dots,t_{n-1}^{-1})$, since the minors $t_1,\dots,t_n$ of the matrix T must equal the products of the corresponding minors of the matrices T^0 and T_0. This factorization might not be defined for all $\alpha\in\mathbb{R}$, since in general, the entries of T^0 and T_0 have denominators that are products of the minors $t_1,\dots,t_{n-1}$ (as we shall see later, the entries of T^0 are polynomials in t_i^j). In a neighbourhood of the point $\alpha=\infty$, such a decomposition is well-defined (see above).

Analogously, set $T = \widetilde{T}_0\widetilde{T}^0$, where $\widetilde{T}_0$ and $\widetilde{T}^0$ are upper and lower triangular respectively, $\operatorname{diag}\widetilde{T}_0 = \operatorname{diag}(\widetilde{t}_0,\widetilde{t}_1,\dots,\widetilde{t}_{n-1})$ and $\operatorname{diag}\widetilde{T}^0 = \operatorname{diag}(\widetilde{t}_1^{-1},\dots,\widetilde{t}_{n-1}^{-1},\widetilde{t}_n^{-1})$. Such a decomposition, called the *conjugate triangular factorization*, exists in a neighbourhood of $\alpha=\infty$.

Let us construct two new solutions of (2.2) for $\alpha\in\mathbb{R}$ from $E_\pm$:

$$\Phi = E_+T^0 = E_-(T_0)^{-1},$$
$$\Psi = E_+\widetilde{T}^0 = E_-(\widetilde{T}_0)^{-1}.$$

Set $\widehat{\Phi} = \Phi R_0^{-1}$, $\widehat{\Psi} = \Psi R_0^{-1}$, $\widehat{\varphi}^k = \varphi^k\exp(-\alpha\mu_k x)$, and $\widehat{\psi}^k = \psi^k\exp(-\alpha\mu_k x)$, where φ^k and ψ^k are the k-th columns of Φ and Ψ, respectively.

Theorem2.1. a) *The minors t_p and $\widetilde{t}_{-p}$ for $1\leqq p\leqq n$ of Φ are continuous functions of $\alpha\in\mathbb{R}$ and have analytic continuations to the upper half-plane $\operatorname{Im}\alpha>0$.*

Minors $\widetilde{t}_p$ and t_{-p} of Ψ are continuous functions of $\alpha \in \mathbb{R}$ and admit analytic continuations to the lower half-plane $\operatorname{Im}\alpha < 0$.

b) If we denote the analytic continuations by the same letters, then for $1 \leqq p \leqq n$, as $|\alpha| \to \infty$

$$t_p \to 1, \qquad \widehat{\Phi} \to I, \qquad (\operatorname{Im}\alpha \geqq 0),$$
$$\widetilde{t}_p \to 1, \qquad \Psi \to I, \qquad (\operatorname{Im}\alpha \leqq 0).$$

Proof. By construction, $\varphi^n = e^n_+$, $\varphi^1 = e^1_-$, $\psi^1 = e^1_+$, $\psi^n = e^n_-$ and $t_1 = t^1_1$, $\widetilde{t}_{1-n} = t^n_{-n}$, $\widetilde{t}_{n-1} = t^n_n$, $t_{-1} = t^1_{-1}$. Hence the statement of the theorem for the first and the last columns of Φ and Ψ and for the first and last coefficients of the matrices T and T^{-1} follows immediately from Lemma2.2. Note also that by formulae (2.7) $\widetilde{t}_{1-n} = t_{n-1}$ and $t_{-1} = \widetilde{t}_1$. Now we proceed to the proof of the general case.

Let us recall that we denote the p-th exterior power of $\mathbb{C}^n$ by $\bigwedge^p \mathbb{C}^n$. This is the subspace of $(\mathbb{C}^n)^{\otimes p}$ of dimension $\binom{n}{p}$ generated by vectors $z_1 \wedge \ldots \wedge z_p$, $\{z_j\} \subset \mathbb{C}^n$. Given an $n \times n$-matrix $A = (a^j_i)$, we associate to it two square matrices $\bigwedge^n A$ and $\lambda^p A$ of order $\binom{n}{p}$ by the formulae

$$\lambda^p A(z_1 \wedge \ldots \wedge z_p) =$$
$$Az_1 \wedge z_2 \wedge \ldots \wedge z_p + z_1 \wedge Az_2 \wedge \ldots \wedge z_p + z_1 \wedge z_2 \wedge \ldots \wedge z_{p-1} \wedge Az_p,$$

$$\bigwedge^p A(z_1 \wedge \ldots \wedge z_p) = Az_1 \wedge Az_2 \wedge \ldots \wedge Az_p. \tag{2.8}$$

We number the coordinates of $\bigwedge^p \mathbb{C}^n$ in the natural lexicographic order by the sets $1 \leqq i_1 < i_2 < \ldots < i_p \leqq n$. Then at the crossing of row $(i_1, \ldots, i_p)$ and column $(j_1, \ldots, j_p)$ of matrix $\bigwedge^p A$ stands the corresponding minor which is equal to the determinant of the matrix $(\widetilde{a}^m_l) \stackrel{\text{def}}{=} (a^{j_m}_{i_l})$ of order $p \times p$ (we remind that minors in this book mean the determinants of the minors). If $(i_1, \ldots, i_p) = (j_1, \ldots, j_p)$, then the corresponding diagonal element of $\lambda^p A$ is the trace of the matrix $(\widetilde{a}^m_l) = (a^{i_m}_{i_l})$.

It is easy to check that $\lambda^p U_0$ is again diagonal. Its $(i_1, \ldots, i_p)$-th entry is equal to $\mu_{i_1} + \cdots + \mu_{i_p}$. There is another obvious relation: $\bigwedge^p R_0 = \exp(\alpha x(\lambda^p U_0))$. If the first coefficient $\mu_1 + \cdots + \mu_p$ of matrix $\lambda^p U_0$ is equal to the coefficient $\mu_{i_1} + \cdots + \mu_{i_p}$, then $i_1 = 1$, $i_2 = 2$, ..., and $i_s = s$ for $s = \max\{j, \mu_j \neq \mu_p, j < p\}$ and $\mu_{s+1} = \ldots \mu_p = \mu_{i_{s+1}} = \ldots = \mu_{i_p}$. We assume the condition $Q' = Q$ in this section. This implies that the entry of $\lambda^p Q$ at the crossing of the first column with the index $(1, 2, \ldots, p)$ and row $(i_1, \ldots, i_p)$ for which $\mu_{i_1} + \cdots + \mu_{i_p} = \mu_1 + \cdots + \mu_p$

is equal to zero. An analogous statement holds for the first row. We can check by the same method that for the $(i_1, \dots, i_p)$-th element of $\lambda^p U_0$ which coincides with its last element $\mu_n + \mu_{n-1} + \dots + \mu_{n-p+1}$, the entries of $\lambda^p Q$ with indices $((n-p+1, \dots, n), (i_1, \dots, i_p))$ and $((i_1, \dots, i_p), (n-p+1, \dots, n))$ are zero.

Differentiating $\bigwedge^p E$ for an arbitrary solution E of equation (2.2), we obtain the relation

$$(\bigwedge^p E)_x + (\lambda^p Q)(\bigwedge^p E) = \alpha(\lambda^p U_0)(\bigwedge^p E).$$

Indeed, for any invertible matrix function $A(x)$, differentiating the second formula of (2.8), we get the equation

$$(\bigwedge^p A)_x (\bigwedge^p A)^{-1} = \lambda^p (A_x A^{-1}).$$

The equation for $\bigwedge^p E$ has the same structure as (2.2). For $p = 1$ we have equation (2.2) and for $p = n$ we get (2.5). The entries of $\lambda^p Q$ are sums of entries of Q and, consequently, are absolutely integrable from $x = -\infty$ to $x = +\infty$. Note that $\bigwedge^p (E_\pm) = (\bigwedge^p E)_\pm$, i.e., $\bigwedge^p (E_\pm) \to \bigwedge^p R_0$ as $x \to \pm\infty$, and $\bigwedge^p E_+ = (\bigwedge^p E_-)(\bigwedge^p T)$. This follows from the multiplicativity of $\bigwedge^p$.

Thus we have seen that we can apply Lemma2.2 to $\lambda^p Q$, $\lambda^p U_0$, $\bigwedge^p E_\pm$ and $\bigwedge^p T$ instead of Q, U_0, $E_\pm$, and T. The above vanishing conditions for the entries of $\bigwedge^p Q$ is, of course, weaker than the relation $(\bigwedge^p Q)' = \bigwedge^p Q$. But even this statement is enough to show that the limits of $\bigwedge^p \hat{e}^s_\pm$ and $(\bigwedge^p T)^s_{\pm r}$ in statement b) of the lemma are equal to unity for $s = 1 = r$ or $s = n = r$. This is obvious from the proof of Lemma2.2. We want to use Lemma2.2 exactly for such indices.

Set

$$\hat{\varepsilon}^p_+ = \hat{e}^{p+1}_+ \wedge \dots \wedge \hat{e}^n_+, \qquad \hat{\varepsilon}^p_- = \hat{e}^{p+1}_- \wedge \dots \wedge \hat{e}^n_-.$$

Let $\hat{\varepsilon}^n_+ = \hat{\varepsilon}^0_- = 1$. Then for $0 \leqq p \leqq n$ the vector fuctions $\hat{\varepsilon}^p_\pm$ and the function t_p (the first coefficient of the matrix $\bigwedge^p T$) are analytic for $\operatorname{Im} \alpha > 0$. Moreover,

$$\hat{\varepsilon}^p_+ \to 1^{p+1} \wedge \dots \wedge 1^n, \qquad \hat{\varepsilon}^p_- \to 1^1 \wedge \dots \wedge 1^p, \qquad t_p \to 1$$

as $|\alpha| \to \infty$, $\operatorname{Im} \alpha \geqq 0$. In particular, we obtain the proof of the statements concerning the t_p. Note the formula

$$t_p = \hat{\varepsilon}^p_- \wedge \hat{\varepsilon}^p_+, \qquad 1 \leqq p \leqq n. \tag{2.9}$$

Now we turn to Φ.

For $\alpha \in \mathbb{R}$, $1 < p < n$, the functions $\hat{\varphi}^p$ satisfy the following linear relations

$$\hat{\varepsilon}^{p-1}_- \wedge \hat{\varphi}^p = t_{p-1} \hat{\varepsilon}^p_-, \tag{2.10a}$$

$$\hat{\varphi}^p \wedge \hat{\varepsilon}^{p-1}_+ = t_p \hat{\varepsilon}^{p-1}_+, \tag{2.10b}$$

which follow from the definition of Φ. We show that for each x and for $1 < p < n$, the function $\widehat{\varphi}^p(x;\alpha)$ is uniquely determined by equation (2.10) for $\alpha \in \mathbb{R}$ with large enough absolute value. Let us suppose that this were not so. Then there exists a solution $z \in \mathbb{C}^n$, $z \neq 0$, of the homogeneous system of linear equations:

$$\widehat{\varepsilon}_-^{p-1} \wedge z = 0 = z \wedge \widehat{\varepsilon}_+^p, \qquad 1 < p < n,$$

for a large enough $|\alpha|$ and $x = x(\alpha)$. But then $\widehat{\varepsilon}_-^{p-1} \wedge \widehat{\varepsilon}_+^p = 0$ (z can be expressed as a linear combination of either $\widehat{e}_-^1, \dots, \widehat{e}_-^{p-1}$, or $\widehat{e}_+^{p+1}, \dots, \widehat{e}_+^n$). Consequently, for an appropriate sequence $|\alpha_j| \to \infty$, $x_j = x(\alpha_j)$, the vectors $\widehat{e}_-^1, \dots, \widehat{e}_-^{p-1}, \widehat{e}_+^{p+1}, \dots, \widehat{e}_+^n$ are linearly dependent and

$$t_p = \widehat{\varepsilon}_-^p \wedge \widehat{\varepsilon}_+^p = 0.$$

As the minor t_p does not depend on x, we see that $t_p \to 0$ when $|\alpha| \to 0$. This is a contradiction since we know that $t_p \to 1$ when $|\alpha| \to 0$.

Now let us replace t and $\widehat{\varepsilon}_\pm$ in (2.10) by their analytic continuations in the upper half-plane. We will show that this inhomogeneous system of linear equations on $\widehat{\varphi}^p$ remains solvable and even uniquely solvable for almost all α, $\operatorname{Im}\alpha \geqq 0$. The solvability condition is given by stating that a certain polynomial function which is defined in terms of the coefficients of system (2.10), is equal to zero. These coefficients are the components of the vectors $\widehat{\varepsilon}_-^{p-1}$, $\widehat{\varepsilon}_+^p$, $t_{p-1}\widehat{\varepsilon}_-^p$, and $t_p\widehat{\varepsilon}_+^{p-1}$. This function is analytically continued to the upper half-plane as shown above, continuous at the point $\alpha = \infty$, and equal to zero on the real axis. Therefore it is identically zero for all α, $\operatorname{Im}\alpha \geqq 0$ by the Maximum Principle. The proof of the uniqueness is the same as that for $\alpha \in \mathbb{R}$. We need only to remove a finite number of zeros of t_p on the upper half-plane.

So, let $\widehat{\varphi}^p$ be defined for all α in the upper half-plane (except possibly at a finite number of points) as a solution of system (2.10). Then the components of $\widehat{\varphi}^p$ are expressed rationally in terms of the corresponding components of $\varepsilon_\pm$ and $t\varepsilon_\pm$ (i.e., coefficients of the system) and, as a result, they are meromorphic functions on the half-plane $\operatorname{Im}\alpha > 0$. We will now prove the continuity of $\widehat{\varphi}^p$ at $\alpha \in \mathbb{R}$ and analyticity on the upper half-plane.

Lemma2.4. *Let us suppose that for certain analytic vector functions $\widehat{\varepsilon}_\pm^p(\alpha) \in \bigwedge^p \mathbb{C}^n$ $(1 \leqq p \leqq n)$ of parameter $\alpha \in \mathbb{C}$, system (2.10) with t_p from (2.9) is uniquely solvable in a punctured neighbourhood of a point α_0 and $\widehat{\varepsilon}_\pm^p(\alpha_0) \neq 0$ for all p ($\widehat{\varepsilon}_\pm^0 \overset{\text{def}}{=} 1$). Then*

a) there exist vector functions which are holomorphic at α_0, $f_\pm^1(\alpha), \dots, f_\pm^n(\alpha)$ such that

$$\widehat{\varepsilon}_-^p = f_-^1 \wedge \dots \wedge f_-^p,$$

$$\widehat{\varepsilon}_+^{p-1} = f_+^p \wedge \dots \wedge f_+^n,$$

$(1 \leqq p \leqq n)$;

b) the solution $\widehat{\varphi}^p$ of system (2.10) is analytically continued to α_0.

Proof. Statement a) will be proved by induction. We can take $f^1_- = \widehat{\varepsilon}^1_-$ and $f^n_+ = \widehat{\varepsilon}^{n-1}_+$. The claims for $\widehat{\varepsilon}_-$ and $\widehat{\varepsilon}_+$ are analogous, so we prove only the first of them. Let us assume that $f^1_-, \dots, f^{p-1}_-$ are already constructed. Then

$$t_{p-1}\widehat{\varepsilon}^p_- = f^1_- \wedge f^2_- \wedge \dots \wedge f^{p-1}_- \wedge \widehat{\varphi}^p$$

(see (2.10a)), where $\widehat{\varphi}^p$ is meromorphic at α_0. If the function $f^p \stackrel{\text{def}}{=} \widehat{\varphi}^p t^{-1}_{p-1}$ is holomorphic at α_0, then we can just take $f^p_- = \widehat{f}^p$. In the contrary case, we represent $\widehat{f}^p$ in the form $\widehat{f}^p = (\alpha - \alpha_0)^{-m}\widehat{g}^p$, where $m \geqq 1$, $\widehat{g}^p(\alpha_0) \neq 0$ (we use here that $\widehat{\varphi}^p$ is meromorphic, which follows from linear system (2.10)). The vector function $\widehat{g}^p$ admits the following expansion:

$$\widehat{g}^p = \sum_{i=1}^{p-1} c_i f^i_- + (\alpha - \alpha_0)^m g, \tag{2.11}$$

where $g(\alpha)$ and functions $c_i(\alpha)$ for $1 \leqq i \leqq p-1$ are holomorphic at α_0. Really, the exterior product $f^1_- \wedge \dots \wedge f^{p-1}_- \wedge \widehat{g}^p = \widehat{\varepsilon}^{p-1}_- \wedge \widehat{g}^p$ has a zero of order m at α_0 (equation (2.10a)), and $f^1_-, \dots, f^{p-1}_-$ are linearly independent at α_0, since $f^1_- \wedge \dots \wedge f^{p-1}_- = \widehat{\varepsilon}^{p-1}_-$ is not zero at that point. Therefore $\widehat{g}(\alpha_0)$ is linearly expressed in terms of $f^i_-(\alpha_0)$:

$$\widehat{g}^p(\alpha) = \sum_{i=1}^{p-1} c'_i f^i_-(\alpha) + (\alpha - \alpha_0)\widehat{g}'(\alpha), \tag{2.11'}$$

where $c'_i \in \mathbb{C}$, $\widehat{g}'$ is holomorphic in a neighbourhood of α_0. The order of the zero of $f^1_- \wedge \dots \wedge f^{p-1}_- \wedge \widehat{g}'$ at α_0 is $m-1$. Analogously we can represent $\widehat{g}'$ in the form (2.11'), constructing $\widehat{g}''$ and so on. Finally we come to $\widehat{g}^{(m)}$ for which

$$f^1_- \wedge \dots \wedge f^{p-1}_- \wedge \widehat{g}^{(m)}(\alpha_0) \neq 0.$$

If we set $g = \widehat{g}^{(m)}$, then relation (2.11) is fulfilled for suitable functions $c_i(\alpha)$. Hence we can take g as f^p_-. This proves statement a).

Keeping the same notation, we use (2.11) for $\widehat{g}^p = (\alpha - \alpha_0)^m \widehat{\varphi}^p t^{-1}_{p-1}$, where m is the order of pole of $\widehat{\varphi}^p t^{-1}_{p-1}$ at α_0 (i.e., the maximum order of the pole of the components of $\widehat{\varphi}^p t_{p-1}^{-1}$ at α_0). If $m \geqq 1$, at least one of the coefficients $\{c_i(\alpha)\}$ is not equal to zero at α_0; denote it by c_{i_0}. Equation (2.10b) implies that

$$\widehat{g}^p \wedge \widehat{\varepsilon}_+^p = t_p t_{p-1}^{-1} \widehat{\varepsilon}_+^{p-1} (\alpha - \alpha_0)^m.$$

Multiply by $f_+^p \wedge f_-^1 \wedge \ldots \wedge f_-^{i_0-1} \wedge f_-^{i_0+1} \wedge \ldots \wedge f_-^{p-1}$ on both sides of this equation. Then the right hand side is identically zero as α varies ($f_+^p \wedge \widehat{\varepsilon}_+^{p-1} = 0$ by statement a)), and the left hand side can be rewritten as

$$\pm c_{i_0} \widehat{\varepsilon}_-^{p-1} \wedge \widehat{\varepsilon}_+^{p-1} + (\alpha - \alpha_0)^m g',$$

where g' is holomorphic at α_0. Thus $t_{p-1} = \widehat{\varepsilon}_-^{p-1} \wedge \widehat{\varepsilon}_+^{p-1}$ has a zero of order not less than m at α_0 (see (2.9)). Therefore $\widehat{\varphi}^p$ is holomorphic at α_0. □

Lemma2.5. *For an arbitrary invertible matrix $T = (t_i^j)$, the entries of the matrices T^0, T_0^{-1}, $\widetilde{T}_0$ and $(\widetilde{T}^0)^{-1}$ constructed above and of the matrices $\mathcal{D}T_0$, $\mathcal{D}(T^0)^{-1}$, $\widetilde{\mathcal{D}}\widetilde{T}^0$ and $\widetilde{\mathcal{D}}\widetilde{T}_0^{-1}$ are polynomials in t_i^j, where*

$$\mathcal{D} = \mathrm{diag}(t_0 t_1, \ldots, t_{i-1} t_i, \ldots, t_{n-1} t_n),$$
$$\widetilde{\mathcal{D}} = \mathrm{diag}(\widetilde{t}_0 \widetilde{t}_1, \ldots, \widetilde{t}_{i-1} \widetilde{t}_i, \ldots, \widetilde{t}_{n-1} \widetilde{t}_n).$$

Proof. Set $\widehat{E}_+ = I$ and $\widehat{E}_- = T$ in Lemma2.4 and define $\widehat{\varepsilon}_\pm^p$ from $\widehat{E}_\pm$ by the same formulae as above. Then $\widehat{\Phi} = T^0$ is a solution of system (2.10). Let $\alpha \in \mathbb{C}$, $\widetilde{T} = (\widetilde{t}_i^j)$, where $\widetilde{t}_i^j = t_i^j + \alpha f_{ij}$, where the f_{ij} are linear combinations of entries of T. We can always take f_{ij} such that the entries of $\widetilde{T}^0$ are analytic in a certain punctured neighbourhood of $\alpha = 0$. Applying Lemma2.4 to $\widehat{\Phi} = \widetilde{T}^0$, we obtain that all entries of $\widetilde{T}^0$ are analytically continued to $\alpha = 0$ for suitable $\{f_{ij}\}$. Thus entries of T^0 (which are *a priori* rational functions of t_i^j) are regular for all concrete T and must be polynomials of t_i^j. Indeed, singularities of T^0 (if any) form a hypersurface $\mathcal{D}$ (of codimension 1) in the space of all invertible T. Therefore there exists a matrix $T \in \mathcal{D}$ and an entry of $\widetilde{T}^0$ with a pole at $\alpha = 0$ for any choice of $\{f_{ij}\}$, which contradicts the fact proved above.

We can apply an analogous argument to T_0^{-1} (set $\widehat{E}_- = I$, $\widehat{E}_+ = T^{-1}$). Transposing and inverting the decomposition $T = T^0 T_0$, we show that the entries of the other six matrices are polynomial. □

Lemma2.4 for $\widehat{\varphi}^p(x; \alpha)$ above assures that $\widehat{\varphi}^p$ is holomorphic for $1 \leqq p \leqq n$, if $\mathrm{Im}\, \alpha > 0$.

The continuity of $\widehat{\varphi}^p$ for $\operatorname{Im}\alpha \geqq 0$ follows from Lemma2.5. To complete the proof of the theorem concerning Φ, we have only to check that $\widehat{\varphi}^p \to 1^p$ when $|\alpha| \to \infty$, $\operatorname{Im}\alpha \geqq 0$. This is easy to see from relation (2.10), if we replace $\widehat{\varepsilon}_\pm$ and t by their asymptotics as $|\alpha| \to \infty$.

The statements of the theorem for Ψ are proved in exactly the same way. We leave the proof of the analogues of formulae (2.9), (2.10) and Lemma2.4 to the reader as an exercise (see formula (2.13) below). We can also use the following method. Set $\Pi = (\pi_i^j)$, where $\pi_i^j = \delta_{i-1}^{n-j}$ and $A(\alpha)^\omega \overset{\text{def}}{=} \Pi A(-\alpha)\Pi$. Then $E_\mp^\omega$ and T^ω correspond to the matrices $\Pi Q \Pi$ and $-\Pi U_0 \Pi$ in the sense of §2.1. Moreover, the involution ω transforms the decomposition $T = T^0 T_0$ into the conjugate triangular decomposition of T^ω. Consequently, Φ^ω (which is analytic on $\operatorname{Im}\alpha < 0$) is Ψ constructed for $\Pi Q \Pi$ and $-\Pi U_0 \Pi$ and all statements for Ψ follow formally from the properties of Φ. □

Corollary2.1. a) *The multiplicity of each zero α_0 of vector functions $\widehat{\varphi}^p$ ($1 \leqq p \leqq n$), i.e., the minimum order of the zero among all the components of $\widehat{\varphi}^p$ at α_0 ($\operatorname{Im}\alpha_0 \geqq 0$), does not exceed the multiplicity of the zero of the minor t_{p-1} and the minor t_p at that point. The zeros of $\widehat{\varphi}^1 \wedge \ldots \wedge \widehat{\varphi}^p$ (counted with multiplicity) coincide with the zeros of $t_0 t_1 \ldots t_{p-1}$, and the zeros of $\widehat{\varphi}^p \wedge \ldots \wedge \widehat{\varphi}^n$ coincide with the zeros of the product of the minors $t_p \ldots t_n$ ($t_0 = 1 = t_n$).*

b) For $\operatorname{Im}\alpha \geqq 0$, $1 \leqq p, q \leqq n$ we have:

$$\begin{aligned} x \to +\infty &\Rightarrow \widehat{\varphi}_p^p \to t_p \text{ and } \widehat{\varphi}_q^p \to 0 && \text{for } q < p, \\ x \to -\infty &\Rightarrow \widehat{\varphi}_p^p \to t_{p-1} \text{ and } \widehat{\varphi}_q^p \to 0 && \text{for } q > p, \\ \operatorname{Im}\alpha > 0, x \to \pm\infty &\Rightarrow \widehat{\varphi}_q^p \to 0 && \text{if } \mu_q \neq \mu_p, \end{aligned}$$

where $\widehat{\varphi}_q^p$ is the q-th component of $\widehat{\varphi}^p$.

c) At points in the lower half-plane the multiplicities of the zeros of $\widehat{\psi}^p$ ($1 \leqq p \leqq n$) do not exceed the multiplicities of the zeros of $\widetilde{t}_{p-1}$ and $\widetilde{t}_p$ (cf. a)). The zeros of $\widehat{\psi}^1 \wedge \ldots \wedge \widehat{\psi}^p$ (with multiplicities) coincide with the zeros of $\widetilde{t}_0 \widetilde{t}_1 \ldots \widetilde{t}_{p-1}$ and the zeros of $\widehat{\psi}^p \wedge \ldots \widehat{\psi}^n$ coincide with the zeros of $\widetilde{t}_p \widetilde{t}_{p+1} \ldots \widetilde{t}_n$. For $\operatorname{Im}\alpha \leqq 0$ we have:

$$\begin{aligned} x \to +\infty &\Rightarrow \widehat{\psi}_p^p \to \widetilde{t}_{p-1} \text{ and } \widehat{\psi}_q^p \to 0 && \text{for } q > p, \\ x \to -\infty &\Rightarrow \widehat{\psi}_p^p \to \widetilde{t}_p \text{ and } \widehat{\psi}_q^p \to 0 && \text{for } q < p, \\ \operatorname{Im}\alpha < 0, x \to \pm\infty &\Rightarrow \widehat{\psi}_q^p \to 0 && \text{if } \mu_q \neq \mu_p. \end{aligned}$$

Proof. By Lemma2.3 the vector functions $\widehat{\varepsilon}^p_\pm$ do not vanish for $\operatorname{Im}\alpha \geqq 0$. Thus the zeros of $\widehat{\varphi}^p$ are contained (with multiplicities) in the set of zeros of t_{p-1} and t_p (see formula (2.10)). Applying (2.10) step by step, we obtain the relations

$$\widehat{\varphi}^1 \wedge \ldots \wedge \widehat{\varphi}^p = (t_0 t_1 \ldots t_{p-1})\widehat{\varepsilon}^p_-,$$
$$\widehat{\varphi}^p \wedge \ldots \wedge \widehat{\varphi}^n = (t_p \ldots t_n)\widehat{\varepsilon}^{p-1}_+,$$

which prove statement a).

From the definition of $\widehat{\varepsilon}_\pm$ (and Lemma2.2) it follows that for $\operatorname{Im}\alpha \geqq 0$,

$$\widehat{\varepsilon}^p_+ \to 1^{p+1} \wedge \ldots \wedge 1^n, \qquad (x \to +\infty)$$
$$\widehat{\varepsilon}^p_- \to 1^1 \wedge \ldots \wedge 1^p, \qquad (x \to -\infty).$$

The first component of $\lim_{x\to+\infty} \widehat{\varepsilon}^p_-$ (the coefficient of $1^1 \wedge \ldots \wedge 1^p$) is equal to t_p, while the last component of $\lim_{x\to-\infty} \widehat{\varepsilon}^p_-$ (the coefficient of $1^{p+1} \wedge \ldots \wedge 1^n$) is equal to $\tilde{t}_{-p} = t_p$. For $\operatorname{Im}\alpha > 0$ and $\mu_q \neq \mu_p$, $q > p$ the coefficient of $1^q \wedge (\cdot)$ in the expansion of $\lim_{x\to+\infty} \widehat{\varepsilon}^p_-$ is equal to zero, where $(\cdot)$ is an exterior product of certain basis vectors of the form 1^j $(j \neq q)$. Correspondingly, for $q < p$, the coefficients of $1^q \wedge (\cdot)$ of $\lim_{x\to-\infty} \widehat{\varepsilon}^p_+$ are zero (see Lemma2.2) if $\operatorname{Im}\alpha > 0$, $\mu_q \neq \mu_p$.

Substituting the limits $\lim_{x\to\pm\infty} \widehat{\varepsilon}^p_\pm$ into relation (2.10) we get the asymptotics of $\widehat{\varphi}^p_q$ for $\operatorname{Im}\alpha \geqq 0$. The discussion for the region $\operatorname{Im}\alpha > 0$ is analogous. Let us assume, for example, that $\operatorname{Im}\alpha > 0$, $\mu_q \neq \mu_p$ and $q > p$. Then

$$\widehat{\varepsilon}^{p-1}_-(x = \infty) \wedge \widehat{\varphi}^p(x = \infty) = t_{p-1}\widehat{\varepsilon}^p_-(x = \infty).$$

Vectors of the form $1^q \wedge (\cdot)$ (see above) do not appear in the expansion of $\widehat{\varepsilon}^{p-1}_-(x = \infty) \neq 0$ with respect to the standard basis vectors of $\bigwedge^{p-1} \mathbb{C}^n$. Consequently, if $\widehat{\varphi}^p_q(x = \infty) \neq 0$, such vectors must appear in the expansion of $\widehat{\varepsilon}^p_-(x = \infty)$, which is also impossible. Therefore $\widehat{\varphi}^p_q(x = \infty) = 0$. In the case of $\operatorname{Im}\alpha \geqq 0$, $q \leqq p$ we have:

$$\widehat{\varphi}^p(x = \infty) \wedge \widehat{\varepsilon}^p_+(x = \infty) = t_p \widehat{\varepsilon}^{p-1}_+(x = \infty)$$
$$\Rightarrow \widehat{\varphi}^p(x = \infty) \wedge 1^{p+1} \wedge \ldots \wedge 1^n = t_p 1^p \wedge \ldots \wedge 1^n$$
$$\Rightarrow \widehat{\varphi}^p_q(x = \infty) = \delta^p_q t_p, \text{ where } \infty = +\infty.$$

Asymptotics of the vector function $\widehat{\psi}^p$ as $x \to \pm\infty$ are calculated analogously. We can also use the involution ω from the proof of Theorem2.1. □

Exercise2.1. *Using Lemma2.2, compute the limit of* $\widehat{\varphi}^p$ *for* $\operatorname{Im}\alpha > 0$ *and of* $\widehat{\psi}^p$ *for* $\operatorname{Im}\alpha < 0$ *as* $x \to \pm\infty$. (Corollary2.1 allows us to do this computation in the case when the eigenvalues μ_p have multiplicity 1). □

Exercise2.2. *a) Using estimate (2.4), show that the minors* $t_1, \dots, t_{n-1}$ *and* $\widetilde{t}_1, \dots, \widetilde{t}_{n-1}$ *do not have zeros for* $\operatorname{Im}\alpha \geqq 0$ *and* $\operatorname{Im}\alpha \leqq 0$ *respectively, if* $\exp(nc_Q) < 2$ (cf. [ZMaNP], Ch.I §10).

b) Check that for compactly supported Q *(zero outside of a finite interval), the functions* $\widehat{E}_\pm$ *are analytic for all* $\alpha \in \mathbb{C}$ *except at* $\alpha = \infty$. □

2.3. Variants.

When some of the eigenvalues $\{\mu_j\}$ coincide, the entries of T behave better analytically than in the case of pairwise distinct $\{\mu_j\}$. For instance, if $\mu_1 = \mu_2$, then not only t_1^1 but also the entries t_1^2, t_2^1 and t_2^2 are analytically continued to the upper half-plane (Lemma2.2). Theorem2.1 only guarantees that t_1^1 and $t_2 = t_1^1 t_2^2 - t_1^2 t_2^1$ are analytic. It is exactly the same for $E_\pm$ — more entries admit analytic continuation to the upper and lower half-plane. This means that the procedure of constructing analytic solutions of equation (2.2) from $E_\pm$ can be refined and, in fact, simplified in the case when $\{\mu_j\}$ have multiplicities.

Our first goal in this section is to modify Theorem2.1 in the case when some of the μ_j coincide. The reader can skip this section ignoring all dots in the indices when reading the next sections.

First of all, we modify the decomposition $T = T^0 T_0$ (a variant of the Bruhat decomposition with respect to the Borel subgroup of lower triangular matrices). Keeping the notation and conventions of the previous section, set

$$i^* = \max\{j \mid \mu_j = \mu_i\}, \qquad i_* = \min\{j \mid \mu_j = \mu_i\} - 1$$

for $1 \leqq i \leqq n$; the number of distinct values which i^* and i_* might take is the same as the number of pairwise distinct μ_i among $\{\mu_j\}$. If $\mu_i \neq \mu_j$ for all $j \neq i$, then $i^* = i$, $i_* = i - 1$.

Let us introduce the factorization $T = T^{\bullet 0} T_0^\bullet$, where $T^{\bullet 0} \overset{\text{def}}{=} (\tau_i^{0j})$ and $T_0^\bullet \overset{\text{def}}{=} (\tau_{0i}^j)$ and

$$\begin{aligned} i < j, \mu_i \neq \mu_j &\Rightarrow \tau_i^{0j} = 0, \\ i > j &\Rightarrow \tau_{0i}^j = 0, \\ \mu_i = \mu_j, i \neq j &\Rightarrow \tau_{0i}^j = 0. \end{aligned}$$

This decomposition is uniquely determined if we fix the diagonal of $T_0^\bullet$. Let

$$\operatorname{diag} T_0^\bullet = \operatorname{diag}(1, t_{2_\bullet}^{-1}, \dots, t_{i_\bullet}^{-1}, \dots, t_{n_\bullet}^{-1}),$$

where the t_i are the "upper" principal minors (see §2.2). Then the entry τ_i^{0j} for which $\mu_i = \mu_j$ ($\Leftrightarrow i_* = j_*$) is equal to the determinant of the matrix of order $i_* + 1$ consisting of the entries t_l^m of T for $l = 1, 2, \dots, i_*, i$ and $m = 1, 2, \dots, i_*, j$. In particular, $\tau_i^{0j} = t_i^j$ for $\mu_i = \mu_1 = \mu_j$. Similarly, we can express τ_i^{0j} (for $\mu_i = \mu_j$) in terms of the matrices $\bigwedge^{i_*+1} T$. Note that for $p = p^*$, the minor $\det(\tau_i^{0j}, i^* = j^* - p)$ of the matrix $T^{\bullet 0}$ is $t_p t_{p_*}^{(p-p_*-1)}$ and $\tau_{p^*+1}^{0\ p+1} = t_{p+1}$.

In an analogous way we define a decomposition $T = \widetilde{T}_0^\bullet \widetilde{T}^{\bullet 0}$, where zero entries of $\widetilde{T}_0^\bullet$ and $\widetilde{T}^{\bullet 0}$ are arranged as the zeros in the transposed matrices ${}^tT^{\bullet 0}$ and ${}^tT_0^\bullet$ respectively. We fix

$$\operatorname{diag} \widetilde{T}^{\bullet 0} = \operatorname{diag}(\widetilde{t}_{1^\bullet}^{-1}, \widetilde{t}_{2^\bullet}^{-1}, \dots, \widetilde{t}_{i^\bullet}^{-1}, \dots, 1),$$

where $\widetilde{t}_i$ are "lower" minors of T (see §2.2). The minor $\det(\widetilde{\tau}_{0i}^j, i^* = j^* = p^*)$ of the matrix $\widetilde{T}_0^\bullet$ is equal to $\widetilde{t}_{p_*} \widetilde{t}_{p^*}^{(p^*-p_*-1)}$.

The minors $\{t_{i_*}, t_{i^\bullet}\}$ cannot be zero for $\alpha \in \mathbb{R}$ with big enough absolute values. This ensures the existence of the above decomposition "almost everywhere" for $\alpha \in \mathbb{R}$.

Theorem2.2. a) *Functions* $\Phi^\bullet \overset{\text{def}}{=} E_+ T^{\bullet 0} = E_-(T_0^\bullet)^{-1}$ *and* $\Psi^\bullet = E_+ \widetilde{T}_0^\bullet = E_-(\widetilde{T}^{\bullet 0})^{-1}$ *are continuous functions of* $\alpha \in \mathbb{R}$ *and can be analytically continued for* α *in the upper and the lower half-plane, respectively. This is also true with respect to the functions* τ_i^{0j} *and* $\widetilde{\tau}_{0i}^j$ *for* $\mu_i = \mu_j$, *where* $T^{\bullet 0} = (\tau_i^{0j})$ *and* $\widetilde{T}_0^\bullet = (\widetilde{\tau}_{0i}^j)$.

b) If we denote the analytic continuations by the same letters and set $\widehat{\Phi}^\bullet = \Phi^\bullet R_0^{-1}$ *and* $\widehat{\Psi}^\bullet = \Psi^\bullet R_0^{-1}$, *then*

$$\tau_i^{0j} \to \delta_i^j, \qquad \widehat{\Phi}^\bullet \to I,$$
$$\widetilde{\tau}_{0i}^j \to \delta_i^j, \qquad \widehat{\Psi}^\bullet \to I$$

as $|\alpha| \to \infty$, $\operatorname{Im} \alpha \geqq 0$ *or* $\operatorname{Im} \alpha \leqq 0$, $\mu_i = \mu_j$.

c) For $p = p^*$, $p < p_1 < \dots < p_k \leqq (p+1)^* \overset{\text{def}}{=} p'$, *set*

$$\widehat{\varphi}_-^{\bullet\{p_i\}} = \widehat{\varphi}^{\bullet 1} \wedge \dots \wedge \widehat{\varphi}^{\bullet p} \wedge \widehat{\varphi}^{\bullet p_1} \wedge \dots \wedge \widehat{\varphi}^{\bullet p_k},$$
$$\widehat{\varphi}_+^{\bullet\{p_i\}} = \widehat{\varphi}^{\bullet p_1} \wedge \dots \wedge \widehat{\varphi}^{\bullet p_k} \wedge \widehat{\varphi}^{\bullet p'+1} \wedge \dots \wedge \widehat{\varphi}^{\bullet n},$$

and analogously define $\widehat{e}_-^{\{p_i\}}$ *and* $\widehat{e}_+^{\{p_i\}}$ *via* e_- *and* e_+ *respectively. Then,*

$$\widehat{\varphi}_-^{\bullet\{p_i\}} = d_- \widehat{e}_-^{\{p_i\}},$$
$$\widehat{\varphi}_+^{\bullet\{p_i\}} = \sum_{\{q_i\}} d_+^{\{q_i\}} \widehat{e}_+^{\{q_i\}},$$

where

$$d_- = (t_p)^k \prod_{i=1}^{p} t_{i_*},$$

$$d_+^{\{q_i\}} = \det(\tau_{q_i}^{0\ p_j};\ 1 \leqq i,j \leqq k) t_{p'}^{-1} \prod_{i=p'+1}^{n} t_{i_*},$$

$\{q_i\}$ runs over the sets $p < q_1 < \ldots < q_k \leqq p'$. The zeros of $\widehat{\varphi}_-^{\bullet\{p_i\}}$ coincide with the zeros of d_- together with multiplicities; the multiplicity of an arbitrary zero of the vector function $\widehat{\varphi}_+^{\bullet\{p_i\}}$ is equal to the minimum of the multiplicities of the zeros of $d_+^{\{q_i\}}$ for all possible $\{q_i\}$.

d) Set $[T^{\bullet 0}] = (\tau_i^{0j}\delta_{i_*}^{j^*})$ (this matrix is obtained by letting entries τ_i^{0j} be zero for $\mu_i \neq \mu_j \Leftrightarrow i^* = j^*$). Then for $\operatorname{Im}\alpha > 0$,

$$\lim_{x\to-\infty} \widehat{\Phi}^\bullet = \operatorname{diag}(1, \ldots, t_{i_*}, \ldots, t_{n_*}),$$

$$\lim_{x\to+\infty} \widehat{\Phi}^\bullet = [T^{\bullet 0}].$$

Analogously for $\operatorname{Im}\alpha < 0$,

$$\lim_{x\to-\infty} \widehat{\Psi}^\bullet = \operatorname{diag}(\widetilde{t}_{1^\bullet}, \ldots, \widetilde{t}_{i^\bullet}, \ldots, 1),$$

$$\lim_{x\to+\infty} \widehat{\Psi}^\bullet = [\widetilde{T}_0^\bullet].$$

The same relations also hold for $[\widehat{\Phi}^\bullet]$ and $[\widehat{\Psi}^\bullet]$ instead of $\widehat{\Phi}^\bullet$ and $\widehat{\Psi}^\bullet$ when $\operatorname{Im}\alpha \gtreqless 0$.

Proof. The only difference in the proofs here from those of Theorem2.1 and Corollary2.1 is the extended use of Lemma2.2. Lemma2.4 and Lemma2.5 are transferred to the case of analytic functions with dots without difficulty.

Set

$$\widehat{\varepsilon}_-^{\bullet p}(x;\alpha) = \widehat{e}_-^1 \wedge \ldots \wedge \widehat{e}_-^{p_*} \wedge \widehat{e}_-^p \in \bigwedge^{p_*+1} \mathbb{C}^n,$$

$$\widehat{\varepsilon}_+^{\bullet p-1}(x;\alpha) = \widehat{e}_+^p \wedge \widehat{e}_+^{p^*+1} \wedge \ldots \wedge \widehat{e}_-^n \in \bigwedge^{n-p_*+1} \mathbb{C}^n,$$

keeping the notation $\widehat{e}_\pm^p$ of the previous section. Then for $\alpha \in \mathbb{R}$ the following generalizations of (2.10) hold

$$\widehat{\varepsilon}_-^{p_*} \wedge \widehat{\varphi}^{\bullet p} = t_{p_*} \widehat{\varepsilon}_-^{\bullet p}, \tag{2.10a$^\bullet$}$$

$$\widehat{\varphi}^{\bullet p} \wedge \widehat{\varepsilon}_+^{p^*} = \sum_{i=n-p^*+1}^{p^*} \tau_i^{0p} \widehat{\varepsilon}_+^{\bullet i-1}, \tag{2.10b$^\bullet$}$$

where for $1 \leqq p \leqq n$, $\widehat{\varphi}^{\bullet p}$ is the p-th column of $\widehat{\Phi}^{\bullet}$. The functions $\widehat{\varepsilon}^{p}_{\pm}$, $\widehat{\varepsilon}^{\bullet p}_{\pm}$, $t_{p\bullet}$, and τ^{0j}_{i} for $i^* = j^*$ ($\Leftrightarrow \mu_i = \mu_j$) are analytically continued to the upper half-plane. Hence $\widehat{\varphi}^{\bullet p}$ can be defined as solutions of (2.10$^\bullet$). The uniqueness is proven in the same way as it was done in Theorem2.1. The functions $\widehat{\psi}^{\bullet p}$ are constructed similarly. Corollary2.1 implies statements c) and d). (We leave the proof of the counterpart of c) in the case of $\widehat{\Psi}^{\bullet}$ to the reader as an exercise.) □

We can easily generalize Theorem2.2 to the case of an arbitrary semisimple or reductive algebraic group G, embedding G into the general linear group GL_n for a suitable n.

Exercise2.3*. *a) Let G be a semisimple algebraic group over $\mathbb{C}$ with Lie algebra $\mathfrak{g}$. Suppose that the matrix $\mathrm{ad}_{U_0} = [U_0, (\cdot)]$ for $U_0 \in \mathfrak{g}$ is diagonizable and has purely imaginary eigenvalues (i.e., $\exp U_0$ belongs to a compact Lie subgroup of G, where exp is the exponential map from $\mathfrak{g}$ to G). Assume that the entries of $Q(x) \in \mathfrak{g}$ for $x \in \mathbb{R}$ are absolutely integrable for $-\infty < x < \infty$ with respect to an embedding of $\mathfrak{g}$ into (suitable) $\mathfrak{gl}_n$. Then Proposition2.1 about the existence of $E_{\pm}(x;\alpha)$ for $\alpha \in \mathbb{R}$ holds, where $E_{\pm}$ take their values in G. This allows us to introduce the monodromy matrix $T(\alpha) = E_{+}^{-1}E_{-} \in G$.*

b) Let us denote by $\mathfrak{u}_{\pm}$ (respectively, $\mathfrak{u}_0$) the Lie subalgebra of $\mathfrak{g}$ generated by elements $z \in \mathfrak{g}$ over $\mathbb{C}$ such that $[iU_0, z] = rz$ for $r = r(z) \in \mathbb{R}$ and $\pm r > 0$ (respectively, $r = 0$). Let $\mathcal{P}_{\pm} = \exp(\mathfrak{u}_{\pm} \oplus \mathfrak{u}_0)$, $\mathcal{U}_{\pm} = \exp(\mathfrak{u}_{\pm})$. Represent T in the form $T = T^{\bullet 0}T_0^{\bullet} = \widetilde{T}_0^{\bullet}\widetilde{T}^{\bullet 0}$ for "almost all" $\alpha \in \mathbb{R}$ and suitable $T^{\bullet 0}(\alpha) \in \mathcal{P}_{-}$, where $T_0^{\bullet}(\alpha) \in \mathcal{U}_{+}$, $\widetilde{T}_0^{\bullet}(\alpha) \in \mathcal{P}_{+}$, $\widetilde{T}^{\bullet 0}(\alpha) \in \mathcal{U}_{-}$. Then the functions $\widehat{\Phi}^{\bullet}$ and $\widehat{\Psi}^{\bullet}$ defined by the formulae of Theorem2.2 are meromorphically continued to the upper and lower half-plane of α respectively. If $Q(x) \in \mathfrak{u}_{+} \oplus \mathfrak{u}_{-}$, then $\widehat{\Phi}^{\bullet} \to I$ and $\widehat{\Psi}^{\bullet} \to I$ as $|\alpha| \to \infty$ ($\mathrm{Im}\,\alpha \geqq 0$ or $\mathrm{Im}\,\alpha \leqq 0$). □

Unitarity conditions. Let Ω be a diagonal matrix consisting of ± 1. Set $A^* = \Omega A^{+}\Omega$, where $A^{+} = {}^t(\overline{A})$ is the hermitian conjugate of the matrix A. In the most important applications, $Q^* + Q = 0$ for a suitable matrix Ω. We assume here that this condition, called the $*$-anti-hermitian condition, is fulfilled. All the previous conventions and notation remain valid. Then

$$(E_{\pm}(\alpha))^* = E_{\pm}(\alpha), \qquad (T(\alpha))^* = T^{-1}(\alpha)$$

for $\alpha \in \mathbb{R}$, i.e., $E_{\pm}$, T are $*$-unitary matrix functions. This is a direct consequence of the identity $R_0^* = R_0^{-1}$ and the definition of $E_{\pm}$. Using (2.7), the following relations are easily obtained:

$$\widetilde{T}_0 = ((T^0)^*)^{-1}\mathcal{D}^*, \quad \widetilde{T}^0 = (\mathcal{D}^*)^{-1}((T_0)^*)^{-1},$$

where $\mathcal{D} \overset{\text{def}}{=} \operatorname{diag}(t_0 t_1, t_1 t_2, \dots, t_{n-1} t_n)$, $\alpha \in \mathbb{R}$.

If $\alpha \in \mathbb{C}$, then we set $A^*(\alpha) \overset{\text{def}}{=} (A(\overline{\alpha}))^*$ for any matrix function $A(\alpha)$. In this notation $\Psi(\alpha) = (\Phi^*(\alpha))^{-1}\mathcal{D}^*(\alpha)$ for $\operatorname{Im}(\alpha) \leqq 0$. Indeed, the last equation is true for $\alpha \in \mathbb{R}$ and by the Maximum Principle, is also true for all α ($\operatorname{Im}\alpha \leqq 0$). An analogous relation for $\Psi^\bullet$ also holds as well (see Theorem2.2):

$$\Psi^\bullet(\alpha) = (\Phi^{\bullet *}(\alpha))^{-1}\mathcal{D}^{\bullet *}(\alpha),$$
$$\mathcal{D}^\bullet \overset{\text{def}}{=} \operatorname{diag}(t_{1^\bullet}, \dots, t_{i^\bullet} t_{i_\bullet}, \dots, t_{n_\bullet}). \tag{2.12}$$

The case of an arbitrary matrix U_0. We will briefly sketch the necessary changes in the construction of analytic solutions of (2.2) for matrices U_0 (diagonal) which are not purely imaginary. Though the approach is almost the same, the construction becomes complicated. First of all, we no longer have the concept of a (total) monodromy matrix. The entries of $Q(x)$ are assumed to be absolutely integrable for x from $-\infty$ to ∞ as before. For the sake of simplicity, let $Q' = Q$.

We associate to $U_0 = \operatorname{diag}(\mu_1, \dots, \mu_n)$, $\mu_j \in \mathbb{C}$ two sectors $M_\pm = M_\pm(U_0) \subset \mathbb{C}$:

$$M_\pm = \pm\{\alpha \in \mathbb{C} \mid k < l \Rightarrow \operatorname{Re}(\mu_k \alpha) \geqq \operatorname{Re}(\mu_l \alpha)\}.$$

For $\{\mu_j\} \subset i\mathbb{R}$ with the standard ordering (see above), $M_\pm$ are the upper and lower half-planes. Sectors $M_\pm$ may consist of only zero or of half-lines of the form $\mathbb{R}_\pm \alpha$, where $\mathbb{R}_\pm = \{\pm r^2, r \in \mathbb{R}\}$, $\alpha \in \mathbb{C}$. Let us assume that the interiors $M_\pm^\circ$ of sectors $M_\pm$ are not void. Then $M_\pm^\circ$ are (connected) simply connected domains (intersections of half-planes).

For $\alpha \in M_+$, we define vector functions $e_-^1(x;\alpha)$, $e_+^n(x;\alpha)$ as the first and last column of series (2.3) respectively. Estimate (2.4) for $|k_+^{(p+1)}|_j^n$ and $|k_-^{(p+1)}|_j^1$ remains valid without change. Thus these series converge. Moreover, estimates (2.4a) and (2.4b) also hold. In an analogous way we can construct vector functions $e_\pm^s$ ($\alpha \in M_+$) for $\mu_s = \mu_n$ or $\mu_s = \mu_1$ respectively, and $e_\mp^s$ ($\alpha \in M_-$) for $\mu_s = \mu_1$ or $\mu_s = \mu_n$ respectively. With all these we can rewrite Lemma2.2 and 2.3 for vector functions $e_\pm^s$ defined by series (2.3) and boundary coefficients T defined by series (2.5) for $\alpha \in M_+$ (instead of the upper half-plane) and $\alpha \in M_-$ (instead of the lower half-plane) respectively.

Note that in §2.1 the series which define $E_\pm$ and T were meaningful for all $\alpha \in \mathbb{R}$. This allowed us to pose a question about analytic continuations of the entries of these matrices (or a linear combination of them). If such continuations are found, then they are uniquely determined. Now the situation is different. We need to define the entries of analytic solutions of (2.2) constructively.

We construct the vector functions $\varepsilon_\pm^p(x;\alpha) \in \bigwedge^p \mathbb{C}^n$ from the boundary columns of formulae (2.3) for $\lambda^p Q$ and $\lambda^p U_0$ instead of Q and U_0. These functions are continuous for $\alpha \in M_+$ and analytic for $\alpha \in M_+^0$. As in the proof of Theorem2.1, ε_+^p is constructed from the last column of the "E_+-formula" and ε_-^p from the first column of the "E_--formula", $1 \leqq p \leqq n$. Analogously we can introduce $t_p(\alpha)$ (the first coefficient of the $\bigwedge^p$-analogue of the formula for T). In particular, $\varepsilon_+^1 = e_+^n$, $\varepsilon_-^1 = e_-^1$, and $t_1 = t_1^1$. By the same formulae as in §§2.1, 2.2 we define $\widehat{\varepsilon}_\pm^p$ ($\widehat{e}_\pm^k = \exp(-\alpha\mu_k x)e_\pm^k$ and so on). Analogously for $\alpha \in M_-$ (starting with $\gamma_+^1 \stackrel{\text{def}}{=} e_+^1$, $\gamma_-^1 \stackrel{\text{def}}{=} e_-^n$ and $\widetilde{t}_{n-1} = t_n^n$), let us construct $\gamma_\pm^p(x;\alpha) \in \bigwedge^p \mathbb{C}^n$, $\widetilde{t}_p(\alpha)$ and $\widehat{\gamma}_\pm^p$, where $1 \leqq p \leqq n$. Note that $t_n = 1 = \widetilde{t}_0$ and $\widehat{\varepsilon}_+^n = \widehat{\varepsilon}_-^n = 1 = \widehat{\gamma}_+^1 = \widehat{\gamma}_-^n$. We also set $t_0 = 1 = \widetilde{t}_n$ and $\widehat{\varepsilon}_\pm^0 = 1 = \widehat{\gamma}_\pm^0$.

Proposition2.2. *For any $x \in \mathbb{R}$ and for each $1 \leqq p \leqq n$, there exists a unique solution $\widehat{\varphi}^p(x;\alpha) \in \mathbb{C}^n$ of system (2.10) which is continuous on M_+ and analytic on M_+°. Analogously there exists a unique analytic (continuous) solution $\widehat{\psi}^0$ of the system of equations*

$$\begin{aligned} \widehat{\gamma}_+^{p-1} \wedge \widehat{\psi}^p &= \widetilde{t}_{p-1}\widehat{\gamma}_+^p, \\ \widehat{\psi}^p \wedge \widehat{\gamma}_-^p &= \widetilde{t}_p \widehat{\varepsilon}_-^{p-1}, \end{aligned} \tag{2.13}$$

for $\alpha \in M_-^\circ$ ($\alpha \in M_-$). We construct the matrices $\widehat{\Phi}$ and $\widehat{\Psi}$ from $\{\widehat{\varphi}^p\}$ and $\{\widehat{\psi}^p\}$ as the columns and set $\Phi = \widehat{\Phi}R_0$ and $\Psi = \widehat{\Psi}R_0$, $R_0 = \exp(\alpha U_0 x)$. Then $\widehat{\Phi} \to I$ and $\widehat{\Psi} \to I$ when $|\alpha| \to \infty$ ($\alpha \in M_\pm$), and Φ and Ψ satisfy equation (2.2).

Proof. The solvability of (2.10) (or (2.13)) is an equation of the form $\pi_p = 0$ (or $\widetilde{\pi}_p = 0$), where π_p (or $\widetilde{\pi}_p$) is a polynomial of the components of $\widehat{\varepsilon}_\pm^p$, $\widehat{\varepsilon}_\pm^{p-1}$ t_p and t_{p-1} (or $\widehat{\gamma}_\pm^{p,p-1}$ and $\widetilde{t}_{p,p-1}$). Let us extend the definition of the functions $\varepsilon_\pm^p$, $\gamma_\pm^p$, t_p and t_{p-1}. Set $\alpha = 1$ in formulae (2.3), (2.5), (2.6) as well as in all other formulae containing α and consider these functions as functions of the eigenvalues $\mu_1, \ldots, \mu_n$. We get immediately that $\{\widehat{\varepsilon}_\pm^p, t_p\}$ and $\{\widehat{\gamma}_\pm^p, \widetilde{t}_p\}$ depend analytically on $\mu_1, \ldots, \mu_n$ in the domains $\mathcal{M}_+^0 \stackrel{\text{def}}{=} \{\operatorname{Re}\mu_k > \operatorname{Re}\mu_{k+1}, k = 1, \ldots, n-1\}$ and $\mathcal{M}_-^0 \stackrel{\text{def}}{=} -\mathcal{M}_+^0$ respectively and continuously in their closures $\mathcal{M}_\pm$. Now let us prove, for example, that system (2.10) is solvable for $p = 2$ for $(\mu_1, \ldots, \mu_n) \in \mathcal{M}_+$. General p and (2.13) are completely analogous.

Set $\mathcal{F} = \mathcal{M}_+ \cap \{\operatorname{Re}\mu_1 = \operatorname{Re}\mu_2\}$. For $(\mu_j) \in \mathcal{F}$, more terms in series (2.3) and in the series for T converge than in all the domain $\mathcal{M}_+$. Actually we can handle those series after restricting to $\mathcal{F}$ as we did in the case $\mu_1 = \mu_2$ (see §2.1, Theorem2.2). This allows us to apply Lemma2.2 to its full extent. In addition to $\widehat{\varepsilon}_-^1 = \widehat{e}_-^1$ and $\widehat{\varepsilon}_-^2$ we can define a continuous $\mathbb{C}^n$-valued vector functions $\widehat{e}_-^2$ of $(\mu_j) \in \mathcal{F}$, such that $\widehat{e}_-^1 \wedge \widehat{e}_-^2 = \widehat{\varepsilon}_-^2$. Analogously we define a continuous function t_1^2 of $(\mu_j) \in \mathcal{F}$. Then $\widehat{\varphi}^2 \stackrel{\text{def}}{=} t_1\widehat{e}_-^2 - t_1^2\widehat{e}_-^1$ satisfies system (2.10).

Thus we have proven the solvability of system (2.10) for $p = 2$ at $(\mu_j) \in \mathcal{F}$. Hence, $\pi_2|_{\mathcal{F}} = 0$. The real codimension of $\mathcal{F}$ in $\mathcal{M}_+$ is one. Hence $\pi_2 \equiv 0$ on $\mathcal{M}_+$ (π_2 is analytic). Similarly, $\pi_p \equiv 0$ and $\widetilde{\pi}_p \equiv 0$ for all p on $\mathcal{M}_+$ and $\mathcal{M}_-$ respectively. Therefore systems (2.10) and (2.13) are solvable for any p, $x \in \mathbb{R}$ and for arbitrary $(\mu_j) \in \mathcal{M}_+$. Of course, this also holds for functions of $\alpha \in M_\pm$ at fixed $\mu_1, \dots, \mu_n$.

Moreover, using the solvability of (2.10), (2.13) proven above, it is easy to show the uniqueness for these systems in a neighbourhood of the point $\alpha = \infty$ (i.e., for big enough $|\alpha|$, $\alpha \in M_\pm$), to derive relation (2.2), and to find the asymptotics of $\widehat{\Phi}$ and $\widehat{\Psi}$ when $|\alpha| \to \infty$. Finally Lemma2.4 implies that $\widehat{\Phi}$ and $\widehat{\Psi}$ have unique analytic continuations from a neighbourhood of $\alpha = \infty$ to corresponding sectors $\alpha \in M_+^0, M_-^0$ which are continuous for all $\alpha \in M_\pm$. $\square$

It remains to cover the complex plane $\mathbb{C}$ by sectors, where the above construction can be applied. For an arbitrary permutation σ of n elements we denote the corresponding matrix by P_σ:

$$P_\sigma \begin{pmatrix} x_1 \\ \vdots \\ x_n \end{pmatrix} \stackrel{\text{def}}{=} \begin{pmatrix} x_{\sigma(1)} \\ \vdots \\ x_{\sigma(n)} \end{pmatrix}, \qquad P_\sigma(a_i^j)P_\sigma^{-1} = (a_{\sigma(i)}^{\sigma(j)}).$$

Set $U_0^\sigma = P_\sigma U_0 P_\sigma^{-1}$ and $M_\pm^\sigma = M_\pm(U_0^\sigma)$. We call permutation σ and sector $M_\pm^\sigma$ *admissible*, if $(M_\pm^\sigma)^\circ \neq \emptyset$. Admissible sectors $\{M_+^\sigma\}$ cover the whole complex plane and any two of them can either coincide, or intersect in a half-line, or intersect at the origin (and analogously for $\{M_-^\sigma\}$). When $\mu_1, \dots, \mu_n$ are in a generic position (i.e., lines $\mu_j\mathbb{R}$ are pairwise distinct), the number of admissible sectors is maximal and equal to $n(n-1)$. In each admissible sector M_+^σ we construct a solution of equation (2.2) for U_0^σ and $Q^\sigma \stackrel{\text{def}}{=} P_\sigma Q P_\sigma^{-1}$ by Propostion2.2. Set $\Phi^{(\sigma)} = P_\sigma^{-1}\Phi_\sigma P_\sigma$. Then $\Phi^{(\sigma)}$ is a solution of (2.2) for U_0 and Q which is analytic for $\alpha \in M_+^\sigma$ and continuous up to the boundary.

When some of $\{\mu_j\}$ coincide, there is one deficiency in the above construction. It is easy to see that

$$M_+^{\sigma'} = M_+^\sigma \Leftrightarrow [P_{\sigma'\sigma^{-1}}, U_0] = 0$$

for admissible σ and σ'. Therefore, generally speaking, one admissible sector M_+^σ might correspond to several different permutations σ' and functions $\Phi^{(\sigma')}$ (their number is equal to $\prod_{i^*}(i^* - i_*)!$, the order of the centralizer of U_0 in the permutation group). The best way to define Φ^σ on M_+^σ canonically is to turn to the construction of the functions $\Phi^\bullet$ instead of Φ (see Theorem2.2) for arbitrary (not purely imaginary) U_0:

Exercise2.4. a) *Following the method of Proposition2.2, construct analogues of* $\Phi^\bullet$ *and* $\Psi^\bullet$ *of Theorem2.2 in sectors* $M_\pm(U_0)$. (Rewrite system (2.13) for $\Psi^\bullet$ and then generalize it and (2.10•) to the case of an arbitrary matrix U_0).

b) Show that the function $\Phi^{\bullet(\sigma)}$ *defined in the same manner as* $\Phi^{(\sigma)}$ *depends only on its admissible sector* M_+^σ, *i.e.,*

$$[P_{\sigma'}, U_0^\sigma] = 0 \Rightarrow [P_{\sigma'}, \Phi^{\bullet(\sigma)}] = 0.$$

In particular, $\Psi^\bullet = \Phi^{\bullet(w_0)}$ *for* $w_0(1,\dots,n) = (n, n-1,\dots,1)$ (cf. the end of the proof of Theorem2.1). □

2.4. Riemann-Hilbert problem.

We begin with a "spectral" interpretation of the degenerations of the functions $\Phi^\bullet$ and $\Psi^\bullet$ of Theorem1.1,2, i.e., points α in the upper or the lower half-plane at which these functions are not invertible. We keep all assumptions and notation of those theorems.

We can naturally regard the zeros of $\det \Phi^\bullet = \prod_{i=1}^n t_{i_\bullet}$ and $\det \Psi^\bullet = \prod_{i=1}^n \widetilde{t}_{i^\bullet} = \prod_{i=1}^n \widetilde{t}_{i_\bullet}$ as points of the *discrete spectrum* of equation (2.2) in the upper and the lower half-plane respectively. Let us recall the usual definition of a point of the discrete spectrum α_0. For such points, equation (2.2) (or any other similar differential equation with parameter) has vector solution ψ which is normally integrable from $-\infty$ to $+\infty$. The concept of normal integrability depends on concrete problems. Note that $\{i_*\} \cup n = 0 \cup \{i^*\}$.

Proposition2.3. a) *For each zero* $\alpha_0 \notin \mathbb{R}$ *of functions* $\prod_{i_\bullet} t_{i_\bullet}$ *or* $\prod_{i_\bullet} \widetilde{t}_{i_\bullet}$, *we can find a vector solution* $\varphi(x;\alpha_0) = {}^t(\varphi_k) \neq 0$ *of equation (2.2) and two numbers* $a, b \in \mathbb{R}$, $b > 0$ *for which*

$$|e^{ax}\varphi_k(x;\alpha_0)| < e^{-b|x|} \text{ when } |x| \to \infty, \qquad 1 \leqq k \leqq n.$$

b) If $\prod_{i=1}^n t_{i_\bullet}(\alpha_0) \neq 0 \neq \prod_{i=1}^n \widetilde{t}_{i_\bullet}(\alpha_0)$ *for* $\operatorname{Im}\alpha_0 > 0$ *or, respectively,* $\operatorname{Im}\alpha_0 < 0$, *then for any* $a \in \mathbb{R}$ *and an arbitrary solution* $\varphi(x;\alpha_0) \in \mathbb{C}^n$ *of equation (2.2), at least one component of the vector function* $e^{ax}\varphi(x;\alpha_0)$ *does not vanish when* $|x| \to +\infty$ *or* $|x| \to -\infty$.

Proof. Let $t_{p_\bullet}(\alpha_0) = 0$, $\operatorname{Im}\alpha_0 > 0$ for a certain p. We use the notation of Theorem2.2. In a neighbourhood of α_0, set $\widehat{g}^q = (\alpha-\alpha_0)^{k_q}\widehat{\varphi}^{\bullet q} \neq 0$, where k_q is the order of a zero of the vector function $\widehat{\varphi}^{\bullet q}$. Since $\varphi^{\bullet q}$ satisfies equation (2.2), the order k_q is locally constant with respect to x and, consequently, independent of x in general (cf., for example, Lemma2.3). Let us show that there is an index q such that $\widehat{\varepsilon}_-^{q\bullet} \wedge \widehat{g}^q(\alpha_0) = 0$. We suppose that this were false.

If $t_{p_*}(\alpha_0) = 0$, then (see equation (2.10a$^\bullet$))

$$(\alpha - \alpha_0)^{-k_q} \widehat{\varepsilon}_-^{p_*} \wedge \widehat{\varphi}^{\bullet q} = (\alpha - \alpha_0)^{-k_q} t_{p_*} \widehat{\varepsilon}_-^{\bullet q} = \widehat{\varepsilon}_-^{p_*} \wedge \widehat{g}^q \neq 0$$

for all q with the condition $q^* = p^*$ ($\Leftrightarrow p_* < q \leqq p^*$). Thus t_{p_*} has a zero of order k_q at α_0 since $\widehat{\varepsilon}_-^{\bullet q}$ does not vanish by Lemma2.3. Moreover, all k_q coincide with each other (and with $k_{p^\bullet}$). Using (2.10a$^\bullet$) directly or statement c) of Theorem2.2, we get that

$$\Big(\bigwedge_{q^*=p^*} \widehat{\varphi}^{\bullet q}\Big) \wedge \widehat{\varepsilon}_+^{p^*} = t_{p^\bullet} t_{p_*}^{(p^* - p_* - 1)} \widehat{\varepsilon}_+^{p_*}, \qquad \widehat{\varepsilon}_+^{p_*}(\alpha_0) \neq 0.$$

The order of the zero of the left hand side at point α_0 is greater than or equal to $(p^* - p_*)k_{p^\bullet}$. Therefore $t_{p_*}(\alpha_0) = 0 \Rightarrow t_{p^\bullet}(\alpha_0) = 0$. Applying this argument to $t_{p^\bullet}$ (instead of t_{p_*}) and so on, we finally get a contradiction since $t_n = 1$.

Let p be the minimal index for which $\widehat{\varepsilon}_-^{p_*} \wedge \widehat{g}^p(\alpha_0) = 0$ ($p > 1$, since $\widehat{\varepsilon}_-^{1_*} \wedge \widehat{g}^1 = \widehat{\varepsilon}_-^1 \neq 0$). It is easy to check by using induction and (2.10a$^\bullet$) that

$$\widehat{g}^1 \wedge \ldots \wedge \widehat{g}^{p_*}(\alpha_0) = c\widehat{\varepsilon}_-^{p_*}(\alpha_0)$$

for $c \neq 0$. We can also use formulae c) of Theorem2.2. As a result, $\widehat{g}^1, \ldots, \widehat{g}^{p_*}$ are linearly independent at α_0 and $\widehat{g}^p(\alpha_0)$ is expressed linearly in terms of them. Finally we obtain the identity:

$$g^p(\alpha_0) = \sum_{q=1}^{p_*} c_i g^q(\alpha_0),$$

where $c_i \in \mathbb{C}$ and $g^i = \exp(\alpha\mu_i x)\widehat{g}^i$. The coefficients c_i do not depend on x, since g^p, $\{g^q\}$ are solutions of equation (2.2).

Let $a_p > a' > a_{p_*}$ (recall that $\mu_s = ia_s$, $q > r \Rightarrow a_q > a_r$), $0 < b' < a_p - a'$ and $b' < a' - a_{p_*}$ for $a', b' \in \mathbb{R}$. Set $a = \operatorname{Im} \alpha_0 a'$, $b = \operatorname{Im} \alpha_0 b'$ and $\varphi = g^p$. Letting $|x| \to \infty$, we obtain the estimates:

$$|\varphi_k(x;\alpha_0)| < O(1)e^{-|x|a_p \operatorname{Im} \alpha_0} \qquad \text{as } x \to +\infty,$$
$$|\varphi_k(x;\alpha_0)| < O(1)e^{|x|a_{p_*} \operatorname{Im} \alpha_0} \qquad \text{as } x \to -\infty.$$

Here we use the definition of $\varphi = g^p$ and the relations obtained above (together with formulae d) of Theorem2.2). Thus statement a) is proven for $\prod t_{i_*}$. For $\prod \widetilde{t}_{i_*}$ the proof is the same.

Now let us suppose that $\prod t_{i_\bullet}$ does not have a zero in the upper half-plane. Then at any point $\alpha = \alpha_0$ with $\operatorname{Im}\alpha_0 > 0$, the matrix $\Phi^\bullet$ is invertible and $\varphi^{\bullet 1}, \dots, \varphi^{\bullet n}$ are linearly independent. Any vector solution φ of equation (2.2) for $\alpha = \alpha_0$ is represented in the form

$$\varphi(\alpha_0) = \sum_{i=1}^{n} c_i \varphi^{\bullet i}(\alpha_0), \qquad c_i \in \mathbb{C}$$

by the uniqueness theorem of differential equations. Let $i_0 = \min\{i, c_i \neq 0\}$, $i^0 = \max\{i, c_i \neq 0\}$. Finally we obtain the following estimates with the help of Theorem2.2 (cf. above):

$$|\varphi_{k_1}(\alpha_0)| > |C_1| e^{-|x| a_{i_0} \operatorname{Im}\alpha_0} \qquad \text{as } x \to +\infty,$$

$$|\varphi_{k_2}(\alpha_0)| > |C_2| e^{|x| a_{i^0} \operatorname{Im}\alpha_0} \qquad \text{as } x \to -\infty,$$

where $C_1 \neq 0 \neq C_2$, for certain constants k_1, k_2. Analogous estimates for $\Psi^\bullet$ hold when $\operatorname{Im}\alpha_0 < 0$. Applying the inequality $i^0 \geqq i_0$, we get the proof of statement b). □

Now let $\widetilde{\Phi}$ be an arbitrary analytic solution of equation (2.2) in a punctured neighbourhood of the point $\alpha = \alpha_0$ for $-\infty < x < +\infty$. We assume that $\det \widetilde{\Phi}(\alpha) \neq 0$ for $\alpha \neq \alpha_0$ and the entries of $\widetilde{\Phi}(\alpha)$ are meromorphic at α_0. For scalar ($\mathbb{C}$-valued) analytic functions of α, the most important invariant of zeros and poles is their order. For matrix functions, the degenerate points (the poles of the entries of $\widetilde{\Phi}$ or $\widetilde{\Phi}^{-1}$) are described by a set of vector spaces which we now construct for $\widetilde{\Phi}$. Actually we have already encountered an analogous problem in §1 (see Proposition1.2). Let us recall the notation we use there.

Let $\mathcal{O}_0 = \mathcal{O}_{\alpha_0}$ be the ring of formal power series of $(\alpha - \alpha_0)$ whose coefficients depend on x, and let $\mathcal{O}_0^n = \mathcal{O} \otimes_{\mathbb{C}} \mathbb{C}^n$ be the space of vector series. Set $\widetilde{\mathcal{O}}_0 = \{\sum_{i \geqq k}^{\infty} c_i(x)(\alpha - \alpha_0)^i, k \in \mathbb{Z}\}$ and $\widetilde{\mathcal{O}}_0^n = \widetilde{\mathcal{O}}_0 \otimes_{\mathbb{C}} \mathbb{C}^n$ (then $\widetilde{\mathcal{O}}_0$ is the field of fractions of the ring $\mathcal{O}_0$). We call an $\mathcal{O}_0$-module $\mathcal{K} \subset \widetilde{\mathcal{O}}_0^n$ a *lattice* when it generates $\widetilde{\mathcal{O}}_0^n$ as an $\widetilde{\mathcal{O}}_0$-module (i.e., $\mathcal{O}_0\mathcal{K} = \mathcal{K}$ and $\widetilde{\mathcal{O}}_0\mathcal{K} = \widetilde{\mathcal{O}}_0^n$).

It may happen that $\widetilde{\Phi}(\alpha_0)^{-1}$ cannot be defined. However we can always define $\widetilde{\Phi}(\alpha)^{-1}$ as an element of $GL_n(\widetilde{\mathcal{O}}_0)$ (as an invertible matrix with entries in $\widetilde{\mathcal{O}}_0$). Set $\widetilde{\mathcal{K}}_0 = \widetilde{\Phi}(\alpha)^{-1}\mathcal{O}_0^n$. Then $\widetilde{\mathcal{K}}_0$ is a lattice since $\det \widetilde{\Phi} \not\equiv 0$. We will show that $\widetilde{\mathcal{K}}_0$ is generated as an $\mathcal{O}_0$-module by vectors which are constant with respect to x. This independence from x can also be written in the form $(\widetilde{\mathcal{K}}_0)_x \subset \widetilde{\mathcal{K}}_0$, which can be easily checked:

$$(\widetilde{\mathcal{K}}_0)_x = (Q - \alpha U_0)\widetilde{\Phi}^{-1}\mathcal{O}_0^n \subset \widetilde{\mathcal{K}}_0.$$

With the aid of $\widetilde{\mathcal{K}}_0$, we can formulate a trivial but useful criterion for $\widetilde{\Phi}$ to be analytic and invertible at α_0:

$$\widetilde{\Phi} \in GL_n(\mathcal{O}_0) \Leftrightarrow \widetilde{\mathcal{K}}_0 \stackrel{\text{def}}{=} \widetilde{\Phi}^{-1}\mathcal{O}_0^n = \mathcal{O}_0^n. \tag{2.14}$$

Obviously $\mathcal{K}_0$ plays the role of the "vector multiplicity" of α_0.

We will apply the construction $\widetilde{\Phi} \mapsto \widetilde{\mathcal{K}}_0$ to functions $\widetilde{\Phi} = \Phi, \Phi^\bullet, \Psi$ and $\Psi^\bullet$. Let us recall that in the case of pairwise distinct $\{\mu_j\}$, the functions Φ and Ψ coincide with $\Phi^\bullet$ and $\Psi^\bullet$ (analogously for T_0, T^0, $\widetilde{T}_0$ and $\widetilde{T}^0$) and $i^* = i$, $i_* = i-1$.

Lemma2.6. *Define (see Lemma2.5 and (2.12))*

$$\mathcal{D} = \operatorname{diag}(t_i t_{i-1}), \mathcal{D}^\bullet = \operatorname{diag}(t_{i^\bullet} t_{i_\bullet}), 1 \leqq i \leqq n,$$

and analogous diagonal matrices $\widetilde{\mathcal{D}}$ and $\widetilde{\mathcal{D}}^\bullet$ for $\widetilde{t}_i$ instead of t_i. Then the functions $\mathcal{D}\widehat{\Phi}^{-1}, \mathcal{D}^\bullet\widehat{\Phi}^{\bullet-1}, \widetilde{\mathcal{D}}\widehat{\Psi}^{-1}$, and $\widetilde{\mathcal{D}}^\bullet\widehat{\Psi}^{\bullet-1}$ are analytic for $\operatorname{Im}\alpha > 0$, $\operatorname{Im}\alpha < 0$ respectively and continuous up to the boundary $\alpha \in \mathbb{R}$. When $|\alpha| \to \infty$, all these functions converge to I.

Proof. These statements are easily derived from Theorem2.1, 2.2. We introduce an involution $A^\tau(\alpha) \stackrel{\text{def}}{=} ({}^t\!A(-\alpha))^{-1}$. Functions $E_\pm^\tau$ and T^τ are the $E_\pm-$ and T-functions corresponding to the same matrix U_0 and $-{}^t Q$ instead of Q; the decomposition $T^\tau(\alpha) = (T^{0\tau}(\alpha)\mathcal{D}(-\alpha))(\mathcal{D}(-\alpha)^{-1}T_0^\tau(\alpha))$ turns out to be the conjugate factorization of T^τ. Really, this decomposition has the necessary type. Moreover, the i-th diagonal element of $T^{0\tau}(\alpha)\mathcal{D}(-\alpha)$ is equal to

$$t_i^{-1}(-\alpha)t_i(-\alpha)t_{i-1}(-\alpha) = t_{i-1}(-\alpha)$$

and, consequently, coincides with the $\widetilde{t}_i$-minor (lower $(i-1)$-th minor) of the matrix $T^{-1}(-\alpha)$ or $({}^tT(-\alpha))^{-1}$ in view of formula (2.7) (the transposition does not change the principal minors). Therefore ${}^t(\mathcal{D}\Phi^{-1}(-\alpha))$ becomes the Ψ-function for $-{}^tQ$ and has the desired analyticity and continuity by Theorem2.1. The statements of the lemma for the other functions can be checked similarly (see Lemma2.5). □

In the notation of Lemma2.6, set $\widehat{\Phi}_- = \widehat{\Psi}\widetilde{\mathcal{D}}^{-1}$ and $\widehat{\Phi}_-^\bullet = \widehat{\Psi}^\bullet\widetilde{\mathcal{D}}^{\bullet-1}$. For consistency of notation we write $\widehat{\Phi}_+$ and $\widehat{\Phi}_+^\bullet$ instead of $\widehat{\Phi}$ and $\widehat{\Phi}^\bullet$. Analogously, $\Phi_\pm \stackrel{\text{def}}{=} \widehat{\Phi}_\pm R_0$ and $\Phi_\pm^\bullet \stackrel{\text{def}}{=} \widehat{\Phi}_\pm^\bullet R_0$, where $R_0 = \exp(\alpha U_0 x)$. When $\alpha \in \mathbb{R}$ we have

$$\widehat{\Phi}_-^{-1}\widehat{\Phi}_+ = R_0 S R_0^{-1}, \qquad \widetilde{\mathcal{D}}^{-1}S \stackrel{\text{def}}{=} \widetilde{T}_0^{-1}T^0 = \widetilde{T}^0 T_0^{-1}, \tag{2.15}$$

$$\widehat{\Phi}_-^{\bullet-1}\widehat{\Phi}_+^\bullet = R_0 S^\bullet R_0^{-1}, \qquad \widetilde{\mathcal{D}}^{\bullet-1}S^\bullet \stackrel{\text{def}}{=} \widetilde{T}_0^{\bullet-1}T^{\bullet 0} = \widetilde{T}^{\bullet 0}T_0^{\bullet-1}. \tag{2.15•}$$

According to Lemma2.5, S and $S^\bullet$ are polynomials of the entries of T and, in particular, are continuous functions of $\alpha \in \mathbb{R}$. Strictly speaking, this lemma can be applied only to S, but its formulation and proof hold in the case with $^\bullet$. By Lemma2.6 *the functions* $\widehat{\Phi}_-^{-1}$ *and* $\widehat{\Phi}_-^{\bullet-1}$ *have analytic continuations in the lower half-plane.*

Let $\alpha_1^\bullet, \dots, \alpha_{N_+^\bullet}^\bullet$ be the pairwise distinct zeros of $\det \widehat{\Phi}_+^\bullet = \prod_{i=1}^n t_{i_\bullet}$ and let $\alpha_{N_+^\bullet+1}^\bullet, \dots, \alpha_{N^\bullet}^\bullet$ be the zeros of $\det \widehat{\Phi}_-^{\bullet-1} = \prod_{i=1}^n \widetilde{t}_{i_\bullet}$ in the upper ($\operatorname{Im}\alpha > 0$) and the lower ($\operatorname{Im}\alpha < 0$) half-planes respectively. Furthermore for the sake of simplicity we assume that the minors $t_{i_\bullet}$, $\widetilde{t}_{i_\bullet}$ do not have zeros on the real axis $\alpha \in \mathbb{R}$. Set (see (2.14))

$$\mathcal{K}_j^\bullet = \Phi_+^{\bullet-1}(\mathcal{O}_j^n) \qquad (1 \leqq j \leqq N_+^\bullet),$$
$$\mathcal{K}_j^\bullet = \Phi_-^{\bullet-1}(\mathcal{O}_j^n) \qquad (N_+^\bullet < j \leqq N^\bullet),$$

where $\mathcal{O}_j = \mathcal{O}_{\alpha_j^\bullet}$. Note that $\mathcal{K}_j^\bullet \supset \mathcal{O}_j^n$ and $\mathcal{K}_j^\bullet \subset \mathcal{O}_j^n$ for $j \leqq N_+^\bullet$ or $j > N_+^\bullet$ respectively. We also define the analogous lattices for $\Phi_\pm$ (without $^\bullet$). All lattices $\mathcal{K}$ introduced here do not depend on x. We call $\{\alpha_j^\bullet, \mathcal{K}_j^\bullet\}$ *the discrete scattering data.*

Exercise2.5. *Show that we can recover every zero of each minor* $t_{i_\bullet}$ $(1 \leqq j \leqq N_+^*)$ *and the lattices* $[T^{\bullet 0}]^{-1}\mathcal{O}_j^n$ *(in the notation of Theorem2.2) from the set* $\{\alpha_j^\bullet, \mathcal{K}_j^\bullet\}$ $(1 \leqq j \leqq N_+^\bullet)$. *Similar statements hold for* $\{\alpha_j^\bullet, \mathcal{K}_j^\bullet, j > N_+^\bullet\}$, $\widetilde{T}$ *and for* α, $\mathcal{K}$ *without* $^\bullet$. (Use statement c) of Theorem2.2, its analogue for Ψ, and Corollary2.1.) □

Theorem2.3. *a) Let us suppose that a continuous function* $S(\alpha) \in GL_n$, $\alpha \in \mathbb{R}$, *certain sets of pairwise distinct points* $\{\alpha_j, 1 \leqq j \leqq N_+, \operatorname{Im}\alpha_j > 0\}$, $\{\alpha_j, N_+ < j \leqq N, \operatorname{Im}\alpha_j < 0\}$, *and a set of* $\mathcal{O}_j$*-lattices* $\{\mathcal{K}_j\}$ *for* $\mathcal{O}_j = \mathcal{O}_{\alpha_j}$ *are given. Also we suppose that there exist matrix-valued functions* $\widehat{\Phi}_+(x;\alpha)$ *and* $\widehat{\Phi}_-^{-1}(x;\alpha)$ *which are continuous for* $\operatorname{Im}\alpha \geqq 0$ *(respectively,* $\operatorname{Im}\alpha \leqq 0$*) and analytic in* $\operatorname{Im}\alpha > 0$ *(respectively,* $\operatorname{Im}\alpha < 0$*), and satisfy the following properties:*

0) $\{\alpha_j, j \leqq N_+\}$ *are zeros of* $\det \widehat{\Phi}_+$ *and* $\{\alpha_j, j > N_+\}$ *are poles of* $\det \widehat{\Phi}_-$;
i) $\widehat{\Phi}_-^{-1}\widehat{\Phi}_+ = R_0 S R_0^{-1}$;
ii) $\widehat{\Phi}_+, \widehat{\Phi}_- \to \boldsymbol{I}$ *for* $|\alpha| \to \infty$;
iii) $\mathcal{K}_j = \Phi_+^{-1}(\mathcal{O}_j^n)$ $(j \leqq N_+)$, $\mathcal{K}_j = \Phi_-^{-1}(\mathcal{O}_j^n)$ $(j > N_+)$.

Then $\widehat{\Phi}_+$, $\widehat{\Phi}_-$ *are uniquely determined by relations i), ii), and iii). (Here* $R_0 = \exp(\alpha U_0 x)$, $\Phi_\pm = \widehat{\Phi}_\pm R_0$.*)*

b) The functions $\Phi_\pm$ *satisfy relation (2.2) for a suitable matrix function* $Q(x)$. *Let us suppose that functions* $\Phi_\pm^{\mathrm{reg}}$ *are invertible for all* α $(\operatorname{Im}\alpha \geqq 0$ *or* $\operatorname{Im}\alpha \leqq 0)$

and that they fulfill relations i) and ii). Then $\Phi_\pm = B\Phi_\pm^{\rm reg}$ *for a matrix function* $B(x;\alpha)$. *This function* $B(x;\alpha)$ *is rational in* α *and corresponds to the functions* Φ_+ *or* Φ_- *(depending on the sign of* $\operatorname{Im}\alpha_j$*) in the sense of Proposition1.2 of §1. The function* B *is uniquely determined from the relations:*

i) $B(\alpha = \infty) = \boldsymbol{I}$;

ii) $B^{-1}\mathcal{O}_j^n = \Phi_\pm^{\rm reg}\mathcal{K}_j$.

Proof. The proof of this theorem is shorter than its formulation. If $\widehat{\Phi}'_\pm$ satisfy i) – iii) with the same S, and $\{\alpha_j, \mathcal{K}_j\}$, then $\widehat{\Phi}_+(\widehat{\Phi}'_+)^{-1} = \widehat{\Phi}_-^{-1}\widehat{\Phi}'_-$ for $\alpha \in \mathbb{R}$. The left hand side of this equation is analytically continued in α to the upper half-plane and the right hand side to the lower half-plane (use the criterion (2.14)). As both sides tend to $\boldsymbol{I}$ when $|\alpha| \to \infty$, Liouville's theorem gives that $\widehat{\Phi}_\pm(\widehat{\Phi}'_\pm)^{-1} = \boldsymbol{I}$. We can check the statement for $\Phi_\pm(\Phi_\pm^{\rm reg})^{-1}$ analogously. By the same properties of analytic continuations we get

$$(\Phi_+)_x\Phi_+^{-1} = (\Phi_-)_x\Phi_-^{-1},$$

which is obtained by differentiating the relation $\Phi_-^{-1}\Phi_+ = S$. In a neighbourhood of $\alpha = \infty$, both sides are equal to $\alpha U_0 - Q$ for a suitable Q up to $O(\alpha^{-1})$ and, again by Liouville's theorem they coincide with $\alpha U_0 - Q$ for all α. □

According to Theorem2.3, the composition of arrows

$$Q \mapsto E_\pm \mapsto \{\Phi, \Psi\} \mapsto S, \{\alpha_j, \mathcal{K}_j\}$$
$$(\text{or, } Q \mapsto S^\bullet, \{\alpha_j^\bullet, \mathcal{K}_j^\bullet\})$$

is a one-to-one correspondence. Indeed, Φ_+ and Φ_- are uniquely recovered from S and $\{\alpha_j, \mathcal{K}_j\}$ and satisfy differential equation (2.2), which makes it possible to reconstruct $Q(x)$ from Φ_+ or Φ_-. As is obvious from the theorem, the central part of this procedure is the solution of the *regular Riemann problem*, i.e., finding the functions $\Phi_\pm^{\rm reg}$ for $\pm\operatorname{Im}\alpha > 0$ satisfying relations i) and ii) of the theorem. Then $\Phi_\pm$ (or, completely analogously, $\Phi_\pm^\bullet$) are given by the purely algebraic operation of "dressing" (see §1) in terms of $\Phi_\pm^{\rm reg}$ and $\{\alpha_j, \mathcal{K}_j\}$. Note that, in general, there exists exactly one relation for the dimensions of $\{\mathcal{K}_j\}$ and a sequence of inequalities for $\Phi_\pm^{\rm reg}(\mathcal{K}_j)$ which guarantee the dressing procedure. Therefore, if we do not take any symmetries into account, Q could have singularities.

Here we do not touch upon the natural (and difficult) problem of describing all $S(\alpha)$, $\alpha \in \mathbb{R}$, which correspond to the matrix functions $Q(x)$ with absolutely integrable entries. We only mention that the entries of $S - \boldsymbol{I}$ must be Fourier transformations of absolutely integrable functions (which is a simple consequence of Lemma2.1). We refer the reader to the paper [DS] or to classical works (see [GK], [Kre]) for methods of solving this kind of problem, generalizing the Wiener-Hopf method. There is a huge amount of literature on this problem.

We can take an arbitrary closed contour homeomorphic to a circle instead of the real line in Theorem2.3 (on which the function S was defined). The formulation and proof of the theorem do not change. The inner and outer domains bounded by this contour play the role of the upper and the lower half-plane. This contour may have a more complicated topology. For instance, it is easy to carry over Theorem2.3 to the case of an arbitrary matrix U_0 (see the previous section) where S is defined on a collection of half-lines bounding sectors $\{M_+^\sigma\}$. Analogues of $\Phi_\pm$ are defined in each of the sectors. They are the functions $\Phi^{(\sigma)}$. It is worth noting that the level of difficulty of the Wiener-Hopf method increases radically when the class of contours in the problem changes. In particular, the Riemann-Hilbert problem for the interior and the exterior of only one sector in $\mathbb{C}$ came to be studied only recently.

For the readers who are not familiar with the *inverse scattering problem* and the *Riemann-Hilbert problem* (which is the recovery of Q and the corresponding $\Phi_\pm$ from S and $\{\alpha_j, \mathcal{K}_j\}$), the following comments might be useful. When we turn from Φ and Ψ to S, implicit dependence of Φ and Ψ on x (defined by equation (2.2)) is replaced by a simple evolutionary relation

$$(\widehat{\Phi}_-)^{-1}\widehat{\Phi}_+(x;\alpha) = R_0(x;\alpha)S(\alpha)R_0^{-1}(x;\alpha)$$

without loss of information. If we know values of $\widehat{\Phi}_+(x_0;\alpha)$ and $\widehat{\Phi}_-(x_0;\alpha)$ for a point $x = x_0$, then after conjugation by the matrix $R_0(x - x_0;\alpha)$ we obtain functions depending on x which satisfy relation (2.15) like $\widehat{\Phi}_\pm$. They are, of course, analytically continued to a suitable half-plane for all x. However, in contrast to the "true" functions $\widehat{\Phi}_+(x)$ and $\widehat{\Phi}_-(x)$, the functions $R_0(x - x_0)\widehat{\Phi}_\pm(x_0)R_0(x - x_0)^{-1}$ have essential singularities when $|\alpha| \to \infty$, $\alpha \notin \mathbb{R}$. Hence such a naive attempt to recover directly $\widehat{\Phi}_+(x)$ and $\widehat{\Phi}_-(x)$ from $\widehat{\Phi}_+(x_0)$ and $\widehat{\Phi}_-(x_0)$ turns out to be inconsistent and, actually, we have to solve the Riemann problem for each x and the corresponding matrix $R_0(x)S(\alpha)R_0^{-1}(x)$ again and again. This makes the problem of describing the dependence of Q on x highly complicated even for a simple matrix S. For $\alpha \in \mathbb{R}$, conjugation by the matrix $R_0(x)$ is harmless and no singularities arise when $\alpha \to \pm\infty$.

As the function $Q(x)$ can be recovered from S and $S^\bullet$ and discrete scattering data, the matrix T is uniquely determined by S and $\{\alpha_j, \mathcal{K}_j\}$ (similarly for $S^\bullet$ and $\{\alpha_j^\bullet, \mathcal{K}_j^\bullet\}$). Let us rewrite the corresponding procedure without passing through Q. We will examine only the arrow $S^\bullet \mapsto T$ and leave the case $S \mapsto T$ to the reader (omit the index $^\bullet$ and set $i^* = i$, $i_* = i - 1$). We keep the notation of Theorem2.2. We assume that the minors $\{t_{i_*}, \widetilde{t}_{i_*}\}$ do not have zeros on the real axis, as before.

Proposition2.4. *Let us assume that for the function $T(\alpha)$, $\alpha \in \mathbb{R}$, the entries τ_i^{0j} and τ_{0i}^j of the matrices $T^{\bullet 0}$ and $T_0^\bullet$ are analytically continued to the upper and the lower half-plane for $\mu_i = \mu_j$ ($\Leftrightarrow i^* = j^*$), respectively. Let $\tau_i^{0j}, \widetilde{\tau}_{0i}^j \to \delta_i^j$ as $|\alpha| \to \infty$. Then T is uniquely recovered from the matrix $S^\bullet = \widetilde{\mathcal{D}}^\bullet \widetilde{T}_0^{\bullet -1} T^{\bullet 0} = \widetilde{\mathcal{D}}^\bullet \widetilde{T}^{\bullet 0} T_0^{\bullet -1}$ (see (2.15•)), the zeros of the minors t_{i_*}, $\widetilde{t}_{i_*}$ $(1 \leqq i \leqq n)$ on $\operatorname{Im}\alpha > 0$, $\operatorname{Im}\alpha < 0$ (with multiplicities) and corresponding lattices $[T^{\bullet 0}]^{-1}\mathcal{O}_j^n$ and $[\widetilde{T}_0^\bullet]^{-1}\mathcal{O}_j^n$ for the matrices $[T^{\bullet 0}]$ and $[\widetilde{T}_0^\bullet]$ defined in Theorem2.2.*

Proof. We recall that $\operatorname{diag} T_0^\bullet = \operatorname{diag}(t_{i_*}^{-1})$, $\operatorname{diag} \widetilde{T}^{\bullet 0} = \operatorname{diag}(\widetilde{t}_{i^*}^{-1})$, and $\widetilde{\mathcal{D}}^\bullet = \operatorname{diag}(\widetilde{t}_{i_*} \widetilde{t}_{i^*})$ $(1 \leqq i \leqq n)$. Consequently,

$$s_j^\bullet \stackrel{\text{def}}{=} \det(s_p^{\bullet q}, 1 \leqq p, q \leqq j) = \prod_{i=1}^{j} \widetilde{t}_{i_*} t_{i_*}, \qquad 1 \leqq j \leqq n. \tag{2.16}$$

The minors t_{j_*} and $\widetilde{t}_{j_*}$ are analytically continued to the corresponding half-planes and have zeros α_{s_j}, $\widetilde{\alpha}_{s_j}$ with multiplicities κ_{s_j}, $\widetilde{\kappa}_{s_j}$ (which are constant for $j_* < j \leqq j^*$) as prescribed by the conditions of the proposition. Moreover $t_{j_*}, \widetilde{t}_{j_*} \to 1$ when $|\alpha| \to \infty$. Hence we come to the scalar Riemann problem and can solve it explicitly:

$$\log t_{j_*}(\alpha) = \frac{1}{2\pi i}\int_{-\infty}^{+\infty} \frac{\log(s_j^\bullet s_{j-1}^{\bullet -1}(z))}{z-\alpha} + \log \prod_s \frac{(\alpha - \alpha_{s_j})^{\kappa_{s_j}}}{(\alpha - \widetilde{\alpha}_{s_j})^{\widetilde{\alpha}_{s_j}}},$$

$$\log \widetilde{t}_{j_*}(\alpha) = \frac{1}{2\pi i}\int_{-\infty}^{+\infty} \frac{\log(s_j^\bullet s_{j-1}^{\bullet -1}(z))}{\alpha - z} + \log \prod_s \frac{(\alpha - \widetilde{\alpha}_{s_j})^{\widetilde{\alpha}_{s_j}}}{(\alpha - \alpha_{s_j})^{\kappa_{s_j}}},$$

where $\operatorname{Im}\alpha > 0$ for t, $\operatorname{Im}\alpha < 0$ for $\widetilde{t}$, $j \leqq j^*$ and log is a suitable branch of the natural logarithm.

Note that given only the lattice $[T^{\bullet 0}]^{-1}\mathcal{O}_j$ at the point α_j, we can determine the minors t_{i_*} whose zero is α_j and calculate the multiplicities (cf. Exercise2.5). The same is true for $\widetilde{t}_{i_*}$ and $[\widetilde{T}_0^\bullet]^{-1}\mathcal{O}_j$.

As we have recovered t_{j_*}, $\widetilde{t}_{j_*}$, we can express the matrix $\widetilde{\mathcal{D}}^\bullet$ via the minors $s_j^\bullet$. Thus the matrix $\widetilde{T}_0^{\bullet -1} T^{\bullet 0} = \widetilde{T}^{\bullet 0} T_0^{\bullet -1}$ can be expressed via $S^\bullet$ and, consequently, this holds for $\widetilde{T}^{\bullet 0}$ and $T_0^\bullet$ separately. Indeed, the triangular matrices $\widetilde{T}^{\bullet 0}$ and $T_0^{\bullet -1}$ have their diagonals expressed in terms of the minors of $S^\bullet$ and using the triangular factorization of $\widetilde{\mathcal{D}}^{\bullet -1} S^\bullet$, we can determine $T_0^{\bullet -1}$ uniquely. Let us show that $\widetilde{T}_0^\bullet$ and $T^{\bullet 0}$ are also expressed in terms of $\widetilde{\mathcal{D}}^{\bullet -1} S^\bullet$ (they are not triangular, and the algebraic arguments are not enough).

Let $\widetilde{T}_0^{\bullet -1}T^{\bullet 0} = \widetilde{T}_{10}^{\bullet -1}T_1^{\bullet 0}$ for the matrices $\widetilde{T}_{10}^{\bullet}$ and $T_1^{\bullet 0}$ which satisfy the same conditions as $\widetilde{T}_0^{\bullet}$ and $T^{\bullet 0}$. Then the matrix $\widetilde{T}_{10}^{\bullet}\widetilde{T}_0^{\bullet -1} = A \stackrel{\text{def}}{=} T_1^{\bullet 0}T_1^{\bullet 0\,-1}$ commutes with U_0 (in order to check this, examine the left hand side and the right hand side of the above equality). Using the notation of Theorem2.2, we have:

$$[\widetilde{T}_{10}^{\bullet}][\widetilde{T}_0^{\bullet}]^{-1} = A = [T_1^{\bullet 0}][T_1^{\bullet 0}]^{-1}.$$

As the lattices for $[\widetilde{T}_{10}^{\bullet}]$ and $[T_1^{\bullet 0}]$ coincide with the lattices for $[\widetilde{T}_0^{\bullet}]$ and $[T^{\bullet 0}]$ at the corresponding zeros $\widetilde{\alpha}_j$, α_j, the function A can be analytically continued to both half-planes by the criterion (2.14) (cf. Theorem2.3). Moreover, $A \to \boldsymbol{I}$ when $|\alpha| \to \infty$. Thus, $A \equiv \boldsymbol{I}$. □

Before ending this section, let us show the constraints on the scattering data which should be imposed in the case of the $*$-anti-hermitian matrix Q (see §2.3). Equation (2.12) implies that $(\widehat{\Phi}_-)^{-1} = \widehat{\Phi}^*$, $(\widehat{\Phi}_-^{\bullet})^{-1} = (\widehat{\Phi}^{\bullet})^*$. Therefore (2.15) is rewritten in the form

$$\begin{aligned} \widehat{\Phi}^*\widehat{\Phi} &= R_0 S R_0^{-1}, \\ (\widehat{\Phi}^{\bullet})^*\widehat{\Phi}^{\bullet} &= R_0 S^{\bullet} R_0^{-1}, \end{aligned} \tag{2.17}$$

where $S = (T^0)^*T^0 = (T_0^{-1})^*T_0^{-1}$ and $S^{\bullet} = (T^{\bullet 0})^*T^{\bullet 0} = (T_0^{\bullet -1})^*T_0^{\bullet -1}$. The matrices S and $S^{\bullet}$ are $*$-anti-hermitian. We need only $\alpha_1, \dots, \alpha_{N_+}$ and corresponding $\mathcal{K}_j$ (or $\alpha_1^{\bullet}, \dots, \alpha_{N_+^{\bullet}}^{\bullet}$, $\{\mathcal{K}_j^{\bullet}\}$) from the discrete scattering data. The remaining points $\{\alpha_{N_++1}, \dots, \alpha_N\}$ are complex conjugate to $\{\alpha_1, \dots, \alpha_N\}$ ($N = 2N_+$). It is also easy to connect the sets $\{\mathcal{K}_j, j \leqq N_+\}$ and $\{\mathcal{K}_j, j > N_+\}$ by means of the involution $*$. The matrix B of Theorem2.3 satisfies the relation $B^*(\alpha) \stackrel{\text{def}}{=} B(\overline{\alpha})^* = B^{-1}(\alpha)$ for $\alpha \in \mathbb{C}$ (analogously for $B^{\bullet}$).

2.5. Variational derivatives of entries of T (Poisson brackets).

Let us recall the notation: $\boldsymbol{I}_q^p = (\delta_i^q \delta_p^j)$ is the matrix with the unity as the (q,p) element and zeros as the other entries, $A' = (a_i^j(1 - \delta_{\mu_i}^{\mu_j}))$ for a matrix $A = (a_i^j)$, ${}^t\!A$ is the transpose of A. Set (see §0.5 of Introduction):

$$[\boldsymbol{I}]_p = \sum_{q=1}^{p} \boldsymbol{I}_q^q, \qquad [\boldsymbol{I}]^p = \sum_{q=p+1}^{n} \boldsymbol{I}_q^q, \qquad [\boldsymbol{I}]_0 = 0 = [\boldsymbol{I}]^n,$$
$$[A]_p = [\boldsymbol{I}]_p A [\boldsymbol{I}]_p, \qquad [A]^p = [\boldsymbol{I}]^p A [\boldsymbol{I}]^p.$$

We keep the notation of the previous section. For an arbitrary functional f of entries of Q we define a matrix $\delta f/\delta Q = (\delta f/\delta q_i^j)$ whose entry at the crossing of the j-th column and the i-th row is $\delta f/\delta q_i^j$. When $Q' = Q$, the entries $\delta f/\delta q_i^j$ are identically zero for $\mu_i = \mu_j$. Unless otherwise stipulated, we assume that $Q' = Q$.

Proposition2.5. *a) For $E_\pm(x;\alpha)$ of §2.1, we have*

$$\frac{\delta t^p_q}{\delta\, {}^t Q(x)} = -(E_- I^q_p E_+^{-1})', \qquad 1 \leqq q, p \leqq n,$$

when $\alpha \in \mathbb{R}$.

b) For $\operatorname{Im}\alpha \geqq 0$, $\operatorname{Im}\alpha \leqq 0$ respectively,

$$\frac{\delta \log t_p}{\delta\, {}^t Q(x)} = -(\Phi[I]_p \Phi^{-1})',$$
$$\frac{\delta \log \widetilde{t}_p}{\delta\, {}^t Q(x)} = -(\Psi[I]_p \Psi^{-1})',$$

$(1 \leqq p \leqq n)$.

c) In the case of $Q' \neq Q$, the above formulae remain valid if the primes are dropped.

Proof. Let us suppose that E^δ_- and T^δ correspond to another matrix $Q^\delta(x)$. Then recalling the definition of T-matrices and using equation (2.2), we obtain the formula

$$\begin{aligned}\Delta T &= E_+^{-1} E^\delta_- \big|_{x=-\infty}^{x=+\infty} \\ &= \int_{-\infty}^{\infty} (E_+^{-1} E^\delta_-)_x dx \\ &= -\int_{-\infty}^{\infty} E_+^{-1}(\Delta Q) E^\delta_- dx,\end{aligned}$$

where $\Delta T = T^\delta - T$, $\Delta Q = Q^\delta - Q$. Therefore

$$\begin{aligned}\Delta t^p_q &= t^{\delta p}_q - t^p_q \\ &= \operatorname{Sp}(\Delta T I^q_p) \\ &= \int_{-\infty}^{\infty} \operatorname{Sp}(\Delta Q E^\delta_- I^q_p E_+^{-1})\, dx.\end{aligned}$$

By definition of the variational derivation of $f = f(u)$, we have as $\Delta u \to 0$:

$$\Delta f(x) = \int_{-\infty}^{\infty} \frac{\delta f}{\delta u}(y)\, \Delta u(y)\, dy + o(\Delta u).$$

Hence, changing the difference Δ to the variation δ, we establish statement a) and its formulation without primes (when $Q' \neq Q$).

We use the formula $\delta \log\det A = \mathrm{Sp}(\delta A A^{-1})$ which follows directly from the definition of the determinant of the matrix A to get:

$$\frac{\delta t_p}{\delta\, {}^tQ}(x) = -(\Phi T_0[T]_p^{-1} T^0 \Phi^{-1})' t_p$$

(we expressed $E_\pm$ via Φ). Here $[T]_p^{-1}$ is the inverse of $[T]_p$ as a $p\times p$-matrix. Since $[T]_p = [T^0]_p[T_0]_p$, $T_0[I]_p = [T_0]_p$ and $[I]_pT^0 = [T^0]_p$, we get to relation b) for t_p, $\alpha \in \mathbb{R}$. Both sides of the above formula are defined on the real axis and can be continued analytically to meromorphic functions on $\mathrm{Im}\,\alpha > 0$. Hence they coincide (where defined), by the uniqueness of analytic continuations. This argument can be also applied to the "lower" principal minors $\widetilde{t}_p$. □

Exercise2.6. *Show that*

$$\frac{\delta \log t_p}{\delta\, {}^tQ(x)} = -(\Phi^\bullet [I]_p \Phi^{\bullet -1})',$$
$$\frac{\delta \log \widetilde{t}_p}{\delta\, {}^tQ(x)} = -(\Psi^\bullet [I]_p \Psi^{\bullet -1})'$$

for $p = p^*$ *(in the notation of §2.3). Change statement b) of Proposition2.5 to include* τ_i^{0j}, $\widetilde{\tau}_{0i}^{j}$ *when* $\mu_i = \mu_j$ *(see Theorem2.2).* □

We will apply the results of Proposition2.5 to prove the infinitesimal invertibility of the mapping $Q(x) \mapsto S^\bullet(\alpha)$ defined in the previous section. First of all, completing the analysis started in §2.4, we shall find algebraic relations among the entries of $S^\bullet$.

According to formula (2.16), we can calculate the "upper" principal minors $s_j^\bullet$ of the matrix $S^\bullet$. We deduce an analogous formula for the "lower" principal minors under the assumption $j^* = j$. We get

$$\widetilde{s}_j^\bullet \stackrel{\mathrm{def}}{=} \det(s_p^{\bullet q}, j < p, q \leqq n) = \prod_{i=j+2}^{n} t_{i_\bullet} \widetilde{t}_{i_\bullet}, \qquad (2.18)$$
$$\widetilde{s}_{n-1}^\bullet = 1.$$

To prove this, we use the representation $S^\bullet = \widetilde{\mathcal{D}}\widetilde{T}_0^{\bullet -1} T^{\bullet 0}$. The lower minors of the matrices $\widetilde{T}_0^{\bullet -1}$ and $T^{\bullet 0}$ are easily calculated with the help of the formulae in §2.3 for $\det(\tau_i^{0j}, i^* = p^* = j^*)$ and $\det(\widetilde{\tau}_{0i}^{j}, i^* = p^* = j^*)$; $\widetilde{\mathcal{D}}^\bullet = \mathrm{diag}(\widetilde{t}_{i_\bullet} \widetilde{t}_{i\bullet})$. Using this, we can define $\widetilde{s}_j^\bullet$ which are equal to products of the corresponding lower minors of $\widetilde{\mathcal{D}}^\bullet$, $\widetilde{T}_0^{\bullet -1}$ and $T^{\bullet 0}$.

Comparing with formula (2.16), we come to the relations

$$s_1^\bullet = 1, \qquad \widetilde{s}_{j_\bullet}^\bullet s_{j_\bullet+1}^\bullet = s_n^\bullet = \det S^\bullet, \quad (1 \leqq j \leqq n). \tag{2.19}$$

In the case of S (see (2.15)) and the minors s_j and $\widetilde{s}_j$ defined by S instead of $S^\bullet$, or for pairwise distinct eigenvalues μ_j (then $j_* = j-1$, $S = S^\bullet$), we obtain the equations $\widetilde{s}_{j-1}s_j = s_n$, $1 \leqq j \leqq n$. Relations on S are exhausted by these formulae. If there is coincidence among $\{\mu_j\}$, then there are additional analytic constraints on the entries of S, as we already know. These constraints reflect the existence of the analytic continuations of proper polynomial combination of them into the upper and the lower half-planes of α. The matrix $S^\bullet$ is, in principle, free from analytic constraints since all the relations on its entries are algebraic. This statement can be easily made strict, if $S^\bullet$ is "near enough" to the unit matrix. First we check purely algebraically, not considering the dependence of T and $S^\bullet$ on α.

Proposition2.6. a) *Besides relation (2.19), the matrix $S^\bullet$ defined by formula (2.15) via T satisfies the following additional constraints. If $i_* = j_*$, we denote by σ_i^j, the determinant of the matrix of order $i_* + 1$ formed by the $s_l^{\bullet m}$ where l runs over the indices $1, 2, \dots, i_*, i$ and m over $1, 2, \dots, i_*, j$. Then $\sigma_i^j = \delta_i^j s_{i_\bullet+1}^\bullet$ (see (2.16)).*

b) *If $S^\bullet$ satisfies all of the constraints of* a), *$\prod_{i=1}^n s_{i_\bullet}^\bullet \neq 0$ and $\prod_{i=1}^n \widetilde{s}_{i_\bullet}^\bullet \neq 0$, then there is an invertible matrix T (for which $\prod_{i=1}^n t_{i_\bullet} \neq 0 \neq \prod_{i=1}^n \widetilde{t}_{i_\bullet}$) which is related to $S^\bullet$ by (2.15).*

c) *Let $T = 1 + \delta A$ and $S^\bullet = 1 + \delta B$ for $A = (a_i^j)$ and $B = (b_i^j)$. Then*

$$b_i^j = \operatorname{sgn}(i^* - j^*)a_i^j + o(\delta)$$

for $i^* = j^*$ (here, sgn means the sign and $\operatorname{sgn} 0 \stackrel{\text{def}}{=} 0$).

Proof. The decomposition $S^\bullet = (\widetilde{\mathcal{D}}^\bullet \widetilde{T}^{\bullet 0})T_0^{\bullet -1}$ is of the same type as the decomposition $T = T^{\bullet 0}T_0^\bullet$. In the beginning of §2.3 we calculate the entries of $[T^{\bullet 0}]$ (the diagonal of $T_0^\bullet$ is fixed) via the minors of T. We apply this result here noting that, in contrast to $T^{\bullet 0}$, the matrix $\widetilde{\mathcal{D}}^\bullet \widetilde{T}^{\bullet 0}$ is more special: $[\widetilde{\mathcal{D}}^\bullet \widetilde{T}^{\bullet 0}] = (\delta_i^j d_{i_\bullet})$ for certain $d_{i_\bullet} \in \mathbb{C}$. This implies a).

The proof of b) is similar to that of Proposition2.4. Let $t_{i_\bullet}$ and $\widetilde{t}_{i_\bullet}$ satisfy (2.16) and (2.18). We represent $S^\bullet = S^{\bullet 0}S_0^\bullet = \widetilde{S}_0^\bullet \widetilde{S}^{\bullet 0}$ where $S^{\bullet 0}$, $S_0^\bullet$, $\widetilde{S}_0^\bullet$ and $\widetilde{S}^{\bullet 0}$ are of the same type as $\widetilde{T}^{\bullet 0}$, $T_0^\bullet$, $\widetilde{T}_0^\bullet$ and $T^{\bullet 0}$, respectively. Such a representation exists since $s_{i_\bullet} \neq 0 \neq \widetilde{s}_{i_\bullet}^\bullet$ $(1 \leqq i \leqq n)$. We may assume that $\operatorname{diag} S_0^\bullet = \operatorname{diag}(t_{i_\bullet})$ and $\det(\widetilde{\sigma}_i^j, i^* = p = j^*) = t_p t_{p_\bullet}^{(p-p_\bullet-1)}$, where $\widetilde{S}^{\bullet 0} = (\widetilde{\sigma}_i^j)$. Then the upper and the lower principal minors (with the index $j = j^*$) of the matrix $T_1 \stackrel{\text{def}}{=} \widetilde{S}^{\bullet 0} S_0^{\bullet -1}$ are equal to t_j and $\widetilde{t}_j$ respectively. The upper minors of T_1 are computed directly from the constraints on diag $S_0^\bullet$ and $[\widetilde{S}^{\bullet 0}]$. To determine the lower minors, we apply the decomposition $T_1 = \widetilde{S}_0^{\bullet -1} S^{\bullet 0}$ together with formulae which express diag $S^{\bullet 0}$, the minors of $S^\bullet$ and $[\widetilde{S}^{\bullet 0}]$ via $\{t_{j_\bullet}, \widetilde{t}_{j_\bullet}\}$. Furthermore, $S^\bullet = \mathcal{D}' \widetilde{T}_1^{\bullet 0} T_{10}^{\bullet -1}$ for a suitable diagonal matrix $\mathcal{D}' = \operatorname{diag}(d_{i_\bullet})$ and the matrices $\widetilde{T}_1^{\bullet 0}$ and $T_{10}^\bullet$ constructed from the matrix T_1 defined above. Using the formulae for the minors of T_1, we find that $\mathcal{D}'$ coincides with the matrix $\widetilde{\mathcal{D}}_1^\bullet$ corresponding to T_1. Thus $T = T_1$ is the desired matrix.

Let us check statement c). The triangular factorization of $T = \boldsymbol{I} + \delta A$ gives the decomposition of A (up to $o(\delta)$) into a sum (instead of a product) of the matrices of the same type. In particular, $\widetilde{T}^{\bullet 0} = \boldsymbol{I} + \delta \widetilde{A}^0$ and $T_0^\bullet = \boldsymbol{I} + \delta A_0$, where

$$\widetilde{A}^0 - A_0 = B = (\operatorname{sgn}(i^* - j^*) a_i^j) + \Delta + o(\delta)$$

for a suitable diagonal matrix Δ (recall that $S^\bullet = \widetilde{\mathcal{D}}^\bullet \widetilde{T}^{\bullet 0} T_0^{\bullet -1}$). Hence to prove statement c) it is enough to show that diag $B = 0$. We have:

$$\begin{aligned} \operatorname{diag} S^\bullet &= \operatorname{diag}(t_{i_\bullet} \widetilde{t}_{i_\bullet}) = \operatorname{diag}(t_{i_\bullet} t_{-i_\bullet}) \\ &= \operatorname{diag}(t_{i_\bullet} - t_{i_\bullet}) + \boldsymbol{I} = \boldsymbol{I} \mod o(\delta). \end{aligned}$$

Here we use formula (2.7) and the trivial relation $(\boldsymbol{I} + A)^{-1} = \boldsymbol{I} - \delta A + o(\delta)$. □

To complete the analysis of the algebraic structure of $S^\bullet$, let us return to the functions $T(\alpha)$ and $S^\bullet(\alpha)$ for $\alpha \in \mathbb{R}$ defined from $Q(x)$. We remind (Lemma2.1) that the entries of $T - \boldsymbol{I}$ belong to the space $F(\mathcal{L}_1)$ of Fourier transformations of functions from the space $\mathcal{L}_1$ of absolutely integrable complex valued functions of x from $-\infty$ to $+\infty$. This is also true for entries of $S^\bullet - \boldsymbol{I}$ (which are expressed as polynomials of t_i^j). Indeed, the space $F(\mathcal{L}_1)$ is a ring with respect to the multiplication of functions (which corresponds to the convolution in $\mathcal{L}_1$). Let us introduce a norm in $F(\mathcal{L}_1)$ which is induced from $\mathcal{L}_1$. Each variational derivation $\delta s_i^{\bullet j} / \delta q_r^s$ is a linear operator from $\mathcal{L}_1$ to $F(\mathcal{L}_1)$. We will compute them at the "point" $Q(x) \equiv 0$.

We have: $E_\pm = R_0(x; \alpha) = \exp(\alpha U_0 x)$ and $T(\alpha) \equiv \boldsymbol{I} \equiv S^\bullet(\alpha)$ for $Q \equiv 0$. Combining Proposition2.5, a) and Proposition2.6, c), we obtain

Corollary2.2. *For $Q(x) \equiv 0$ the variational derivation $\delta s_l^{\bullet m} / \delta q_r^s$ maps $f \in \mathcal{L}_1$ to*

$$\operatorname{sgn}(m^* - l^*) \int_{-\infty}^{\infty} \exp(i\alpha(a_m - a_l)x) f(x)\, dx \in F(\mathcal{L}_1)$$

for $l = r$ and $m = s$, and to zero if either $l \neq r$ or $m \neq s$ (recall that $\mu_m = i a_m \in i\mathbb{R}$, $m^* = l^* \Leftrightarrow a_m = a_l$, $\operatorname{sgn} 0 = 0$). □

We obtain that the map $Q(x) \mapsto S^\bullet(\alpha) - \boldsymbol{I}$ is an isomorphism of the topological vector spaces of the matrices with entries in $\mathcal{L}_1$ or $F(\mathcal{L}_1)$ in a neighbourhood of

$Q \equiv 0$, $S^{\bullet} \equiv \boldsymbol{I}$. Note that for a matrix $Q(x)$ sufficiently close to zero, the minors t_i, $\widetilde{t}_i$ do not have zeros in the corresponding half-planes and those discrete scattering data are trivial (see Exercise2.2). Hence the inverse scattering problem for the matrices $S^{\bullet}(\alpha)$ close enough to $\boldsymbol{I}$ admits a unique solution, if Q is taken from a neighbourhood of zero even if we take into consideration the discrete spectrum.

Poisson bracket. Another application of the formulae in Proposition2.5 is the computation of Poisson brackets for the entries of T. In order to avoid the problem of convergence, we assume that $Q(x) = 0$ if $x \leqq a$ or $x \geqq b$ $(a < b)$. Let us call the following integral (assuming the existence) the *Poisson bracket* of two given functionals f, g of the entries of Q:

$$\{f,g\} = \int_a^b \mathrm{Sp}\left(\left(\frac{\delta f}{\delta Q}(x)\right)\left[U_0, \frac{\delta g}{\delta Q}(x)\right]\right) dx \tag{2.20}$$

(Sp is the trace). In particular, for $a < z$ and $y < b$ we obtain the relations

$$\begin{aligned}\{q_i^j(z), q_r^s(y)\} &= \int_a^b \mathrm{Sp}(\boldsymbol{I}_i^j\delta(z-x)[U_0, \boldsymbol{I}_r^s\delta(y-x)])\,dx \\ &= (\mu_j - \mu_i)\delta_r^j\delta_i^s\delta(z-y).\end{aligned}$$

Here $\delta(x)$ is the delta function ($\int_a^b f(x)\delta(x-y)\,dx = f(y)$ and $\delta(x) = \delta(-x)$) and the variational derivatives and Poisson brackets are considered as generalized functions. The original definition (2.20) follows from the last formula since by the chain rule

$$\begin{aligned}\{f,g\} &= \sum \int_a^b \frac{\delta f}{\delta q_i^j}(x)\{q_i^j(x), g\}\,dx \\ &= \sum \int_a^b \int_a^b \frac{\delta f}{\delta q_i^j}(x)\{q_i^j(x), q_r^s(y)\}\frac{\delta g}{\delta q_r^s}(y)\,dx\,dy \\ &= \sum \int_a^b \frac{\delta f}{\delta q_i^j}(x)(\mu_j - \mu_i)\frac{\delta g}{\delta q_j^i}(x)\,dx\end{aligned}$$

for any functionals f, g, where the summations are taken over all indices. We also have the following trivial identities:

$$\begin{aligned}\{q_i^j(z), q_r^s(y)\} &= -\{q_r^s(y), q_i^j(z)\}, \\ \{\{q_i^j(z), q_r^s(y)\}, q_l^m(x)\} &= 0,\end{aligned}$$

which implies that bracket (2.20) is skewsymmetric ($\{f,g\} = -\{g,f\}$) and satisfies the *Jacobi identity*:

$$\{\{f,g\},h\} + \{\{g,h\},f\} + \{\{h,f\},g\} = 0.$$

Exercise2.7. *Check these two properties of $\{,\}$, using (2.20). Assume that functionals f, g and h are twice continuously differentiable with respect to Q (see [Ch11]).* □

Proposition2.7. a) *For $\alpha, \beta \in \mathbb{R}$ $(\alpha \neq \beta)$, $1 \leqq i, j, k, l \leqq n$,*

$$\{t_i^j(\alpha), t_k^l(\beta)\} = \frac{1}{\alpha - \beta}(e^{(\mu_k - \mu_i)(\alpha-\beta)b} t_k^j(\alpha) t_i^l(\beta) - e^{(\mu_j - \mu_l)(\alpha-\beta)a} t_i^l(\alpha) t_k^j(\beta)),$$

$$\{\check{t}_i^j(\alpha), \check{t}_k^l(\beta)\} = \frac{1}{\alpha - \beta}(\check{t}_k^j(\alpha)\check{t}_i^l(\beta) - \check{t}_i^l(\alpha)\check{t}_k^j(\beta)),$$

where $(\check{t}_i^j) = \check{T} \stackrel{\text{def}}{=} R_0(b) T R_0(a)^{-1}$.

b) For finite or infinite α, β the brackets $\{t_k(\alpha), t_l(\beta)\}$ are well-defined and identically zero for $\operatorname{Im}\alpha, \beta \geqq 0$.

Proof. We use (2.20) and Proposition2.5 and get

$$\begin{aligned}\{t_i^j(\alpha), t_k^l(\beta)\} &= \int_a^b \operatorname{Sp}({}^t(F_j^i(\alpha)'[U_0, {}^tF_l^k(\beta)']) \, dx \\ &= \int_a^b \operatorname{Sp}([F_l^k(\beta)', U_0] F_j^i(\alpha)') \, dx \\ &= \int_a^b \operatorname{Sp}([F_l^k(\beta), U_0] F_j^i(\alpha)) \, dx,\end{aligned}$$

where we set $F_j^i(\alpha) = E_-(x;\alpha) I_j^i E_+(x;\alpha)^{-1}$, applied the transposition before taking the trace and used the identity $\operatorname{Sp}(A'B) = \operatorname{Sp}(A'B')$. As a consequence of the fundamental equation (2.2)

$$\operatorname{Sp}(F_j^i(\alpha) F_l^k(\beta))_x = (\alpha - \beta) \operatorname{Sp}([U_0, F_j^i(\alpha)] F_l^k(\beta)). \tag{2.21}$$

Really, $\operatorname{Sp}([F_j^i(\alpha), Q] F_l^k(\beta) + F_j^i(\alpha)[F_l^k(\beta), Q]) = 0$, since $\operatorname{Sp}(AB) = \operatorname{Sp}(BA)$. Using (2.21), we get

$$(\alpha - \beta)\{t_i^j(\alpha), t_k^l(\beta)\} = \operatorname{Sp}(F_j^i(\alpha) F_l^k(\beta))|_{x=a}^{x=b}$$

$$= \operatorname{Sp}(R_0(b) T I_j^i R_0^{-1}(b) \widetilde{R}_0(b) \widetilde{T} I_l^k \widetilde{R}_0^{-1}(b) - R_0(a) I_j^i T R_0^{-1}(a) \widetilde{R}_0(a) I_l^k \widetilde{T} \widetilde{R}_0^{-1}(a)),$$

where we set $T(\alpha) = T$, $T(\beta) = \widetilde{T}$, $R_0(x) = R_0(x;\alpha)$ and $\widetilde{R}_0(x) = R_0(x;\beta)$. The last expression can be rewritten as the trace of

$$R_0(b) t_k^j I_k^i R_0^{-1}(b) \widetilde{R}_0(b) \widetilde{t}_i^l I_i^k \widetilde{R}_0^{-1}(b) - R_0(a) t_i^l I_j^l R_0^{-1}(a) \widetilde{R}_0(a) \widetilde{t}_k^j I_l^j \widetilde{R}_0^{-1}(a)).$$

Conjugating this by R_0 and $\widetilde{R}_0$, we get the formulae of a).

Analogously we can establish the relation

$$(\alpha-\beta)\{t_k(\alpha),t_k(\beta)\}=\mathrm{Sp}(F_k(\alpha)F_l(\beta))|_{x=a}^{x=b}$$

for the functions $F_j=\Phi[I]_j\Phi^{-1}$ which satisfy the same equations (2.21) (see Proposition2.5). At $x=a$ or $x=b$, the functions Φ, F_k and F_l are simultaneously upper or lower triangular matrices. Therefore,

$$\mathrm{Sp}(F_k(\alpha)F_l(\beta))|_{x=1}^{x=b}=\mathrm{Sp}(\mathrm{diag}\,F_k(\alpha)\,\mathrm{diag}\,F_l(\beta))|_{x=a}^{x=b}.$$

As $\mathrm{diag}\,F_j=[I]_j$ for $x=a,b$ and $j=k,l$, we get statement b). □

At the end of this section we give three examples which illustrate the coincidence of the number of entries of $Q=Q'$ and which is independent of the number of entries of $S^\bullet$. The most popular case in the inverse problem method is $n=2$, $\mu_1\neq\mu_2$. In this case:

$$\begin{aligned}T&=T^0T_0 &&=\widetilde{T}_0\widetilde{T}^0\\ &=\begin{pmatrix}a&b\\c&d\end{pmatrix}=\begin{pmatrix}a&0\\c&1\end{pmatrix}\begin{pmatrix}1&ba^{-1}\\0&a^{-1}\end{pmatrix}&&=\begin{pmatrix}1&b\\0&d\end{pmatrix}\begin{pmatrix}d^{-1}&0\\cd^{-1}&1\end{pmatrix},\end{aligned}$$

since $\det T=t_2=\widetilde{t}_0=1=t_0=\widetilde{t}_2$, $t_1=a$ and $\widetilde{t}_1=d$. Hence (see (2.15)),

$$\begin{aligned}S&=\mathrm{diag}(\widetilde{t}_i\widetilde{t}_{i-1})\widetilde{T}^0T_0^{-1}\\ &=\begin{pmatrix}d&0\\0&d\end{pmatrix}\begin{pmatrix}d^{-1}&0\\cd^{-1}&1\end{pmatrix}\begin{pmatrix}1&-b\\0&a\end{pmatrix}=\begin{pmatrix}1&-b\\c&1\end{pmatrix}.\end{aligned}$$

Relations (2.19) are reduced to the equations $s_1=\widetilde{s}_1=1$.

The case $n=3$, $\mu_1=\mu_2\neq\mu_3$. One has:

$$\begin{aligned}T&=T^{\bullet0}T_0^\bullet=\widetilde{T}_0^\bullet\widetilde{T}^{\bullet0}\\ &=\begin{pmatrix}a&b&e\\c&d&f\\h&k&g\end{pmatrix}=\begin{pmatrix}a&b&0\\c&d&0\\h&k&1\end{pmatrix}\begin{pmatrix}1&0&(de-bf)t_2^{-1}\\0&1&(af-ce)t_2^{-1}\\0&0&t_2^{-1}\end{pmatrix}\\ &=\begin{pmatrix}ag-he&bg-ke&e\\cg-hf&dg-kf&f\\0&0&g\end{pmatrix}\begin{pmatrix}g^{-1}&0&0\\0&g^{-1}&0\\hg^{-1}&kg^{-1}&1\end{pmatrix},\end{aligned}$$

$t_2 = ad - bc$, $\widetilde{t}_{1\bullet} = \widetilde{t}_{2\bullet} = g$, $\widetilde{t}_0 = \widetilde{t}_3 = 1$ and $\widetilde{\mathcal{D}}^\bullet = \mathrm{diag}(\widetilde{t}_{i_\bullet} \widetilde{t}_{i\bullet}) = \mathrm{diag}(g, g, g)$,

$$S^\bullet = \begin{pmatrix} 1 & 0 & 0 \\ 0 & 1 & 0 \\ h & k & g \end{pmatrix} \begin{pmatrix} 1 & 0 & bf - de \\ 0 & 1 & ce - af \\ 0 & 0 & t_2 \end{pmatrix} = \begin{pmatrix} 1 & 0 & bf - de \\ 0 & 1 & ce - af \\ h & k & 1 \end{pmatrix}.$$

Let us assume now that T is $*$-unitary and $n = 3$; μ_1, μ_2, and μ_3 will be arbitrary purely imaginary numbers. We have:

$$T = T^0 T_0 = \begin{pmatrix} a & 0 & 0 \\ c & t_2 & 0 \\ h & ak - hb & 1 \end{pmatrix} \begin{pmatrix} 1 & ba^{-1} & ea^{-1} \\ 0 & a^{-1} & af - ec \\ 0 & 0 & t_2^{-1} \end{pmatrix}$$

in the notation above, and

$$\begin{aligned} S &= (T^0)^* T^0 = \Omega (T^0)^* \Omega T^0 \\ &= \begin{pmatrix} 1 & -b & \overline{h}\omega_1\omega_2 \\ -\omega_1\omega_2\overline{b} & a\overline{a} + \omega_1\omega_2 b\overline{b} & -f \\ h & -\omega_2\omega_3\overline{f} & 1 \end{pmatrix}. \end{aligned}$$

The entries of S satisfy the unique (real) relation $s_2\widetilde{s}_1 = \det S$. We can express $a\overline{a}$ in terms of the entries b, f and h which are independent. The quadratic equation for $a\overline{a}$ has one or two non-negative solutions for $\omega_2\omega_3 = \pm 1$ respectively, if it has real solutions at all. The number of real parameters determining S is equal to 6 and coincides with the number of real (functional) parameters for the $*$-anti-hermitian matrix Q with zeros on the diagonal.

2.6. Comments.

The aim of this section is to give a systematic introduction to the inverse scattering problem method in the formulation by A. B. Shabat (see [Sh1,2]). Actually we are concerned only with the direct scattering problem. This includes the construction of the scattering matrix and solutions of the original linear problem (in our case (2.2)) which are analytic in the spectral parameter. Since we always consider a one-dimensional variable x, we prefer the term "monodromy matrix" to the term "scattering matrix" and use the notation "T". We understand the inverse problem (from T to Q) as the Riemann problem. In a more traditional approach, one usually includes in it the reformulation of the Riemann problem in terms of integral equations and the proper existence theorems. For instance, the inverse problem for the KdV equation can be reduced to the Gelfand-Levitan-Marchenko equations.

In the case of matrices of order 2, a systematic exposition of the inverse scattering problem can be found in [TF2] (see also [F2], [DT]). As to arbitrary matrices, the corresponding methods have not yet been worked out completely. There are a number of theoretical problems. For instance, the space of T-matrices corresponding to the matrices Q with absolutely integrable entries have not yet been fully described. The construction of the analytic solutions Φ and Ψ of equation (2.2) by means of the triangular factorization is due to M. G. Krein (see, for example, [GK], [Kre]). For $n = 2$ the triangular factorization is trivial. We can construct analytic continuations for the columns of matrices $E_{\pm}$ directly. In this case the study of the discrete spectrum is essentially simple. Note that we do not examine real points where Φ and Ψ are degenerate (i.e., the zeros of the minors t_i, $\widetilde{t}_i$ on the real axis). For $n = 2$ such points are studied in detail in [TF2].

We include this "analytic" section because of the following reasons. First of all, its content is closely related to more algebraic material in other sections. In general, the inverse problem method is used in the soliton theory more algebraically than analytically (though, of course, such a statement should be understood with reservations). Another reason is that, in spite of the popularity of the inverse problem method, the direct scattering problem for $n > 2$ has not been discussed in the literature in a sufficiently complete and rigorous way. Usually the analyticity of Φ and Ψ is explained by the limiting procedure from comactly supported $Q(x)$ (equal to zero apart from a finite interval) to absolutely integrable functions (see, for example, [ZMaNP], [Sh1]). This procedure is not simple and still uses a kind of reduction as in §2.2 to prove the analyticity of Φ and Ψ. This way or that way, we need exterior powers of the matrices $E_{\pm}$ which play a central role in the present section. Our approach sits next to the methods in other sections and has explicit invariant meanings from the viewpoint of the representation theory. The analytic part (§2.1) consists of elementary direct estimates.

We follow paper [Ch2] and drop the usual condition of pairwise distinctness of numbers μ_j $(1 \leqq j \leqq n)$. Our "algebraized" approach to the direct problem makes the necessary refinement easy. We need to take possible coincidence of certain μ_j into account for concrete equations (S^{n-1} fields, the GHM and the VNS for $n > 2$). From the theoretical viewpoint this refinement is necessary to handle the case of an arbitrary reductive group instead of GL_n (Exercise2.3).

Another demonstration of the universality of our procedure is the construction of the system of analytic solutions of equation (2.2) for arbitrary complex μ_j (see Proposition2.2). We give this construction as an illustration only without going into detail. However, one can easily complete the story and prove the counterparts of Theorem2.3, Proposition2.6 and Corollary2.2. The procedure of dividing the complex plane into sectors and constructing solutions of (2.2) in each sector is not new. For a series of soliton equations, such an approach was applied by R. Beals, R. Coifman, P. Deift, D. Kaup, A. V. Mikhailov, E. Trubowitz and others (see e.g. [BDZ]). As we noted above, there are difficulties in working out the analytic methods of solving the associated Riemann problem.

We hope that our constructions simplify the technique of the inverse scattering problem especially the study of the analytic continuations. Even in the well-known case of the purely imaginary matrix U_0 with pairwise distinct μ_j, when the results in this section are mostly classical, it seems to be useful.

As is obvious from the proof of Theorem2.3, we can apply the Riemann problem in order to solve equation (2.2) for functions $Q(x)$ of other types (not only for normally integrable ones) if we choose suitable contours and matrix functions $S(\alpha)$ with appropriate properties. This was pointed out in [ZMaNP], [Kri5] (see also [TF2]). The corresponding analytic technique can be rather involved (except for the case we studied here and for algebraic-geometric Q).

The formulae for the variational derivatives of the entries of T is universal, too. Note that the results of Proposition2.5 are rather well-known (see, for example, [ZMaNP], [Sh2], [TF2], and also [Ch2] for the formulae for principal minors) and often used in soliton theory. The proof of the infinitesimal solvability of the direct problem by means of the computation of variational derivatives (for pairwise distinct μ_j) can be found in [Sh2]. The Poisson bracket (2.20) turns out to be a special case of the Gelfand-Dickey bracket (see [GD3]), if μ_j are in a generic position. Such brackets play an important role in mathematical physics and the soliton theory. For instance, for $n = 2$ bracket (2.20) is one of the fundamental objects studied in the book [TF2]. The results of Proposition2.7 (and the variants) are also known (see [ZMaNP], [TF2]).

Analysis of examples of the Riemann problems connected to soliton equations is the subject of the next section. The reader can also refer to books [ZMaNP], [TF2], [AS], [CD]. Note that Ch.III of [ZMaNP] contains main formulations related to the matrix Riemann problem for pairwise distinct $\{\mu_j\}$. See also papers [BC1, BC2, BDZ] on more exact and systematic theory. We also mention the paper of A. Fordy (J. Phys. A, **17**, (1984)) devoted to an analogous problem.

§3. Applications of the inverse problem method

In §3.1 we construct "stabilizing" $*$-unitary solutions of the basic equations by means of the Riemann problem (and the results of §2). In §3.2 we calculate the asymptotic expansions of the principal minors of the T-matrix. We also deduce the trace formulae, making it possible to compute the integrals of the basic equations (§1) from the scattering data. The next §3.3 is devoted to the study of special cases. We consider O_n-fields, S^{n-1}-fields, and equations with $n = 2$. In §3.4 we examine the nonlinear Schrödinger equation (NS) in more detail, and compute the entries of T (scattering coefficients) for several solutions. In §3.5 we study the derivative nonlinear Schrödinger equation (DNS), which is an analogue of the NS and an example of a more general associated inverse problem. In §§3.2, 3.4 and 3.5 we use the formulae for variational derivatives from §2 proving the involutivity of local conservation laws and computing the variations of the discrete scattering data in the case of the NS and DNS.

We remind the reader of all necessary constructions as needed. When reading §§3.4 and 3.5 it is useful (but not necessary) to be familiar with basic facts about the inverse problem method for the NS equation presented in the books [ZMaNP] and [TF2].

3.1. Inverse problem for basic equations.

Let $U(x,t)$, $V(x,t)$ be a solution of the PCF system (0.2) (Introduction) or let U be a solution of the GHM equation (0.9). As in Ch.I,

$$(3.1) \qquad U = F_0 U F_0^{-1}, \qquad Q \stackrel{\text{def}}{=} F_0^{-1}(F_0)_x = Q',$$

where $U_0 = \operatorname{diag}(\mu_1, \dots, \mu_n)$ for constant $\{\mu_j\} \subset \mathbb{C}$ and a suitable invertible matrix function $F_0(x,t)$. Let us recall that

$$Q = Q' \stackrel{\text{def}}{\Longleftrightarrow} Q(x,t) \in \mathfrak{gl}'_n = [U_0, \mathfrak{gl}_n].$$

For the GHM we impose restriction (0.10) on U_0:

$$\mu_1 = \dots = \mu_p = c_1, \qquad \mu_{p+1} = \dots = \mu_n = c_2,$$
$$c = c_1 - c_2 \neq 0, \ 1 \leqq p < n.$$

The functions F_0 and Q are not uniquely determined by U. By Corollary1.1 of Ch.I we can choose the function F_0 so that it is subject to the relations

$$(3.2\text{A}) \qquad 2(F_0)_t F_0^{-1} = V \Leftrightarrow 2F_0^{-1}(F_0)_t = F_0^{-1} V F_0 \stackrel{\text{def}}{=} W,$$

$$(3.2\text{B}) \qquad F_0^{-1}(F_0)_t = [U_0, Q_x] - \tfrac{1}{2}[Q, [U_0, Q]],$$

where we denote the formulae corresponding to the PCF by the letter "A" and to the GHM by the letter "B", as in Ch.I.

If Φ satisfies system (1.1), (1.2A,B) of Ch.I for $k = \alpha$ (see also (1.1,2) of the present chapter and (0.3) and (0.11) of Introduction), then $E = F_0^{-1}\Phi$ is a solution of the system of equations

$$E_x + QE = \alpha U_0 E, \tag{3.3}$$

$$E_t E^{-1} = \frac{1}{2(2\alpha - 1)} W, \tag{3.4A}$$

$$E_t E^{-1} = c^2\alpha^2 U_0 - c^2\alpha Q + [Q_x, U_0] - \tfrac{1}{2}[Q, [Q, U_0]], \tag{3.4B}$$

where W is defined in (3.2A) and $c = c_1 - c_2$ (see §1.2, Ch.I). It is proved using constraints (3.2), the relation $[U_0, [Q, U_0]] = -c^2 Q$ for the GHM (see (0.11)) and the following differential equation which may be checked directly:

$$E_t E^{-1} = \frac{\alpha}{2\alpha - 1} W - F_0^{-1}(F_0)_t,$$
$$E_t E^{-1} = c^2\alpha^2 U_0 + \alpha[U_0, [Q, U_0]] - F_0^{-1}(F_0)_t,$$

in the cases A and B respectively.

Conversely, suppose that an invertible matrix function $E(x, t; \alpha)$ together with the functions Q and W satisfy (3.3) and (3.4) for the PCF (A) or the GHM (B). Then F_0 is determined from Q and W by (3.1) and (3.2) up to left multiplication by a constant matrix, and U and V are uniquely determined by (3.1) and (3.2A) modulo conjugation by constant matrices. As $\Phi \stackrel{\text{def}}{=} F_0 E$, U, V are connected by formulae (1.1) and (1.2) of Ch.I for $k = \alpha$, it is easily checked that the U and V constructed above are solutions of system (0.2) or U is a solution of (0.9) together with (0.10) respectively.

Let us recall that $E = F_0^{-1}\Phi$ is analogous to the formal series $\check{\Psi} F_0^{-1}\check{\Phi}$ from Ch.I (see Corollary1.1 and the proof of Proposition1.2). Using the results obtained there, or by direct computation, it is easy to show that the compatibility condition of equations (3.3) and (3.4A) is given by the following system:

$$\begin{aligned} W_x &= \tfrac{1}{2}[U_0, W] - [Q, W], \\ Q_t &= \tfrac{1}{4}[U_0, W]. \end{aligned} \tag{3.5}$$

Analogously, the compatibility of (3.3) and (3.4B) leads to the equations

$$\frac{1}{c} Q_t^{(+)} = Q_{xx}^{(+)} + [[Q^{(-)}, Q^{(+)}], Q^{(+)}], \tag{3.6a}$$

$$-\frac{1}{c} Q_t^{(-)} = Q_{xx}^{(-)} + [[Q^{(+)}, Q^{(-)}], Q^{(-)}], \tag{3.6b}$$

where $Q = Q^{(+)} + Q^{(-)}$ and $[U_0, Q^{(\pm)}] = \pm cQ^{(\pm)}$ (see equation (0.13) of Introduction). Note that $\frac{1}{2}[Q,[Q,U_0]] = c[Q^{(+)}, Q^{(-)}]$ (see (3.4B)). Thus *equation (0.2) and (0.9) together with (0.10) can be transformed into (3.5) or (3.6), and vice versa.*

Further in this section we assume that

$$U^* + U = 0 = V^* + V$$

for an anti-involution $A^* = \Omega A^+ \Omega$ with $\Omega = \mathrm{diag}(\omega_j)$ and $\omega_j = \pm 1$ (where $A^+ = {}^t(\overline{A})$ is the hermitian conjugation). Then we can assume for such U and V that F_0 is a $*$-unitary matrix (that is, $F_0 F_0^* = I$) and, consequently, $Q^* + Q = 0 = W^* + W$. Note that the matrix U_0 in this case must be purely imaginary: $\{\mu_j\} \subset i\mathbb{R}, c, c_1, c_2 \in i\mathbb{R}$. For the GHM equation, the second equation of (3.6) is obtained from the first by applying $*$, since $Q^{(-)} = -(Q^{(+)})^*$.

We leave the general complex case to the reader. Almost always, the most interesting applications are for soliton equations with certain reality conditions (for instance, for $*$-anti-hermitian or $*$-unitary solutions).

We further assume that the derivatives W_x, Q_t (PCF), Q_x, Q_{xx} and Q_t (GHM) exist and that the entries of Q, $[U_0, W]$ and of Q, Q_x and Q_{xx} for the PCF and GHM, respectively, are absolutely integrable with respect to x on the real axis. Such $*$-anti-hermitian W and Q are called *stabilizing* (as always, $Q' = Q$). In this case there exist the limits $\widetilde{W}_+$ and $\widetilde{W}_-$ of the function $W(x,t)$ for $x \to \pm\infty$ (which possibly depend on t). By equation (3.5),

$$[U_0, \widetilde{W}_\pm] = (Q(\pm\infty, t))_t = 0.$$

We may assume that the variable t belongs to an interval containing the initial point t'.

The above conditions on the solutions Q and W are equivalent to the existence of the derivative U_t and the existence and the absolute integrability of the entries of the matrix functions U_x and V_x (PCF) and of U, U_{xx} and U_{xxx} (GHM) which are constructed from Q and W by the above procedure. Now we are going to check this statement. By (3.1), F_0 depends continuously on x and there exist limits $F(\pm\infty, t)$. Therefore the entries of F_0 and F_0^{-1} are bounded functions of $-\infty \leqq x \leqq +\infty$. Since

$$U_x = F_0[Q, U_0]F_0^{-1}, \qquad V = F_0 W F_0^{-1},$$
$$W_x = F_0([Q,W] + W_x)F_0^{-1} = \tfrac{1}{2}F_0[U_0, W]F_0^{-1},$$

we get the equivalence of absolute integrability of Q and U_x, and that of $[U_0, W]$ and W_x. Analogous reasoning proves the same equivalence for U_{xx}, U_{xxx} and Q_x, Q_{xx} in the case of the GHM. Note that the absolute integrability of the entries of U_x and V_x implies the existence of the limits $U(\pm\infty, t)$ and $V(\pm\infty, t)$. We call functions U and V satisfying all the above conditions *stabilizing*.

Scattering data. Let the numbers μ_j (where $U_0 = \operatorname{diag}(\mu_j)$) be ordered as in §2, i.e., $i\mu_j \geqq i\mu_k$ if $j < k$, and assume that $ic_1 > ic_2$ and $ic > 0$. We construct from $*$-antihermitian stabilizing matrix functions Q and W (satisfying (3.5) or (3.6)) solutions $E_\pm(x,t;\alpha)$ of equation (3.3) and the monodromy matrix $T(t;\alpha)$ (in the notation of §2) for $\alpha \in \mathbb{R}$. By the conditions on Q and W, the entries of Q_t are absolutely integrable (see (3.5) and (3.6)). Therefore we can differentiate the series for $E_\pm$ and T term-wise with respect to t since they converge uniformly and absolutely. The functions $(E_\pm)_t$ and T_t obtained in this way have the same continuity properties as $E_\pm$ and T (see Lemma2.1, b)). Moreover, $(\widehat{E}_\pm)_t, T_t \to 0$ when $|\alpha| \to \infty$. Recall from §2.3 that

$$E_\pm(t;\alpha)^* = E_\pm^{-1}(t;\alpha), \qquad T(t;\alpha)^* = T^{-1}(t;\alpha).$$

In §2 we introduced matrices S and $S^\bullet$ (see formulae (2.15) and (2.15$^\bullet$)). These matrices are $*$-hermitian: $S^* = S$ and $(S^\bullet)^* = S^\bullet$ (see (2.17)). We also constructed solutions Φ and $\Phi^\bullet$ of equation (3.3) for $\operatorname{Im}\alpha \geqq 0$ which depend on x and α (and now also on t) and are analytic with respect to α, as well as the discrete scattering data (which depend only on t):

$$\{\alpha_j, \mathcal{K}_j \stackrel{\text{def}}{=} \Phi^{-1}(\mathcal{O}_j^n)\}, \qquad \{\alpha_j^\bullet, \mathcal{K}_j^\bullet \stackrel{\text{def}}{=} \Phi^{\bullet-1}(\mathcal{O}_j^n)\},$$

where $1 \leqq j \leqq N_+, N_+^\bullet$ and $\alpha_j, \alpha_j^\bullet$ are the zeros of $\det\Phi$ and $\det\Phi^\bullet$ (see §2.4) respectively. If all μ_j are pairwise distinct, then $S = S^\bullet$, $\Phi = \Phi^\bullet$ and $\{\alpha_j, \mathcal{K}_j\} = \{\alpha_j^\bullet, \mathcal{K}_j^\bullet\}$ (and $i^* = i$, $i_* = i - 1$).

As in §2, we assume, restricting the class of Q under consideration, that $\operatorname{Im}\alpha_j > 0$ and $\operatorname{Im}\alpha_j^\bullet > 0$ (i.e., $\det\Phi$ and $\det\Phi^\bullet$ do not have real zeros). Recall that

$$\begin{aligned} T = E_+^{-1}E_- &= R_0^{-1}E_-(x = +\infty) \\ &= E_+^{-1}R_0(x = -\infty), \end{aligned}$$

where $R_0 = \exp(\alpha U_0 x)$. Let us define a function $\check{R}(t;\alpha)$ with the normalization $\check{R}(0;\alpha) = \boldsymbol{I}$ by:

$$\begin{aligned} \check{R}_t &= \frac{\widetilde{W_-}}{2(2\alpha - 1)}\check{R}, \qquad \text{(PCF)}, \\ \check{R} &= \exp(c^2\alpha^2 U_0 t), \qquad \text{(GHM)}. \end{aligned}$$

If $\widetilde{W}_- \stackrel{\text{def}}{=} W(-\infty, t)$ does not depend on t, then $\check{R} = \exp(t\widetilde{W}_-/(4\alpha - 2))$ (for the PCF).

Theorem3.1. *The matrix T depends on the variable t as follows:*

$$T(t;\alpha) = \check{R}(t-t';\alpha)T(t';\alpha)\check{R}^{-1}(t-t';\alpha), \tag{3.7}$$

where the right hand side of (3.7) is defined for all $\alpha \in \mathbb{R} \cup \infty$, since $[\widetilde{W}_-, T(\alpha = 1/2)] = 0$ (PCF) and $[\alpha^2 U_0, T(\alpha)] \to 0$ as $|\alpha| \to \infty$ (GHM). The same formula holds for the matrix $S^\bullet$ defined by formula (2.17) of §2.4 (see also (2.15$^\bullet$)). Analogously, the evolution of discrete scattering data is determined by the relations $(\alpha_j^\bullet)_t = 0$ and

$$\mathcal{K}_j^\bullet(t) = \check{R}(t-t';\alpha_j^\bullet)\mathcal{K}_j^\bullet(t'), \qquad 1 \leqq j \leqq N_+^\bullet. \tag{3.8}$$

Proof. Let us denote the right hand sides of formulae (3.4A) (PCF) and (3.4B) (GHM) by M. Now we temporarily assume that $\alpha \neq 1/2$ (A), $\alpha \neq \infty$ (B). The compatibility of (3.3) and (3.4) given by the equation $[\partial/\partial x + Q - \alpha U_0, \partial/\partial t - M] = 0$ implies the following general property: *For an arbitrary invertible solution E of equation (3.3), the function $E_t - ME$ also satisfies the same equation and is obtained from E by multiplying on the right by a matrix which is constant with respect to x.* Hence, in particular, we can set

$$(E_\pm)_t - ME = -E_\pm C_\pm,$$
$$(C_\pm)_x = 0, \qquad C_\pm^* = C_\pm^{-1}.$$

Letting $x \to \pm\infty$, we get

$$C_+ = M(+\infty, t), \qquad C_- = M(-\infty, t).$$

Here we use the definition of $E_\pm$ ($E_\pm \to R_0$ when $x \to \pm\infty$) and the equation $[U_0, \widetilde{W}_\pm] = 0$ derived above in the case of the PCF. As $E_- = E_+ T$,

$$T_t = C_+ T - T C_-, \qquad \widetilde{W}_+ T(\alpha = 1/2) = T(\alpha = 1/2)\widetilde{W}_-. \tag{3.9}$$

The latter formula is for the PCF only, and follows from the differentiability of T with respect to t and the continuity of T with respect to α for all α. Since $T(\alpha) \to \boldsymbol{I}$, when $|\alpha| \to \infty$, we have that $[\alpha^2 U_0, T(\alpha)] \to 0$ for the GHM when $|\alpha| \to \infty$ by the same reasoning. The differentiability of $E_\pm$ and of T with respect to t follows from the absolute integrability of Q_t, as noted above. In the case of the PCF, the functions $WE_\pm(\alpha = 1/2)$ satisfy (3.3) for $\alpha = 1/2$. It is easy to deduce the second formula of (3.9) from this.

Now we will remind the definition of $S^\bullet$. The factorization $T = T^{\bullet 0}T_0^\bullet$ is uniquely determined by the conditions

$$\tau_i^{0j} = 0 = \delta_{i\bullet}^{j^*}\tau_i^{0j}, \qquad \tau_{0i}^j = 0 = \tau_{0i}^j\delta_{i\bullet}^{j^*}, \qquad \text{for } i < j,$$
$$\operatorname{diag} T_0^\bullet = \operatorname{diag}(1, \ldots, t_{j_*}^{-1}, \ldots, t_{n_*}^{-1}),$$

where $T^{\bullet 0} = (\tau_i^{0j})$, $T_0^\bullet = (\tau_{0i}^j)$, $i^* = \max\{j \mid \mu_j = \mu_i\}$, $i_* = \min\{j \mid \mu_j = \mu_i\} - 1$ (see §2.3) and the t_j are the upper principal minors of T. Then

$$S^\bullet \stackrel{\text{def}}{=} (T^{\bullet 0})^* T^{\bullet 0} = (T_0^{\bullet -1})^* T_0^{\bullet -1},$$

(see (2.17)). Note that, by definition, the minors t_{j_*} do not have real zeros ($\det \widehat{\Phi}^\bullet = \prod_{j=1}^n t_{j_*}$).

Lemma3.1. *The minors $t_{j_*}(\alpha)$ for $\operatorname{Im}\alpha \geqq 0$ do not depend on t ($1 \leqq j \leqq n$).*

Proof. Set $c_{\pm j} = \sum_{i=1}^j c_{\pm j}^i$, where $C_\pm = (c_{\pm i}^j)$ (see (3.9)). As $[U_0, C_\pm] = 0$,

$$(t_j)_t = c_{+j}t_j - t_j c_{-j}, \qquad \text{where } j = j^*. \tag{3.10}$$

In order to check these formulae, we turn from $\mathbb{C}^n$ to the exterior power $\bigwedge^j \mathbb{C}^n$. Only the diagonal entry is non-zero among the entries in the first row and the first column of the matrix $\lambda^j C_\pm$. By the principle of analytic continuations, the relation holds not only for $\alpha \in \mathbb{R}$ but also for all α in the upper half-plane.

It follows from (3.10) that the zeros of t_j (and their multiplicities) for $j = j^*$ do not depend on t (by the uniqueness theorem for differential equations — cf. the proof of Lemma2.3). Since $c_{\pm j}(t) \in i\mathbb{R}$ and $(t_j\bar{t}_j(\alpha))_t = 0$ for $\alpha \in \mathbb{R}$ (here we differentiate $t_j\bar{t}_j$ using (3.10)). But $t_j(\alpha)$ for α such that $\operatorname{Im}\alpha \geqq 0$ is uniquely determined from $t_j\bar{t}_j(\alpha)$, $\alpha \in \mathbb{R}$ and its zeros in the upper half-plane by the Maximum Principle, since $t_j \to 1$ when $|\alpha| \to \infty$ (Theorem2.2). □

The matrices $C_\pm$ belong to the normalizers of the Lie subalgebras of $\mathfrak{gl}_n$ in which the matrices $T^{\bullet 0}$ and $T_0^\bullet$ take their values. Moreover, $C_\pm$ commute with $\operatorname{diag} T_0^\bullet$. The above lemma implies that $\operatorname{diag} T_0^\bullet$ does not depend on t. Consequently, using the uniqueness of the factorization $T = T^{\bullet 0}T_0^\bullet$, we have

$$\begin{aligned}(T^{\bullet 0})_t &= (C_+ T^{\bullet 0} - T^{\bullet 0}C_-),\\ (T_0^\bullet)_t &= [C_-, T_0^\bullet],\end{aligned} \tag{3.11}$$

and $(S^\bullet)_t = ((T^{\bullet -1})^* T_0^{\bullet -1})_t = [C_-, S^\bullet]$. Thus the statement on $S^\bullet$ holds.

Lemma3.2. *a)* $[T^{\bullet 0}] \stackrel{\mathrm{def}}{=} (\delta_{i_\bullet}^{j_\bullet} \tau_i^{0j})$ *(see §2.3) satisfies* $[T^{\bullet 0}]_t = [C_-, [T^{\bullet 0}]]$.
b) $\widetilde{W}_- = W(-\infty, t) = W(+\infty, t) = \widetilde{W}_+$.

Proof. From formula (3.11), we have

$$[T^{\bullet 0}(t)] = \check{R}_+(t-t')[T^{\bullet 0}(t')]\check{R}_-(t-t')^{-1},$$

where we temporarily denoted $\check{R}$ by $\check{R}_-$ and introduced the analogous matrix $\check{R}_+$ for W_+ instead of W_-. This formula holds for $\operatorname{Im}\alpha \geqq 0$ (recall that the entries of $[T^{\bullet 0}]$ are analytically continued to the upper half-plane — Theorem2.2). Set $\mathcal{T}_j = [T^{\bullet 0}]^{-1}(\mathcal{O}_j^n)$ (in the notation of §2) for the same $\alpha_j^\bullet$ $(1 \leqq j \leqq N_+^\bullet)$ as above. Then

$$\mathcal{T}_j(t) = \check{R}_-(t-t'; \alpha_j^\bullet)\mathcal{T}_j(t').$$

Let us define a matrix

$$[T_-^{\bullet 0}] = \check{R}_-(t-t')[T^{\bullet 0}(t')]\check{R}_-(t-t')^{-1},$$

and construct an analogous lattice

$$\mathcal{T}_{-j} \stackrel{\mathrm{def}}{=} [T_-^{\bullet 0}]^{-1}(\mathcal{O}_j^n).$$

Then we have $[T_-^{\bullet 0}]^*[T_-^{\bullet 0}] = [T^{\bullet 0}]^*[T^{\bullet 0}]$ for $\alpha \in \mathbb{R}$ and $\mathcal{T}_j = \mathcal{T}_{-j}$, which follows directly from the definition of $[T_-^{\bullet 0}]$ and the formula for $[T^{\bullet 0}(t)]$. As $[T^{\bullet 0}], [T_-^{\bullet 0}] \to I$ when $|\alpha| \to \infty$ (Theorem2.2), we have that $T^{\bullet 0} = T_-^{\bullet 0}$ by the unique solvability of the Riemann problem of type (2.17) (cf. the proof of Proposition2.4). But then $\check{R}_- = \check{R}_+$ and $\widetilde{W}_- = \widetilde{W}_+$. □

From this lemma and relation (3.9) we get formula (3.7). Note that for the GHM the statements of Lemma3.1 and 3.2 are trivial, since in this case $C_\pm = c^2\alpha^2 U_0$. It remains to check the dependence of $\mathcal{K}_j^\bullet$ on t.

As in the case of $E = E_\pm$, we get for $E = \Phi^\bullet$ that

$$\Phi_t^\bullet - M\Phi^\bullet = -\Phi^\bullet C, \qquad C_x = 0. \tag{3.12}$$

Computing the matrix C by means of Lemma3.1 and Theorem2.2 d), we obtain:

$$x \to +\infty \Rightarrow \widehat{\Phi}^\bullet \stackrel{\mathrm{def}}{=} \Phi^\bullet R_0^{-1} \to \operatorname{diag}(t_{i_\bullet}).$$

Thus, $C = M_-$ since $[U_0, M_-] = 0 = [\operatorname{diag}(t_{i_\bullet}), M_-]$. Note that when $x \to +\infty$, (3.12) tends to the relation for $[T^{\bullet 0}]_t$ derived from (3.11). Set $\check{\Phi}^\bullet = \Phi^\bullet \check{R}$. Then $\check{\Phi}_t^\bullet = M\check{\Phi}^\bullet$ as proved above. The standard argument in §2 or in §1 (see Proposition1.2) leads to the fact that the lattices $\check{\Phi}^{\bullet -1}(\mathcal{O}_j^n) = \check{R}_0^{-1}\mathcal{K}_j^\bullet$ do not depend on t. This is equivalent to relation (3.8). □

Exercise3.1. *Show that formulae (3.7) and (3.8) hold for S and $\{\alpha_j, \mathcal{K}_j\}$, if the matrix $\widetilde{W}_- = \widetilde{W}_+$ is diagonal.* □

Exercise3.2. *Rewrite Theorem3.1 for the Lax equations (2.3) and (2.5) of Ch.I, setting $\check{R} = \exp(\alpha^m t\mathcal{D})$ for $\mathcal{D}$ and m from §2 of Ch.I.* □

Inverse problem. Let us assume that the entries of stabilizing Q (PCF) or of Q, Q_x and Q_{xx} (GHM) are absolutely integrable and $\widetilde{W}(t)$ is continuous in a neighbourhood of a point t'. As above, Q and $\widetilde{W}$ are $*$-anti-hermitian, $Q' = Q$, and $[\widetilde{W}, U_0] = 0$. Let us construct from Q the T-matrix and the scattering data $S^\bullet(\alpha)$ and $\{\mathcal{K}_j^\bullet, \alpha_j^\bullet\}$, where $\operatorname{Im}\alpha_j^\bullet > 0$ for $1 \leqq j \leqq N_+^\bullet$ (see above). We assume that the zeros $\{\alpha_j^\bullet\}$ of the product $\prod_{j=1}^n t_{j\cdot}$ do not lie on the real axis and

$$[\widetilde{W}, T(\alpha = \tfrac{1}{2})] = 0, \qquad \text{(PCF)},$$
$$[\alpha^2 U_0, T(\alpha)] \to 0, \quad \text{when } |\alpha| \to \infty, \qquad \text{(GHM)}.$$

It follows from the previous analysis and Theorem2.3 (see also formula (2.17)), that there exists at most one stabilizing pair of solutions $Q(x,t)$ and $W(x,t)$ of equations (3.5) or (3.6) such that $Q(x,t') = Q(x)$ and $W(\pm\infty, t) = \widetilde{W}(t)$. Hence, the Cauchy problem of this type for differential equations (3.5), (3.6) is uniquely solvable. Note that the function W appears only in system (3.5) (PCF).

We now put the results obtained above and those found in §2 together in order to construct the solutions $Q(x,t)$ and $W(x,t)$ of the corresponding equations in terms of given $Q(x)$ and $\widetilde{W}(t)$. This procedure is called the *inverse problem method.* Below we omit $^\bullet$ for the sake of simplicity of notation. The statements below are also true for "true" S and $\{\alpha_j, \mathcal{K}_j\}$ (defined in §2), if the matrix $\widetilde{W}$ is diagonal (see Exercise3.1). Let $\check{R}$ be constructed by $\widetilde{W}(t)$ for the PCF (see above) and, as before, $\check{R} = \exp(c^2\alpha^2 U_0 t)$ for the GHM. Set

$$\mathcal{K}_j(t) = \check{R}(t - t'; \alpha_j)\mathcal{K}_j, \qquad 1 \leqq j \leqq N_+.$$

Note that $S(t;\alpha)$ is defined for all α including the points $\alpha = 1/2$ (PCF) and $\alpha = \infty$ (GHM) and is continuous with respect to α, since at these points (where $\check{R}$ has essential singularities) S satisfies the same commutativity relation as T (see above).

Let us assume that for all x and t the Riemann problem

$$\text{(3.13)} \qquad \widehat{\Phi}^*\widehat{\Phi} = R_0 S(t;\alpha) R_0^{-1}, \qquad \widehat{\Phi}^{-1}(\mathcal{O}_j^n) = R_0 \mathcal{K}_j(t),$$

is solvable, where $\alpha \in \mathbb{R}$, $1 \leqq j \leqq N_+$, $\widehat{\Phi} = \widehat{\Phi}(x,t;\alpha)$ is analytically continued to the upper half-plane ($\operatorname{Im}\alpha \geqq 0$), and $\widehat{\Phi} \to I$ for $\operatorname{Im}\alpha \geqq 0$, as $|\alpha| \to 0$ (see

(2.17)). Set $\check{\Phi} = \widehat{\Phi} R_0 \check{R}$. Then $\check{\Phi}$ *satisfies equations (3.3) and (3.4) for suitable functions Q and W.* The first of these equations follows from Theorem2.3. $\check{\Phi}_t \check{\Phi}^{-1}$ is calculated analogously. Really, $\check{\Phi}^* \check{\Phi}$ does not depend on t and, consequently, the function $\check{\Phi}_t \check{\Phi}^{-1} = -(\check{\Phi}^*)^{-1} \check{\Phi}_t^*$ is meromorphically continued to the upper and the lower half-planes. Due to the constraints on $\mathcal{K}_j(t)$, the entries of this function do not have singularities at $\alpha = \alpha_j, \overline{\alpha}_j$, $1 \leqq j \leqq N_+$ and may have poles only at $\alpha = 1/2$ (PCF), $\alpha = \infty$ (GHM). Furthermore, using the equation $\widehat{\Phi}(\alpha = \infty) = I$ and algebraic results in Ch.I (for the GHM), it is easy to show that the function $\check{\Phi}_t \check{\Phi}^{-1}$ can be written in the form of (3.4) for a certain $W(x,t)$ in the case of the PCF and (the same) function Q in the case of the GHM.

Thus, from $Q(x)$ and $\widetilde{W}(t)$ we have constructed solutions

$$Q(x,t) = \alpha U_0 - \check{\Phi}_x \check{\Phi}^{-1}, \qquad W(x,t) = (4\alpha - 2)\check{\Phi}_t \check{\Phi}^{-1}$$

of equation (3.5) or equation (3.6). If the Riemann problem (3.13) is solvable and $Q(x,t)$ is a stabilizing function, the pair of functions $Q(x,t)$ and $W(x,t)$ is a stabilizing solution of (3.5) and $\widetilde{W}(t) = W(\pm\infty, t)$. Defining F_0, U and V by formulae (3.1) and (3.2) (recall that U and V are constructed from Q and W uniquely up to conjugation by constant matrices), we get a stabilizing solution $U(x,t)$ and $V(x,t)$ of the system PCF (0.2) or the GHM equation (0.9) of type (0.10). There is an explicit formula for F_0 in terms of $\check{\Phi}$:

$$F_0 = C\check{\Phi}(\alpha = 0)^{-1},$$

where $C_x = C_t = 0$. Recall that $U = F_0 U_0 F_0^{-1}$ and $V = F_0 W F_0^{-1}$. The function $g(x,t) = C\check{\Phi}^{-1}(\alpha = 0)\check{\Phi}(\alpha = 1)C'$ satisfies equation (0.1) for the case of the PCF. We see that $g(x,t)$ is a $*$-unitary principal chiral field for arbitrary constant $*$-unitary matrices C and C'.

If there is no discrete spectrum (i.e., $\prod_{j=1}^{n} t_{j\bullet}(\alpha)$ vanishes nowhere), then (3.13) is called a regular Riemann problem. In the opposite case, when $S \equiv I$, solving (3.13) reduces to a purely algebraic procedure and the corresponding Q, W and U, V are called *multi-soliton solutions.* We study such solutions from an algebraic viewpoint in [Ch11]. Returning to the general case, we can use (3.13) to construct solutions Q^{reg} and W^{reg} from the matrix S, dropping the data of the discrete spectrum. Then Q and W can be obtained from Q^{reg} and W^{reg} by a Bäcklund-Darboux transformation introduced in §1 of this chapter (see Theorem2.3). Analogous statements hold for U and V. Properties of the action of Bäcklund-Darboux transformations on the class of stabilizing solutions strongly depend on the type of the matrix Ω defining the anti-involution $*$.

Proposition3.1. *a) Let $\Omega = I$ (i.e., $* = +$). We suppose that for a given matrix S, the regular Riemann problem (3.13) is solvable and that $Q^{\text{reg}}(x,t)$ is a stabilizing function. Then for arbitrary α_j ($\operatorname{Im}\alpha_j > 0$) and $\mathcal{K}_j$, the problem (3.13) is also solvable. The pair $Q(x,t)$ and $W(x,t)$ is defined everywhere. Moreover, it is a stabilizing solution of (3.5) or (3.6) for sufficiently generic $\{\mathcal{K}_j\}$.*

b) Let $i\mu_1 = \ldots = i\mu_p = 1 = \omega_1 = \ldots = \omega_p$ and $i\mu_{p+1} = \ldots = i\mu_n = -1 = \omega_{p+1} = \ldots = \omega_n$, where $\Omega = \operatorname{diag}(\omega_j)$ and $U_0 = \operatorname{diag}(\mu_j)$, $1 \leqq p < n$. Then for an arbitrary function $Q(x)$ with absolutely integrable entries, the data of the discrete spectrum are absent (i.e., the product $\prod_{j=1}^{n} t_{j*}$ does not vanish in the upper half-plane).

Proof. In the case of an anti-hermitian matrix Q (or U), we can use the results of §1.2 on Bäcklund transformations. Adapting the result of Exercise1.8 to this situation, we get that for a stabilizing $Q^{\text{reg}}(x,t)$, the functions Q and W (or U and V) do not have singularities for all $x, t \in \mathbb{R}$. As a result of a more detailed study of the asymptotics as $|x| \to \infty$, for sufficiently generic $\{\mathcal{K}_j\}$, the entries of Q (or the entries of Q, Q_x and Q_{xx}) are absolutely integrable with respect to x, if the entries of Q^{reg} are.

Now we suppose that assumption b) holds. Then

$$Q = -Q^* = -\Omega Q^+ \Omega = Q^+,$$
$$(U_0^{-1}Q)^+ = -QU_0^{-1} = U_0^{-1}Q.$$

Suppose that $\prod_{j=1}^{n} t_{j*}$ has a zero at α_0 with $\operatorname{Im}\alpha_0 > 0$. By Proposition2.3 there exists a solution φ of equation (3.3) for $\alpha = \alpha_0$ and $a, b \in \mathbb{R}$ with $b > 0$ such that

$$|e^{ax}\varphi_k(x;\alpha)| < e^{-b|x|} \qquad \text{when } |x| \to \infty, \quad k = 1, \ldots, n.$$

Here we can set $a = a' \operatorname{Im}\alpha_0$ for a real number a' such that $-i\mu_1 = 1 > a' > i\mu_n = -1$. Choose $a' = a = 0$. Then we may take any $b = b' \operatorname{Im}\alpha_0$ with $0 < b' < 1$ (cf. proof of Proposition2.3). Let $(\boldsymbol{p}, \boldsymbol{q}) = \boldsymbol{q}^+\boldsymbol{p}$ be the hermitian scalar product of the vectors $\boldsymbol{p}, \boldsymbol{q} \in \mathbb{C}^n$. The operator $L = U_0^{-1}\partial/\partial x + U_0^{-1}Q$ is self-adjoint. Thus we get

$$\alpha_0 \int_{-\infty}^{\infty} (\varphi, \varphi)\, dx = \int_{-\infty}^{\infty} (L\varphi, \varphi)\, dx$$
$$= \int_{-\infty}^{\infty} (\varphi, L\varphi)\, dx = \overline{\alpha}_0 \int_{-\infty}^{\infty} (\varphi, \varphi)\, dx.$$

All of these integrals are finite and nonzero due to the conditions on φ. Therefore $\alpha_0 = \overline{\alpha}_0$; this is a contradiction. □

3.2. Asymptotic expansions and the trace formulae.

The transformation $Q \mapsto T$ has many properties in common with the Fourier transformation. It is well known that a function which is the absolute value of the Fourier transform of an absolutely integrable function such that all its derivatives are also absolutely integrable decays more rapidly than any function $|\alpha|^a$ as $|\alpha| \to \infty$. We will get similar results for the entries of T. For a while we won't be taking the dependence of Q and T on t into account. As before, Q is $*$-anti-hermitian, $Q' = Q$. In this section we assume that *all the entries of Q and their derivatives with respect to x of any order are absolutely integrable from $-\infty$ to ∞.*

Let us introduce the following notation which generalizes (2.3):

$$K_{\pm}^{(p)}(Q^1,\dots,Q^p) = \int_x^{\pm\infty}\int_{x_1}^{\pm\infty}\cdots\int_{x_{p-1}}^{\pm\infty}\prod_{s=1}^{p} R_0^{-1}Q^s R_0(x_s)\,dx_p\dots dx_1,$$

where $p \geqq 1$, and set $K_{\pm}^{(0)} = I$ and $\widehat{K}_{\pm}^{(p)} = R_0(x)K_{\pm}^{(p)}(x)R_0^{-1}(x)$. Here we assume that the matrix functions $Q^1,\dots,Q^p$ have entries which are absolutely integrable from $x = -\infty$ to $+\infty$. The functions $K_{\pm}^{(p)}$ and $\widehat{K}_{\pm}^{(p)}$ depend on $\alpha \in \mathbb{R}$ (let us recall that $R_0 = \exp(\alpha U_0 x)$). In particular, in this notation,

$$\widehat{E}_{\pm} = E_{\pm}R_0^{-1} = \sum_{p=0}^{\infty}\widehat{K}_{\pm}^{(p)}(Q^1,\dots,Q^p),$$

$$T = \sum_{p=0}^{\infty}K_{-}^{(p)}(Q^1,\dots,Q^p)\Big|_{x=+\infty}$$

for the solutions $E_{\pm}(x;\alpha)$ of equation (3.3) constructed in §2.1 and the matrix $T(\alpha)$. Provided that the entries of the derivatives $Q_x^1,\dots Q_x^p$ are also absolutely integrable, we define an operation Δ by the formula:

$$\Delta K_{\pm}^{(p)} = K_{\pm}^{(p)}(Q_x^1,Q^2,\dots,Q^p) + \dots + K_{\pm}^{(p)}(Q^1,\dots,Q^{p-1},Q_x^p).$$

We set $\Delta K_{\pm}^{(0)} = 0$ and $\Delta\sum_{p=0}^{\infty}K_{\pm}^{(p)} = \sum_{p=0}^{\infty}\Delta K_{\pm}^{(p)}$. The last series is term-wise absolutely and uniformly convergent, if the absolute values of all the entries of Q^p and Q_x^p for $p = 1,2,\dots$ are dominated by an absolutely integrable function independent of p.

Proposition3.2. *The entries of the function $[U_0,T]$ decay more rapidly than any function $|\alpha|^a$, $a \in \mathbb{R}$ when $|\alpha| \to \infty$ ($\alpha \in \mathbb{R}$).*

Proof. We will apply the following lemma.

Lemma3.3. *a)* $\Delta T = \alpha[U_0, T]$,
b) $\Delta \widehat{E}_\pm = \alpha[U_0, \widehat{E}_\pm] - Q\widehat{E}_\pm = (\widehat{E}_\pm)_x$.

Proof. We check the lemma only for $\widehat{E}_+$ (for $\widehat{E}_-$ and for T, the argument is similar). The second equality of b) follows immediately from (3.3). Let us prove the first one. Integration by parts for $p \geqq 1$ gives the following formulae for

$$A_l \stackrel{\text{def}}{=} \alpha \widehat{K}_+^{(p)}(Q^1, \dots, Q^{l-1}, [U_0, Q^l], Q^{l+1}, \dots Q^p):$$

$$\begin{aligned} A_l = {}& \widehat{K}_+^{(p)}(Q^1, \dots, Q^{l-1}, Q^l_x, Q^{l+1}, \dots, Q^p) - \widehat{K}_+^{(p-1)}(Q^1, \dots, Q^l Q^{l+1}, \dots, Q^p) \\ &+ \widehat{K}_+^{(p-1)}(Q^1, Q^2, \dots, Q^{l-1}Q^l, \dots, Q^{p-1}, Q^p) \qquad \text{for } 1 < l < n, \\ A_1 = {}& Q^1 \widehat{K}_+^{(p-1)}(Q^2, \dots, Q^p) - \widehat{K}_+^{(p-1)}(Q^1 Q^2, \dots, Q^p), \\ A_n = {}& \widehat{K}_+^{(p-1)}(Q^1, \dots, Q^{p-2}, Q^{p-1}Q^p). \end{aligned}$$

Summing the A_l for $l = 1, \dots, p$, we get the equation

$$\Delta \widehat{K}_+^{(p)} = \alpha[U_0, \widehat{K}_+^{(p)}] - Q^1 K_+^{(p-1)}(Q^2, \dots, Q^p). \tag{3.14}$$

Replacing $Q^l = Q$ in (3.14), we get b). □

Estimates (2.4), (2.4a) and (2.4b) in §2 remain valid for $K^{(p)}$ and $\widehat{K}^{(p)}$ which depend on $Q^1, \dots, Q^p$. Denote the maximum of all the entries of the functions $Q^1, \dots, Q^p$ by $|q(x)|$. In this general situation, the statement of Lemma2.1 holds, when we substitute the corresponding entry of Q^{m+1} for $q_{i_m}^{i_{m+1}}$ in formula (2.6) and define $\widehat{E}_\pm^{\text{gen}} = \widehat{E}_\pm(Q^1, \dots, Q^p, \dots)$, T^{gen} and $(\widehat{E}_\pm^{\text{gen}})^\infty$ in the same way. In particular, the entries of $\Delta^m \widehat{E}_\pm^{\text{gen}}$ and $\Delta^m T^{\text{gen}}$ for $\Delta^m = \Delta \circ \dots \circ \Delta$ tend to zero as $|\alpha| \to \infty$ ($\alpha \in \mathbb{R}$) for arbitrary $m \geqq 1$, since they are Fourier transformations of absolutely integrable functions. Applying formula b) repeatedly, we obtain the statement of the proposition. □

Following the notation of §2, set

$$\widehat{\varepsilon}_+^p = \widehat{e}_+^{p+1} \wedge \dots \wedge \widehat{e}_+^n(x;\alpha) \in \bigwedge^{n-p} \mathbb{C}^n,$$

$$\widehat{\varepsilon}_-^p = \widehat{e}_-^1 \wedge \dots \wedge \widehat{e}_-^p(x;\alpha) \in \bigwedge^{p} \mathbb{C}^n,$$

$$\widehat{\varepsilon}_+^n = \widehat{\varepsilon}_-^0 = 1,$$

where $1 \leqq p \leqq n$, $\hat{e}^s_{\pm}$ is the s-th column of $\widehat{E}_{\pm}$. Let us recall that t_p is the p-th principal minor of the matrix T (see §2.2). By Theorem2.1, $\hat{\varepsilon}^p_{\pm}$ and t_p are analytically continued to the upper half-plane ($\operatorname{Im}\alpha > 0$) and $t_p \to 1$, $\hat{\varepsilon}^p_+ \to 1^{p+1} \wedge \ldots 1^n$ and $\hat{\varepsilon}^p_- \to 1^1 \wedge \ldots \wedge 1^p$ when $|\alpha| \to \infty$. Set

$$\hat{f}^{p^*} = \hat{\varphi}^{\bullet(p_*+1)} \wedge \ldots \wedge \hat{\varphi}^{\bullet p^*},$$

for the function $\widehat{\Phi}^\bullet$ defined in §2.3, i.e., $\hat{f}^{p^*}$ is the exterior product of the columns of the matrix $\widehat{\Phi}^\bullet$ corresponding to $\mu_{p_*+1} = \ldots = \mu_{p^*}$. We assume, as before, that the entries of Q and all their derivatives are absolutely integrable and $Q' = Q$. We also assume that the minors t_{p_*} for all p do not have real zeros.

Proposition3.3. *Suppose that $p^* = p$. Then t_p, $\hat{\varepsilon}^p_{\pm}$, and $\hat{f}^{p^*}$ admit asymptotic expansions, i.e. can be represented by series in α^{-1} when $|\alpha| \to \infty$ and $\operatorname{Im}\alpha \geqq 0$.*

Proof. It is sufficient to show the existence of asymptotic expansions of $\hat{\varepsilon}^p_-$ for $p = 1$ and $\hat{\varepsilon}^p_+$ for $p = n$. Indeed, $\hat{\varepsilon}^p_-$ is the first column of the matrix function $\bigwedge^p \widehat{E}_-$, and t_p is the first entry of $\bigwedge^p T$ (similarly for $\hat{\varepsilon}^p_+$). Therefore the standard reduction argument of the previous section can be applied. Let $p = 1$ (the case $p = n$ is analogous).

If $1^* = 1$ then $\mu_1 = \mu_s$ for $s \neq 1$. Hence by Proposition3.2, the absolute values of the entries t^s_1 for $1 < s \leqq n$ decay more rapidly than any $|\alpha|^a$ as $|\alpha| \to \infty$. Then this also holds for $|t_1|^2 - 1$. Indeed,

$$T^*T = 1 \Rightarrow \sum_{j=1}^{n} \omega_j |t^j_1|^2 = 1,$$

where $\Omega = \operatorname{diag}(\omega_j)$, $\omega_j = \pm 1$.

Let $g(\alpha)$ be a meromorphic function on the upper half-plane which is continuous for $\operatorname{Im}\alpha \geqq 0$. Then $g(\alpha)$ has the form

$$g(\alpha) = \exp\left(\frac{1}{\pi i}\int_{-\infty}^{\infty} \frac{\log|g(z)|}{z-\alpha} dz\right) \frac{G(\alpha)}{\overline{G}(\alpha)}, \tag{3.15}$$

if $g(\alpha) \to 1$ when $|\alpha| \to \infty$. Here we assume that the integral exists. We see that G is the product of the powers of the monomials $(\alpha - \alpha_j)$ over the zeros and poles α_j of the function g. Let us recall that $\overline{G}(\alpha) = \overline{G(\overline{\alpha})}$. If $|g(\alpha)| = 1 + O(|\alpha|^{-p-1})$ as $|\alpha| \to \infty$, then the integrals $\int_{-\infty}^{\infty} |\log|g(z)||z^{j-1}||\,dz$ exist for $j \leqq p$. Expanding $G(\alpha)/\overline{G}(\alpha)$ into a power series and approximating $(z-\alpha)^{-1}$ by $-\alpha^{-1}\sum_{j=0}^{p}(z/\alpha)^j$ for $\operatorname{Im}\alpha > 0$ as $|\alpha| \to \infty$, we obtain

$$g(\alpha) = \sum_{j=0}^{p} c_j \alpha^{-j} + o(|\alpha|^{-p-1}),$$

where $\sum_{j=0}^{p} c_j\alpha^{-j}$ is a partial sum of the Taylor series for g in a neighbourhood of $|\alpha| = \infty$. By continuity, the last formula is also valid for real α.

Using the property of $|t_1|$ checked above and formula (3.15), we prove the existence of an asymptotic expansion of t_1. For $\widehat{e}^1_- = \widehat{e}^1_-$, analogous arguments can be applied due to the algebraic considerations of §1.4, Ch.I. First of all, we rewrite formula b) of Lemma3.3 (for $\operatorname{Im}\alpha \geqq 0$) in the form

$$\Delta\widehat{e}^1_- = \alpha(U_0 - \mu_1)\widehat{e}^1_- - Q\widehat{e}^1_- = (\widehat{e}^1_-)x. \tag{3.16}$$

We will show the existence of an asymptotic expansion of $\widehat{e}^1_-$ by induction.

Let us suppose that

$$\widehat{e}^1_- = 1^1 + \sum_{s=1}^{r} c^s\alpha^{-s} + o(|\alpha|^{-r})$$

when $\alpha \to \infty$ and $\operatorname{Im}\alpha \geqq 0$ for $c^s = c^s(x) \in \mathbb{C}^n$. For $r = 0$ this follows from Lemma2.2. We will now construct c^{r+1}. It follows from (3.16) (see also (3.14)) that

$$\Delta^2\widehat{e}^1_- = (\alpha(U_0 - \mu_1) - Q)^2\widehat{e}^1_- - Q_x\widehat{e}^1_-,$$

where $\Delta^2 = \Delta \circ \Delta$. Analogously,

$$\Delta^{r+1}\widehat{e}^1_- = \alpha^{r+1}(U_0 - \mu_1)^{r+1}\widehat{e}^1_- + \sum_{j=1}^{r} \alpha^j(\cdot) + (-1)^{r+1}(Q)^{r+1}\widehat{e}^1_-,$$

where the coefficients $(\cdot)$ of $\alpha^1, \ldots, \alpha^r$ depend linearly on the components of the vector function $\widehat{e}^1_-$ and polynomially on the entries of $Q, Q_x, \ldots$. As $\Delta^{r+1}\widehat{e}^1_- = o(1)$ when $|\alpha| \to \infty$ (see above), $(U_0 - \mu_1)^{r+1}\widehat{e}^1_-$ and, consequently, the components $\widehat{e}^1_{-\,k}$ for $k > 1$ of the vector function $\widehat{e}^1_-$ admit expansions of the desired type up to α^{-r-1}. Thus it remains to prove the statement for the first component $\widehat{e}^1_{-\,1}$.

Using the fact that $\widehat{E}_-$ is $*$-unitary, for $\alpha \in \mathbb{R}$:

$$\sum_{k=1}^{n} \omega_k|\widehat{e}^1_{-\,k}|^2 = 1 \Rightarrow |\widehat{e}^1_{-\,1}|^2 + \sum_{k=2}^{n} \omega_k\left|\sum_{s=1}^{r+1} c^s_k\alpha^{-s}\right|^2 = 1 + o(|\alpha|^{-r-2}).$$

Therefore, $|\widehat{e}^1_{-\,1}f| = 1 + o(|\alpha|^{-r-1})$ for a suitable polynomial $f(\alpha^{-1})$. Applying (3.15) to $g = \widehat{e}^1_{-\,1}f$, we obtain that $\widehat{e}^1_{-\,1}$ admits an asymptotic expansion up to the term with α^{-r-1}. Thus the induction step from r to $r+1$ is completed. Now we turn to $\widehat{f}^p$.

The vector function $\widehat{f}^p(x;\alpha) \in \bigwedge^{(p-p_*)} \mathbb{C}^n$ is uniquely determined by the relations

$$
\begin{aligned}
\widehat{\varepsilon}_-^{p_*} \wedge \widehat{f}^p &= t_{p_*}^{(p-p_*)} \widehat{\varepsilon}_-^p, \\
\widehat{f}^p \wedge \widehat{\varepsilon}_+^p &= t_p t_{p_*}^{(p-p_*-1)} \widehat{\varepsilon}_+^{p_*}.
\end{aligned}
\tag{3.17}
$$

This follows from formula (2.10•) and relation (2.9). We need the expression for $\widehat{f}^p$ in a "sufficiently small" neighbourhood of $\alpha = \infty$ (where at least t_p and t_{p_*} have no zeros). Replacing $\widehat{\varepsilon}_\mp^{p,p_*}$ and t_{p,p_*} by their asymptotic expansions obtained above, we get the asymptotic expansion of $\widehat{f}^p$ (cf. proof of Theorem2.1, 2.2). □

Let $p = p^* = p_* + 1$, i.e., μ_p is an eigenvalue of U_0 of multiplicity one. Then by Proposition3.3 t_{p-1}, t_p, $\widehat{\varepsilon}_\pm^p$, $\widehat{\varepsilon}_\pm^{p-1}$ and $\widehat{f}^p = \widehat{\varphi}^{\bullet p} = \widehat{\varphi}^p$ have asymptotic expansions as $|\alpha| \to \infty$.

Trace formulae. We keep the previous assumptions and notation. In particular, we assume that $p = p^*$, $1 \leqq p \leqq n$. We rewrite relation (3.15) for $g = t_p$ in a more explicit form. Let $\alpha_{1p}, \dots, \alpha_{rp}$ be the zeros of t_p in the upper half-plane and let $\kappa_{1p}, \dots, \kappa_{rp}$ be their multiplicities. Here $r = r(p)$ (which depends on p), and, as above, t_p does not have zeros on the real axis. Then in a suitable neighbourhood of $\alpha = \infty$

$$
\log t_p = \frac{1}{\pi i} \int_{-\infty}^{\infty} \frac{\log |t_p(z)|}{z - \alpha} dz + \log \prod_{j=1}^{r(p)} \frac{(\alpha - \alpha_{jp})^{\kappa_{jp}}}{(\alpha - \overline{\alpha}_{jp})^{\kappa_{jp}}}.
$$

Expanding with respect to α^{-1} as $|\alpha| \to \infty$ with $\operatorname{Im}\alpha \geqq 0$, we get the following formulae

$$
\begin{aligned}
\log t_p &= \sum_{k=1}^{\infty} c_k^p \alpha^{-k}, \qquad \text{as } |\alpha| \to \infty, \text{ with } \operatorname{Im}\alpha \geqq 0, \\
c_k^p &= \frac{i}{\pi} \int_{-\infty}^{\infty} \alpha^{k-1} \log |t_p(\alpha)|\, d\alpha - \frac{1}{k} \sum_{j=1}^{r(p)} \kappa_{jp} (\alpha_{jp}^k - \overline{\alpha}_{jp}^k).
\end{aligned}
\tag{3.18}
$$

The convergence of the integrals in the expression for c_k^p is guaranteed by the fact that $||t_p| - 1| < |\alpha|^a$ for any $a \in \mathbb{R}$ when $|\alpha| \to \infty$. The last inequality needs to be checked only for $p = 1 = 1^*$, which has been done already (see proof of Proposition3.3).

Now we will compute the expansion of $\log t_p$ in another way. Let us denote the $(p_*+1, \dots, p)$-th component of the vector function $\widehat{f}^p$ by $\widehat{f}_p^p$. Recall that $\widehat{f}^p(x;\alpha) \in$

$\bigwedge^{(p-p_*)}\mathbb{C}^n$ and $\widehat{f}_p^p$ is equal to the minor $\det(\widehat{\varphi}_j^{\bullet i}, p_* < i, j \leqq p)$. Theorem2.2 implies that $\widehat{f}_p^p \to t_{p_*}^{p-p_*}$ when $x \to -\infty$ and $\widehat{f}_p^p \to t_p t_{p_*}^{(p-p_*-1)}$ when $x \to +\infty$, where $\operatorname{Im}\alpha \geqq 0$. Consequently,

$$\log(t_p t_{p_*}^{-1}) = \int_{-\infty}^{\infty} (\log \widehat{f}_p^p)_x dx. \tag{3.19}$$

By the methods presented in Ch.I we can find an asymptotic expansion for $(\log \widehat{f}_p^p)_x$. We will review the necessary results.

Let $\Psi = (\Psi_0 + \Psi_1\alpha^{-1} + \cdots + \Psi_s\alpha^{-s} + \cdots)R_0$ be an invertible formal solution of equation (3.3), $\widehat{\Psi} = \Psi R_0^{-1}$. In §1.2, Ch.I, we denoted the minor $\det(\widehat{\psi}_i^j; p_* < i, j \leqq p)$ by $m_p(\widehat{\Psi})$, where $\widehat{\Psi} = (\widehat{\psi}_i^j)$, and introduced a formal power series $\zeta^p = (\log m_p(\Psi))_x = \sum_{k=1}^{\infty} \zeta_k^p \alpha^{-p}$. The coefficients ζ_k^p do not depend on the choice of a solution Ψ and are polynomials in the entries of $Q, Q_x, Q_{xx}, \ldots$ without constant terms (see Theorem1.2, Ch.I). In §1.4, Ch.I, we discussed methods of computing the coefficients of the series ζ^p. In particular, for $p = 1 = 1^*$ and $\zeta^p = \zeta^1$ we have:

$$\zeta_1^1 = \sum_{s=2}^{n} \frac{q_1^s q_s^1}{\mu_1 - \mu_s} = -\sum_{s=2}^{n} \frac{q_1^s \bar{q}_1^s \omega_s \omega_1}{\mu_1 - \mu_s}.$$

Proposition3.4. *The series ζ^p coincides with the asymptotic expansion of the function $(\log f_p^p)_x$ as $|\alpha| \to \infty$. The following relation (trace formula) holds:*

$$c_k^p - c_k^{p_*} = \int_{-\infty}^{\infty} \zeta_k^p \, dx, \tag{3.20}$$

for $1 \leqq p \leqq n$ and $c_k^0 = 0$.

Proof. For a fixed formal solution Ψ of equation (3.3) (see above), set $\widehat{\pi}_-^p = \widehat{\psi}^1 \wedge \ldots \wedge \widehat{\psi}^p$ and $\widehat{\pi}_+^p = \widehat{\psi}^{p+1} \wedge \ldots \wedge \widehat{\psi}^n$, where $\widehat{\psi}^j$ is the j-th column of $\widehat{\Psi}$. If $p = 1 = 1^*$ (i.e., μ_1 has multiplicity one), then $\widehat{\psi}^1$ is determined by (3.3) up to a scalar factor which is a series constant with respect to x (Proposition1.1, Ch.I). Hence $\widehat{\psi}^1$ is proportional to the asymptotic series for $\widehat{e}_-^1$. Analogously, for $p = p^*$ the asymptotic series for $\widehat{\varepsilon}_\pm^p$ are proportional to $\widehat{\pi}_\pm^p$, as they are uniquely determined by the exterior power ($\bigwedge^p$ and $\bigwedge^{n-p}$ respectively) of equation (3.3) up to a scalar factor. Here all factors of proportionality are invertible power series in α^{-1} which are independent of x. Now let us turn to $\widehat{f}^p$.

The formal series $\widehat{g}^p = \widehat{\psi}^{p_*+1} \wedge \ldots \wedge \widehat{\psi}^p$ is uniquely recovered from $\widehat{\pi}_\pm^{p,p_*}$ by the relations

$$\widehat{\pi}_-^{p_*} \wedge \widehat{g}^p = \widehat{\pi}_-^p, \qquad \widehat{g}^p \wedge \widehat{\pi}_+^p = \widehat{\pi}_+^{p_*}.$$

Moreover, any series in α^{-1} satisfying these equations up to scalar factors is proportional to $\widehat{g}^p$. These claims are immediate consequences of the invertibility of Ψ. Thus, by relation (3.17) and the proportionality of the series $\widehat{\varepsilon}_{\pm}^{p,p_*}$ and $\widehat{\pi}_{\pm}^{p,p_*}$, the asymptotic series for $\widehat{f}^p$ differs from $\widehat{g}^p$ only by a factor which is constant with respect to x. Therefore ζ^p (the logarithmic derivative of the $(p_*+1,\dots,p)$-th component of the series $\widehat{g}^p$) coincides with the asymptotic expansion of $(\log \widehat{f}_p^p)_x$. Use (3.18) and (3.19) to complete the proof. □

We remind that for the proof of the trace formulae we assumed that the entries of Q and all their derivatives are absolutely integrable with respect to x from $x=-\infty$ to $+\infty$. In particular, all the functions ζ_k^p are absolutely integrable from $-\infty$ to ∞. As can be seen from the proofs of Proposition3.3 and 3.4, we can weaken this condition, not requiring integrability of all derivatives of the entries of Q. If one of the integrands in either side of (3.20) is integrable for certain p, k, then both sides of (3.20) are well-defined and coincide for such p, k. We now look at several applications of the trace formulae.

If $p=1=1^*$, then for $k=1$

$$\text{(3.21)} \qquad \frac{i}{\pi}\int_{-\infty}^{\infty} \log|t_1|\,d\alpha - 2i\sum_{j=1}^{r(1)} \kappa_{j1}\,\mathrm{Im}\,\alpha_{j1} = \sum_{s=2}^{n} \frac{\omega_s\omega_1}{\mu_s-\mu_1}\int_{-\infty}^{\infty} |q_1^s|^2 dx.$$

Obviously, both sides of (3.21) are purely imaginary (belong to $-i\mathbb{R}_+$ for $\Omega = I$). In general, we have also $\int_{-\infty}^{\infty} \mathrm{Re}\,\zeta_k^p = 0$, which follows directly from (3.20) and the definition of c_k^p (see (3.18)). Compare this result with Corollary1.2, Ch.I.

In §1 of Ch.I in addition to ζ for $\Omega = I$ we studied the densities χ (formal series in α or in $\overline{\alpha}$) for the conservation laws of the Pohlmeyer type. Note that in Ch.I the series χ was defined only for $p=1=1^*$ (exactly speaking, for $p_*+1=p=p^*$). The formulae, theorems obtained there can be extended easily to the general case. We will now compute the integrals corresponding to the series χ of Ch.I, and to its variants for arbitrary $p=p^*$ using the scattering data.

Let $\Omega = I$ and $\chi^p = (\log(\widehat{g}^p,\widehat{g}^p))_x$, where $\widehat{g}^p$ is the series introduced in the proof of Proposition3.4 and $(,)$ is the standard hermitian form on $\bigwedge^{(p-p_*)}\mathbb{C}^n$. In particular, $\chi^1 = (\log(\widehat{\psi}^1,\widehat{\psi}^1))_x$ for $p=1=1^*$ coincides with χ of Ch.I. The components of the vector function $\widehat{f}^p$ defined above except $\widehat{f}_p^p$ (i.e., with the multi-indices not equal to the multi-index $(p_*+1,\dots,p)$) tend to zero when $x\to\pm\infty$ and $\mathrm{Im}\,\alpha > 0$ (Theorem2.2). Therefore

$$\int_{-\infty}^{\infty} \chi^p dx = \log(t_p t_{p_*}^{-1}) + \overline{\log(t_p t_{p_*}^{-1})},$$

if we substitute the corresponding series in (3.18) for t_p and t_{p_*}. This is consistent with the statement of Theorem1.4, Ch.I on the coincidence of the coefficients of the series χ^p and $\zeta^p + \overline{\zeta}^p$ (for $p = 1$) up to exact derivatives of differential polynomials of the entries of Q.

Let us mention one more simple consequence of formula (3.20). If we consider all the series χ^p for pairwise distinct p^* $(1 \leqq p \leqq n)$, then

$$\sum_{p=p^*} \int_{-\infty}^{\infty} \zeta_k^p dx = 0, \qquad k = 1, 2, \dots, \tag{3.22}$$

since $t_n = \det T = 1$. One can prove this fact purely algebraically. Using the technique in §1, Ch.I, we can check that $\sum_{p=p^*} \zeta_k^p$ for any k is an exact derivative of a certain differential polynomial of the entries of Q. This is the unique relation between integrals $\int_{-\infty}^{\infty} \zeta_k^p dx$ in general.

Corollary3.1. *For an arbitrary finite set of indices k and any numbers $I_k^p \in i\mathbb{R}$ $(p = p^*, 1 \leqq p \leqq n)$ with the condition $\sum_{p=p^*} I_k^p = 0$, there exists a function $Q(x)$ (of the above type) such that $I_k^p = \int_{-\infty}^{\infty} \zeta_k^p dx$.*

Proof. If the numbers I_k^p are sufficiently small, the statement follows directly from (3.20). Indeed, by Corollary2.2 (§2) we can find a matrix function $Q(x)$ which corresponds to any prescribed $*$-anti-hermitian matrix $S^*(\alpha)$ which is sufficiently close to I and satisfies the conditions of Proposition2.6 (§2.5). Multiplying $Q(x)$ by suitable constants $c \in \mathbb{R}_+$ and changing the variable x to $c^{-1}x$, we get arbitrary sets $\{I_k^p\}$ from those close to zero. These $\{I_k^p\}$ also satisfy the conditions of the corollary. Here we use the homogeneity of ζ_k^p, which follows from results in §1, Ch.I (details are left to the reader). □

Now let the pair $Q(x,t)$, $W(x,t)$ be a stabilizing solution of equation (3.5) or (3.6) (or let the pair U, V be a stabilizing solution of (0.2) or (0.9)). As before, the entries of Q (or, equivalently, the entries of U_x) and all their derivatives with respect to x are absolutely integrable. Then Lemma3.1 implies:

Corollary3.2. *For arbitrary $p = p^*$ and k, the integrals $I_k^p = \int_{-\infty}^{\infty} \zeta_k^p dx$ do not depend on t and are integrals of motion of corresponding equations. The number of independent series among $\{I_k^p,\ k = 1, 2, \dots\}$ is equal to the number of pairwise distinct μ_p minus one, i.e. all the integrals I_k^p are functionally independent of each other if $p > 1^*$.* □

Using the Hamiltonian structure introduced in §2.5, we can refine the statement of this corollary. Recall that

$$\{f, g\} \overset{\text{def}}{=} \int_{-\infty}^{\infty} \operatorname{Sp}\left(\frac{\delta f}{\delta Q}(x)\left[U_0, \frac{\delta g}{\delta Q}(x)\right]\right) dx$$

for functionals f and g of the entries of Q (formula (2.20) for $a = -\infty$ and $b = \infty$). If f has the form $\int_{-\infty}^{\infty} F\,dx$, where F is a polynomial of $\{q_i^j\}$ and derivatives $(q_i^j)^{(k)} = \partial^k(q_i^j)/\partial x^k$ of arbitrary orders (i.e., F is a differential polynomial of $\{q_i^j\}$), then

$$\frac{\delta f}{\delta q_i^j}(x) = \sum_{k=0}^{\infty}(-1)^k\left(\frac{d}{dx}\right)^k\left(\frac{\partial F}{\partial (q_i^j)^{(k)}}\right).$$

Therefore we can set $\{I_k^p, I_l^r\} = \int_{-\infty}^{\infty} F_{k,l}^{r,s}dx$, where $p = p^*$, $r = r^*$, $k, l \geqq 1$ and $F_{k,l}^{r,s}$ is a suitable polynomial of the entries of Q.

From Proposition2.4, b) it follows that $\{I_k^p, I_l^r\} = 0$ for any combination of indices. Consequently, the integrals I_k^p are not only independent (for $p > 1^*$), but also pairwise involutive. This statement can be made purely algebraic, since the involutivity means that all the $F_{k,l}^{r,s}$ are exact derivatives of differential polynomials of $\{q_i^j\}$ with respect to x.

In conclusion, note that relation (3.18) and the trace formulae (3.20) also hold without imposing $*$-anti-hermitian conditions on Q, W, U and V. The existence of integrals on both sides of (3.20) is sufficient.

3.3. Examples of reduction (O_n and S^{n-1}-fields; $n = 2$).

We will discuss in this section the constraints imposed on the scattering data for orthogonal PCF's and S^{n-1}-fields. We also study in more detail the above constructions in the special case $n = 2$. We keep all conventions from §3.1.

Let $U(x,t)$ and $V(x,t)$ be $*$-anti-hermitian solutions of system (0.2) which are real-valued (i.e., $U, V \in \mathfrak{gl}_n(\mathbb{R})$). This means that U and V belong to a real Lie algebra $\mathfrak{o}(r, n-r)$, where r is the number of pluses in the matrix Ω defining the anti-involution $*$. All the previous observations can be applied. Denote the matrix (δ_{n+1}^{i+j}) by Π. The diagonal matrix $U_0 = \operatorname{diag}(\mu_j)$ which is equivalent to U is purely imaginary and hence satisfies an additional relation: $U_0 = \Pi U_0 \Pi = \overline{U}_0$. Indeed, the characteristic polynomial of U is real and its zeros (μ_j) are pairwise conjugate (with regard to multiplicities). Since $-i\mu_j \in \mathbb{R}$ are increasing, we get the desired relation for U_0. Set $A^\iota = \Pi\Omega({}^tA)\Omega\Pi$ and assume that $\Pi\Omega = \Omega\Pi$. (This is equivalent to $A^{*\iota} = A^{\iota *}$ for an arbitrary matrix A).

Taking the fact that U and V are real into account, let us take the complex conjugates of (3.1) and (3.2A). We can choose a matrix F_0 of §3.1 such that $\overline{F}_0 = F_0\Pi$ and, hence,

$$\begin{aligned} F_0^\iota &= F_0^{-1}\Pi, \\ Q &= F_0^{-1}(F_0)_x = -Q^\iota, \\ W &= 2F_0^{-1}(F_0)_t = -W^\iota. \end{aligned} \tag{3.23}$$

Here, as above, F_0 is $*$-unitary and Q and W are $*$-anti-hermitian. Hence equations (3.23) are equivalent to the relations

$$\overline{Q} = \Pi Q \Pi, \qquad \overline{W} = \Pi W \Pi.$$

It follows from (3.23) that for any $\alpha \in \mathbb{C}$ and an arbitrary solution E of equation (3.3) (or systems (3.3), (3.4A)), the relation $E^\iota = E^{-1}$ holds identically, which is equivalent to $\overline{E} = \Pi E \Pi$, if it is satisfied for a certain point x' (or a pair (x', t')). Here x' may be equal to $\pm\infty$. In particular, $\overline{E}_\pm = \Pi E_\pm \Pi$, $\overline{T} = \Pi T \Pi$ (recall that $\overline{E}(\alpha) \overset{\text{def}}{=} \overline{E(\overline{\alpha})}$) and

$$E_\pm^\iota = E_\pm^{-1}, \qquad T^\iota = T^{-1} \tag{3.24}$$

for $E_\pm$ and T of §2 (see also §3.1). Let us show that

$$S^\iota = \mathcal{D} S^{-1} \overline{\mathcal{D}}, \qquad (S^\bullet)^\iota = \mathcal{D}^\bullet (S^\bullet)^{-1} \overline{\mathcal{D}}^\bullet, \tag{3.25}$$

where $S(t;\alpha)$ and $S^\bullet(t;\alpha)$ are $*$-hermitian matrix functions defined above (see Theorem3.1 and its proof) and

$$\mathcal{D} = \operatorname{diag}(t_j t_{j-1}), \qquad \mathcal{D}^\bullet = \operatorname{diag}(t_{j^\bullet} t_{j_\bullet}), \qquad 1 \leqq j \leqq n.$$

Note that $t_{-j} = \overline{t}_j = \widetilde{t}_j$ since T is $*$-unitary (see §2.2).

The decomposition $T = (T^{0\iota})^{-1} (T_0^\iota)^{-1}$ has the same type as the triangular factorization $T = T^0 T_0$. Hence the calculation of the matrix $(T_0^\iota)^{-1}$ reduces to the caluculation of its diagonal. We get

$$\begin{aligned} \operatorname{diag}(T_0^\iota)^{-1} &= \operatorname{diag}(\Pi\, {}^t T_0 \Pi)^{-1} \\ &= \operatorname{diag}(t_{n-1}, \ldots, t_{n-j}, \ldots, 1) = \operatorname{diag}(t_j). \end{aligned}$$

Here we used the relation $t_{n-j} = \widetilde{t}_{-j}$ and formulae $\widetilde{t}_{-j} = t_j$ derived from (3.24) (see (2.7) of §2), which imply the identities

$$t_{n-j} = t_j, \qquad 1 \leqq j \leqq n. \tag{3.26}$$

Consequently, $(T_0^\iota)^{-1} = \mathcal{D} T_0$ and $S^\iota = ((T_0^{-1})^* T_0^{-1})^\iota = \mathcal{D} S^{-1} \overline{\mathcal{D}}$. The formula for $S^\bullet$ is proved exactly in the same way.

Analogously we can prove the relations

$$\Phi^\iota = \mathcal{D} \Phi^{-1}, \qquad (\Phi^\bullet)^\iota = \mathcal{D}^\bullet (\Phi^\bullet)^{-1}, \tag{3.27}$$

where $\Phi = E_-(T_0)^{-1}$ and $\Phi^\bullet = E_-(T_0^\bullet)^{-1}$ (see above and §2). It is necessary to use (3.24) and the formulae for T_0^ι and $(T_0^\bullet)^\iota$. These relations hold for analytic continuations of Φ, $\Phi^\bullet$, $\mathcal{D}$ and $\mathcal{D}^\bullet$ in the upper half-plane by the Maximum Principle. The corresponding conditions on the discrete scattering data $\mathcal{K}_j = \Phi^{-1}(\mathcal{O}_j^n)$, where $\{\alpha_j\}$ are the zeros of $\det \Phi$ ($\operatorname{Im} \alpha_j > 0$), have the following form:

$$\mathcal{K}_j = \{z \in \mathcal{O}_j^n \,|\, (z, \mathcal{D}\Pi\Omega\mathcal{K}_j)_s \subset \mathcal{O}_j\}. \tag{3.28}$$

Here $(p, q)_s = {}^t pq$ is the standard scalar product of vectors $p, q \in \mathbb{C}^n, \mathcal{O}_j^n$. Really,

$$\mathcal{K}_j = \{z \,|\, (z, {}^t\Phi\mathcal{O}_j^n)_s \subset \mathcal{O}_j\},$$

where ${}^t\Phi = \mathcal{D}\Pi\Omega\Phi^{-1}\Omega\Pi$. (Note that $\Pi\mathcal{D} = \mathcal{D}\Pi$ by (3.26).) Analogous statements hold for $\{\mathcal{K}_j^\bullet, \alpha_j^\bullet\}$.

Exercise3.3. a) *Drop the $*$-(anti-)hermitian (unitrary) conditions. That is, for ι-skew symmetric U and V, show that (3.23) implies relations (3.24) and (3.25) (for $\widetilde{\mathcal{D}}$ and $\widetilde{\mathcal{D}}^\bullet$ instead of $\overline{\mathcal{D}}$ and $\overline{\mathcal{D}}^\bullet$ — see §2) and (3.26) – (3.28), where U_0 is assumed to be purely imaginary as before. Rewrite equations (3.27) and (3.28) for the functions Ψ and $\Psi^\bullet$ defined on the lower half-plane (see §2.2, 2.3).*

b) *Show that the inverse problem (3.13) (or its variant for the pair Φ and Ψ without the anti-involution $*$ — see §2) reduces to solutions Q, W, U, and V of system (3.5), (0.2) with values in the Lie algebra $\mathfrak{o}(r, n-r)$ (or its complexification), if relations (3.27) and (3.28) are satisfied (together with their analogues for Ψ in the complex case) as well as equation (3.23).* □

Now we will discuss specific properties of the trace formulae of §3.2 for real (and $*$-anti-hermitian) Q. From relations (3.26) it follows that

$$\begin{aligned}\log(t_{p^\bullet} t_{p_\bullet}^{-1}) &= \log(t_{n-p^\bullet} t_{n-p_\bullet}^{-1}) \\ &= \log(t_{(n-p+1)_\bullet} t_{(n-p+1)^\bullet}^{-1}) = -\log(t_{(n-p+1)^\bullet} t_{(n-p+1)_\bullet}^{-1}).\end{aligned}$$

Consequently,

$$\int_{-\infty}^{\infty} \zeta_k^{p^*}\, dx + \int_{-\infty}^{\infty} \zeta_k^{(n-p+1)^*}\, dx = 0. \tag{3.29}$$

In particular, if $\mu_p = 0$, then $p^* = (n-p+1)^*$ and $\int_{-\infty}^{\infty} \zeta^{p^*}\, dx = 0$. We see that for such Q there are exactly $[m/2]$ independent integrals of motion $\{I_k^p, k = 1, 2, \ldots\}$, where $m = \{p = p^*\}$ is the number of pairwise distinct eigenvalues μ_j, $1 \leqq j \leqq n$ (here, $[\cdot]$ denotes the integer part). The proof is similar to the proof of Corollary3.1, 3.2. We leave the proof of the algebraic variant of formula (3.29) as an exercise for the reader.

S^{n-1}**-fields.** Now let $U = 2q \wedge q_x$, $V = 2q \wedge q_\eta$ and $g = 1 - 2P_q$ (where $P_q z = (z, q)q$ and $(a \wedge b)z = (z, a)b - (z, b)a$) for an S^{n-1}-field $q(x,t) \in \mathbb{R}^n$ in the normalized coordinates (see Introduction (0.5) and (0.6)). Recall that $(q, q) = 1 = (q_x, q_x)$ and the eigenvalues of the matrix U are $\mu_1 = -2i$, $\mu_s = 0$ for $1 < s < n$ and $\mu_n = 2i$ (they are ordered to make $-i\mu_j \in \mathbb{R}$ increasing). The eigenvectors of U corresponding to the eigenvalues $\pm 2i$ are proportional to $q_x \pm iq$ respectively (see §1.5, Ch.I) and the matrices U and V are real skew-symmetric. Therefore we can apply all the previous results (for $\Omega = I$, $(\cdot)^* = (\cdot)^+$).

Let us determine the first and the last columns f^1 and f^n of the matrix F_0 defined by relations (3.1), (3.2A) and (3.23). Obviously, $f^n = u(q_x + iq)$ for $u(x,t) \in \mathbb{C}$. As $Q' = Q$ and, in particular, $q_n^n = 0$ (see (3.1)), we have

$$(f^n)^+ f_x^n = 0$$

due to the fact that F_0 is unitary. Thus

$$\begin{aligned} 2u_x + u(q_x - iq, q_{xx} + iq_x) = 0 \Rightarrow 2u_x u^{-1} &= -2i \\ \Rightarrow u = e^{-ix}c, \qquad & c_x = 0. \end{aligned}$$

Here we use the relations

$$\begin{aligned} (q_x, q_{xx}) = (q, q_x) &= 0, \\ (q, q_{xx}) = -(q_x, q_x) &= -1 \end{aligned}$$

where $(,)$ denotes the standard Euclidean scalar product. Now we determine the constant c from (3.2A). Since $Vf^n = u(2iq_t - 2(q_x, q_t)q)$, we see that

$$\begin{aligned} u^{-2}(f^n)^+ V f^n &= 4i(q_x, q_t) \\ &= 2(f^n)^+(F_0)_t F_0^{-1} f^n u^{-2} = 2(f^n)^+(f^n)_t u^{-2} \\ &= (q_{xt} + iq_t, q_x + iq) + 2c_t = 4i(q_x, q_t) + 2c_t \end{aligned}$$

(note that we use equation (0.5) and the relation $(q_t, q_t) = 1$). Consequently, $c_t = 0$. By the symmetry $\overline{F_0} = F_0 \Pi$ (see (3.23)), we obtain the equation $f^1 = \overline{f^n}$. Thus up to a negligible unimodular constant factor c ($c\bar{c} = 1$),

$$\begin{aligned} f^n &= \frac{1}{\sqrt{2}} e^{-ix}(q_x + iq), \\ f^1 &= \frac{1}{\sqrt{2}} e^{ix}(q_x - iq). \end{aligned}$$

It follows from the fact that F_0 is unitary and that the other columns of this matrix are orthogonal to f^1 and f^n, and, hence, orthogonal to q. Therefore

$$F_0^{-1} g F_0 = \exp(U_0 x)\Gamma, \qquad \Gamma^2 = I,$$

where $\Gamma = (\gamma_i^j)$, $\gamma_1^n = \gamma_n^1 = 1$ and $\gamma_j^j = 1$ for $1 < j < n$. Indeed,

$$gf^n = f^n - 2ie^{-ix} q = e^{2ix} f^1,$$
$$gf^1 = e^{-2ix} f^n.$$

We remarked in §1 (see also formula (1.21) of §1, Ch.I) that for an arbitrary solution $\Phi(\alpha)$ of equation (1.1) of §1, the function $g\Phi(1-\alpha)$ also satisfies (1.1) for $\alpha = (1-\lambda)^{-1}$. Consequently, $F_0^{-1} g F_0 E(x; 1-\alpha)$ becomes a solution of equation (3.3) together with $E(x;\alpha)$ (recall that $E = F_0^{-1}\Phi$). Using the above calculation of $F_0^{-1} g F_0$, we obtain the relations (for $\alpha \in \mathbb{R}$):

$$\begin{aligned} \Gamma E_{\pm}(1-\alpha)\Gamma &= \exp(-U_0 x) E_{\pm}(\alpha), \\ \Gamma T(1-\alpha)\Gamma &= T(\alpha). \end{aligned} \tag{3.30}$$

As before, by the uniqueness theorem for differential equations, relation (3.30) can be checked by replacing solutions $E_{\pm}$ by their asymptotics $R_0(x;\alpha)$ for $x \to \pm\infty$. We set $A^{\gamma}(\alpha) = \Gamma A(1-\alpha)\Gamma$ for matrix-valued functions A of α.

In §2 we defined the matrices $T^{\bullet 0}$, $T_0^{\bullet}$, $\widetilde{T}^{\bullet 0}$ and $\widetilde{T}_0^{\bullet}$ and the functions $\Phi^{\bullet} = E_+ T^{\bullet 0}$ and $\Psi^{\bullet} = E_+ \widetilde{T}_0^{\bullet}$. The latter two functions are analytically continued to the upper and the lower half-plane respectively. Let us show that the involution γ maps $T^{\bullet 0}$ and $T_0^{\bullet}$ to $\widetilde{T}_0^{\bullet}$ and $\widetilde{T}^{\bullet 0}$ respectively, and $\Phi^{\bullet}$ to $\exp(-U_0 x)\Psi^{\bullet}$. The matrix $(T_0^{\bullet})^{\gamma}$ is of the same type as $\widetilde{T}^{\bullet 0}$ (we assume now that $\mu_2 = \ldots = \mu_{n-1} = 0$, $1^* = 1$, $2^* = \ldots = (n-1)^* = n-1$ and $n^* = n$). Relations (3.30) and (3.21) imply that

$$\begin{aligned} \widetilde{t}_1(\alpha) &= \bar{t}_1(\alpha) &&= t_1(1-\alpha) \\ = \widetilde{t}_{n-1}(\alpha) &= \bar{t}_{n-1}(\alpha) &&= t_{n-1}(1-\alpha), \end{aligned} \tag{3.31}$$

which results in

$$\operatorname{diag}(T_0^{\bullet})^{\gamma} = \operatorname{diag}(\widetilde{t}_1^{-1}, \ldots, \widetilde{t}_{j^\bullet}^{-1}, \ldots, 1) = \operatorname{diag} \widetilde{T}^{\bullet 0}.$$

Therefore

$$\begin{aligned} (T_0^{\bullet})^{\gamma} = \widetilde{T}^{\bullet 0} &\Rightarrow (T^{\bullet 0})^{\gamma} = \widetilde{T}_0^{\bullet} \\ &\Rightarrow (\Phi^{\bullet})^{\gamma} = \exp(-U_0 x)\Psi^{\bullet}. \end{aligned}$$

Using these formulae, the definition of $S^\bullet$ (see (2.15•) of §2.4), and formula (2.12), we obtain the relations

$$(S^\bullet)^\gamma = \mathcal{D}^\bullet S^{\bullet -1}\overline{\mathcal{D}}^\bullet, \quad \exp(U_0 x)(\Phi^\bullet)^\gamma = (\Phi^{\bullet +})^{-1}\overline{\mathcal{D}}. \tag{3.32}$$

Here $\mathcal{D} = \operatorname{diag}(t_{j^\bullet} t_{j_\bullet}) = \operatorname{diag}(t_1, (t_1)^2, \ldots, (t_1)^2, t_1)$. Formula (3.31) and the second relation of (3.32) hold also for $\operatorname{Im}\alpha \leqq 0$. All these functions are analytically continued to the lower half-plane (by definition, $\overline{F}(\alpha) = \overline{F(\overline{\alpha})}$, $F^+(\alpha) = (F(\overline{\alpha}))^+$ for scalar and matrix functions of $\alpha \in \mathbb{C}$). Putting (3.32) together with (3.25) and (3.27), we establish the connection between the involution γ and the anti-involution ι:

$$(S^\bullet)^\gamma = (S^\bullet)^\iota, \quad \Pi\Gamma S^\bullet(1-\alpha)\Gamma\Pi = {}^t S^\bullet(\alpha) = \overline{S}^\bullet(\alpha), \tag{3.33}$$

$$(\Phi^\bullet)^{\gamma +} = (\Phi^\bullet)^\iota \exp(U_0 x), \quad \Pi\Gamma\overline{\Phi}^\bullet(1-\alpha)\Gamma\Pi = \exp(-U_0 x)\Phi^\bullet(\alpha). \tag{3.34}$$

Applying (3.34) and the relation $t_1(1-\alpha) = \overline{t_1(\overline{\alpha})} \stackrel{\text{def}}{=} \overline{t}_1(\alpha)$ (see (3.31)), we find that the points of the discrete spectrum $\alpha_j^\bullet$ (the zeros of $\det\Phi^\bullet = \prod_{p=1}^n t_{p^\bullet} = (t_1)^{n-1}$) and the lattices $\mathcal{K}_j^\bullet = \Phi^{\bullet -1}(\mathcal{O}_j^n)$ obey the following symmetries:

$$\operatorname{Re}\alpha_j^\bullet = \tfrac{1}{2}, \quad \mathcal{K}_j^\bullet = \{\Pi\Gamma\overline{z}(\alpha_j^\bullet - \alpha) \mid z(\alpha - \alpha_j^\bullet) \in \mathcal{K}_j^\bullet\}. \tag{3.35}$$

Let us remind that $\mathcal{K}_j^\bullet$ is generated by the meromorphic vector series z of the parameter $(\alpha - \alpha_j^\bullet)$. The transformation $z(\alpha - \alpha_j^\bullet) \mapsto \overline{z}(\alpha_j^\bullet - \alpha)$ is the complex conjugation of the coefficients of z and the change of signs of the local parameter.

We leave the converse statement as an exercise to the reader. *Show that: by solving the Riemann problem (3.13) for the hermitian matrix $S^\bullet$ with the discrete spectral data $\{\alpha_j^\bullet, \mathcal{K}_j^\bullet\}$ satisfying (3.33), (3.35), (3.25) and (3.28) and the constraints of Proposition2.6 on $S^\bullet$, we can construct S^{n-1}-fields of stabilizing type (for which q_x and q_t tend to the common limits $q_x(\infty, t)$ and $q_t(\infty, t)$).*

The number of independent local conservation laws considered in §1 reduces much more here than for the orthogonal PCF. Indeed, by (3.29) only the series $\{\zeta_k^1\}$ survives. Moreover, it follows from the formula $t_1(1-\alpha) = \overline{t}_1(\alpha)$ that the coefficients of ζ_k' in the expansion $\zeta^1 = \sum_{k=1}^\infty \zeta_k' \beta^k$ of ζ^1 with respect to $\beta = (1-2\alpha)^{-1} = \alpha^{-1}(\alpha^{-1} - 2)$ satisfy the identity

$$(-1)^k \int_{-\infty}^{\infty} \zeta_k'\, dx = \int_{-\infty}^{\infty} \overline{\zeta}_k'\, dx, \qquad k = 1, 2, \ldots. \tag{3.36}$$

This is consistent with the statement of Corollary1.3 of Ch.I. It is easily shown that (3.36) and the identity $\int_{-\infty}^{\infty}\operatorname{Re}\zeta_k'\,dx = 0$ exhaust all the relations among $I_k' \overset{\text{def}}{=} \int_{-\infty}^{\infty}\zeta_k'$. Thus $\operatorname{Im} I_{2k}'$ are independent integrals of the equation of S^{n-1}-fields (0.5) and linearly generate all the integrals $\{I_k^{p^*}\}$.

Case $n = 2$. Set $n = 2$ and $U_0 = \operatorname{diag}(\mu_1, \mu_2)$ where $\mu = \mu_1 - \mu_2$ and, for the sake of uniformity, $c_1 = \mu_1$, $c_2 = \mu_2$ and $c = \mu_1 - \mu_2$ in the case of the Heisenberg magnet. Let $\omega_1 = 1$ and $\omega_2 = \omega$, where $\Omega = \operatorname{diag}(\omega_1, \omega_2)$ defines the anti-involution $*$. Let us denote (see §2.5):

$$Q = \begin{pmatrix} 0 & \omega\bar{r} \\ -r & 0 \end{pmatrix}, \quad T = \begin{pmatrix} a & -\omega\bar{b} \\ b & \bar{a} \end{pmatrix}, \quad S = \begin{pmatrix} 1 & \omega\bar{b} \\ b & 1 \end{pmatrix}. \tag{3.37}$$

Then

$$E_+T^0 = E_-T_0^{-1} = E_+\begin{pmatrix} a & 0 \\ b & 1 \end{pmatrix} = E_-\begin{pmatrix} 1 & \omega\bar{b} \\ 0 & a \end{pmatrix}$$

$$\overset{\text{def}}{=} \Phi = (ae_+^1 + be_+^2, e_+^2) = (\varphi^1, \varphi^2) = (e_-^1, \omega\bar{b}e_-^1 + ae_-^2). \tag{3.38}$$

The functions a and b depend on $\alpha \in \mathbb{R}$ and Φ depends on α and x; a, φ^1 and φ^2 are analytically continued to the upper half-plane for α.

We denote the pairwise distinct zeros of $a(\alpha) = \det\Phi(\alpha)$ by $\{\alpha_1, \dots, \alpha_N\}$ for $\operatorname{Im}\alpha > 0$. Recall that the α_j are interpreted as points of the discrete spectrum of the operator $\partial/\partial x - iU_0\alpha + Q(x)$ (see Proposition2.3). As usual, we assume that a does not have real zeros. For the sake of simplicity we assume further that all the zeros of a are simple. Then the appearance of the lattice $\mathcal{K}_j = \Phi^{-1}(\mathcal{O}_j^2)$ in the discrete data indicates proportionality of the vector functions φ^1 and φ^2 when $\alpha = \alpha_j$ (they are nonzero by Lemma2.1, b)). Following (3.38), we set:

$$\varphi^1(\alpha_j) = b_j\varphi^2(\alpha_j) \quad \Leftrightarrow \quad e_-^1(\alpha_j) = b_je_+^2(\alpha_j). \tag{3.39}$$

Then $\mathcal{K}_j = {}^t(1, -b_j)(\alpha - \alpha_j)^{-1}\mathbb{C} + \mathcal{O}_j^2$. In particular, the coefficients b_j do not depend on x. We remind that a does not have zeros in the upper half-plane for $\omega = -1$. When $\omega = 1$, arbitrary values of $\{\alpha_j, \operatorname{Im}\alpha_j > 0\}$ and any $b_j \in \mathbb{C}^* = \mathbb{C}\backslash\{0\}$ are allowed (Proposition3.1).

If Q depends on t and is a stabilizing solution of system (3.5) with $W(x,t)$ (for the PCF) or for system (3.6) (in particular, the function $r(x,t)$ is absolutely integrable with respect to x on the real axis), then $a_t = 0$ and $(\lambda_j)_t = 0$ for any j. The matrix $W(+\infty, t) = W(-\infty, t)$ (for (3.6)) is then diagonal. Let $W(\pm\infty, t) = \operatorname{diag}(w_1, w_2)$.

For simplicity, assume that w_1, w_2 and $w \stackrel{\text{def}}{=} w_2 - w_1$ do not depend on t. We know from Theorem3.1 that:

$$\text{(3.40A)} \quad b(t;\alpha) = \exp(wt(4\alpha-2)^{-1})b(0;\alpha) \qquad b_j(t) = \exp(wt(4\alpha_j-2)^{-1})b_j(0)$$

$$\text{(3.40B)} \qquad b(t;\alpha) = \exp(-\mu^3\alpha^2 t)b(0;\alpha) \qquad b_j(t) = \exp(-\mu^3\alpha_j^2 t)b_j(0)$$

for (3.5), (3.6) respectively. The Riemann problem (3.13) becomes

$$\text{(3.41)} \qquad \frac{1}{a}(\varphi^1,\varphi^2)\begin{pmatrix} 1 & \overline{b} \\ -b & -\omega \end{pmatrix} = (\varphi^{2*},\varphi^{1*}), \qquad \alpha \in \mathbb{R},$$

$$\text{(3.41a)} \qquad \widehat{\varphi}^j = \varphi^j \exp(-\mu_j \alpha x) \to 1^j \qquad \text{when } \operatorname{Im}\alpha \geqq 0, |\alpha| \to \infty,$$

and considered together with (3.39) uniquely recovers $\Phi(x,t;\alpha)$, $Q(x,t)$, and $W(x,t)$ from the scattering data $b(0;\alpha)$ and $\{\alpha_j, b_j(0)\}$, $1 \leqq j \leqq N$. Here $\varphi^* \stackrel{\text{def}}{=} {}^t(\overline{\varphi}_2, -\omega\overline{\varphi}_1)$ for $\varphi = {}^t(\varphi_1,\varphi_2)$, $1^1 = {}^t(1,0)$ and $1^2 = {}^t(0,1)$. One needs to express the functions b/a, $\overline{b}/a$, and a^{-1} appearing in (3.41) in terms of b. It can be done easily, since $a(\alpha)$ is uniquely determined by its zeros $\{\alpha_j\}$ and the restriction to $\alpha \in \mathbb{R}$, of the function $|a|^2 = 1 - \omega|b|^2$.

Let us determine r in terms of φ^1:

$$\text{(3.42)} \qquad \varphi^1 = e^{\mu_1 \alpha x}\left(\begin{pmatrix}1\\0\end{pmatrix} + \begin{pmatrix} -\frac{\omega}{\mu}\int_{-\infty}^{x} r\overline{r}(x')\,dx' \\ r/\mu \end{pmatrix}\alpha^{-1} + o(\alpha^{-1})\right).$$

In order to check this formula, we substitute the asymptotic expansion of φ^1 in a neighbourhood of $\alpha = \infty$ up to α^{-2} (see §3.2) into the equation

$$\text{(3.43)} \qquad \varphi_x = \begin{pmatrix} \mu_1 & 0 \\ 0 & \mu_2 \end{pmatrix}\varphi + \begin{pmatrix} 0 & -\omega\overline{r} \\ r & 0 \end{pmatrix}\varphi$$

and compare the constant terms and the coefficients of $\alpha^{-1}1^1$. We can also find the complete asymptotic expansion of φ^1 (provided that it exists) in this way. Moreover, using the method in Ch.I §1, we can compute φ_2^1/φ_1^1 reducing (3.43) to the corresponding Riccati equation. After this it is easy to determine φ_1^1 and the whole asymptotic expansion of φ^1.

Now we will calculate a new solution $r'(x,t)$ of equation (3.6) or (with W) of system (3.5) by a given solution $r(x,t)$ in the following case. Let us assume that $r(x,t)$ and $r'(x,t)$ have the same coefficient $b(\alpha) = b'(\alpha)$ and $\alpha'_j = \alpha_j, b'_j = b_j, 1 \leqq j \leqq N$, but a' for r' has an additional zero $\alpha_0 \neq \alpha_j$ $(1 \leqq j \leqq N)$. To determine such a solution r' we need to know (an arbitrary) α_0 with $\operatorname{Im}\alpha_0 > 0$ and a constant $b_0^0 \in \mathbb{C}^*$:

$$a'(\alpha) = a(\alpha)\frac{\alpha - \alpha_0}{\alpha - \overline{\alpha}_0}, \qquad \operatorname{Im}\alpha \geqq 0,$$
$$b_0'(t = 0) = b_0^0.$$

Here $\omega = 1$, b_0' is constructed by r' using $\varphi'^1(\alpha_0)$ and $\varphi'^2(\alpha_0)$. We will write b_0 instead of b_0'.

According to Theorem2.3, in order to find the function Φ corresponding to $r(x, t)$, it is necessary to find the hermitian projection P satisfying the relations

$$B^{-1}\mathcal{O}_0^2 = \Phi\mathcal{K}_0, \qquad B = I - \frac{\overline{\alpha}_0 - \alpha_0}{\overline{\alpha}_0 - \alpha}P$$

(see the end of §2.4 and §1.2). Since $\mathcal{K}_0 = {}^t(1, -b_0)(\alpha - \alpha_0)\mathbb{C} + \mathcal{O}_0^2$, P is the projection onto $\mathbb{C}\psi$, where

$$\psi = \begin{pmatrix}\psi_1^0\\ \psi_2^0\end{pmatrix} \overset{\text{def}}{=} \varphi^1(\alpha_0) - b_0\varphi^2(\alpha_0), \qquad b_0 = b_0(t)$$

(see above). Following the computations in §1.5 for the NS equation (see (1.29)), we obtain:

$$r' = r - \frac{\mu(\alpha_0 - \overline{\alpha}_0)\varphi}{1 + \omega\varphi\overline{\varphi}}, \qquad \varphi \overset{\text{def}}{=} \psi_2^0/\psi_1^0 = \frac{\varphi_2^1 - b_0\varphi_2^2}{\varphi_1^1 - b_0\varphi_1^2}(\alpha_0). \tag{3.44}$$

Now we will check that it has the required discrete spectrum data. The analogous calculation of W' from (3.5) is left to the reader.

Formula (3.43) coincides with the first equation of (1.28) for $\alpha = \lambda$, where we substituted the values $\mu_j = \mp i$ for μ_1 and μ_2. Strictly speaking, in accordance with (1.29), we should use the vector functions $\check{\varphi}^j = \varphi^j \exp(w_j t(4\alpha - 2)^{-1})$ for (A) ($\check{\varphi}^j = \varphi^j \exp(\mu_j\mu^2\alpha^2 t)$ in the case of (B)) and the constant $b_0(0)$ instead of φ^j and $b_0 = b_0(t)$. However, the exponential factors cancel. The function Φ' corresponding to r' has the form $\Phi' = B\Phi$. Therefore

$$\begin{aligned} a' &= \det\Phi' \\ &= \det B \det\Phi = \left(1 - \frac{\overline{\alpha}_0 - \alpha_0}{\overline{\alpha}_0 - \alpha}\operatorname{Sp}P\right)a = a\frac{\alpha - \alpha_0}{\alpha - \overline{\alpha}_0}, \end{aligned}$$

$\varphi'^1(\alpha_j) = b_j\varphi'^2(\alpha_j) \Rightarrow b'_j = b_j \ (1 \leqq j \leqq N)$ and $b' = b$ by the relations

$$\begin{aligned} S' &= \Phi'^+\Phi' \\ &= \Phi^+B^+B\Phi = \Phi^+\Phi = S. \end{aligned}$$

Finally, from the construction of Φ' we have:

$$(\Phi')^{-1}(\mathcal{O}_0^2) = \Phi^{-1}B^{-1}(\mathcal{O}_0^2) = \mathcal{K}_0.$$

Thus r' from (3.44) indeed has the desired discrete spectrum. This solution is called a *one-soliton solution in the background of the initial solution r (with the parameters α_0 and b_0^0).*

3.4. Scattering data for certain solutions of the NS equation.

We will apply the formula obtained above and describe the scattering data of certain solutions of the NS equation. The solutions in question are those which are obtained from multi-soliton solutions by restricting the domain of $r(x, 0)$ to finite or semi-infinite intervals. This problem is of definite interest since in applied physics the models described by the NS equation always deal with functions $r(x, t)$ on a finite interval. Our approach illustrates the inverse scattering method and makes it possible to trace the way the discrete spectrum appears (or vanishes) when r changes. The results below can be easily transfered to the PCF equation ($n = 2$) and the Sin-Gordon equation.

We keep all notation from the end of the previous section and set $\omega = 1$. We turn to the associated linear problem (1.28) of §1 which is somewhat more traditional for the NS equation. Let $\mu_1 = c_1 = -i$, $\mu_2 = c_2 = +i$ and $\mu = -2i$ in equation (3.3) or equation (3.4B). Equation (3.3) remains unchanged, but (3.4B), which determines the t-dependence, is replaced by the second equation of (1.28), differing a little by the coefficients.

Exercise3.4. *Check that for Q of (3.38) and $\mu_{1,2} = \mp i$, the zero curvature representation (1.28) for the NS equation is obtained by setting $\alpha = -2\lambda$ in formulae (3.3) and (3.4B) and conjugating by the matrix $\Pi = \begin{pmatrix} 0 & 1 \\ 1 & 0 \end{pmatrix}$.* □

We remind that the compatibility condition of system (1.28) is the NS equation (for $\omega = +1$ and in the standard normalization)

$$ir_t = r_{xx} + 2|r|^2 r. \tag{3.45}$$

The evolution of the coefficient b and the discrete scattering data b_j $(1 \leqq j \leqq N)$ with respect to t is now given by formulae

$$\begin{aligned} b(t;\alpha) &= \exp(4i\alpha^2 t)b^0(\alpha), \\ b_j(t) &= \exp(4i\alpha_j^2 t)b_j^0, \end{aligned} \tag{3.46}$$

instead of (3.40B). It is convenient to introduce functions $c_j = b_j/a'(\alpha_j)$, where $a' \stackrel{\text{def}}{=} \partial a/\partial\alpha$. The t-dependence of c_j is given by the same relations (3.46) as for b_j with $c_j^0 = c_j(t=0)$ as the initial values.

We begin with the formulae for N-soliton solutions of (3.45) and the corresponding φ-functions. They are rather well-known. Such solutions are specified by the following defining conditions:

$$b(\alpha) \equiv 0 \Leftrightarrow a(\alpha) = \prod_{j=1}^{N} \frac{\alpha - \alpha_j}{\alpha - \overline{\alpha}_j}. \tag{3.47}$$

The equivalence is a consequence of the relations $a(\infty) = 1$ and $|a|^2 + |b|^2 = 1$ (fulfilled on the real axis). Since $b \equiv 0$, the functions $\widehat{\varphi}^1 = \varphi^1 \exp(i\alpha x)$ and $\widehat{\varphi}^2 = \varphi^2 \exp(-i\alpha x)$ are defined on the entire complex plane by formula (3.41) and are rational with respect to α. For any $\alpha \in \mathbb{C}$, we set

$$\varphi^*(\alpha) = (\varphi(\overline{\alpha}))^* = {}^t(\overline{\varphi_2(\overline{\alpha})}, -\overline{\varphi_1(\overline{\alpha})})$$

for a vector function $\varphi = {}^t(\varphi_1, \varphi_2)$. Note that $\varphi^{**} = -\varphi$. As before, we assume that all of the α_j are pairwise distinct.

Proposition3.5. a) *The N-soliton function φ^1 (which corresponds to $b \equiv 0$) can be represented as follows:*

$$\begin{aligned} a^{-1}\varphi^1(\alpha)e^{i\alpha x} &= \begin{pmatrix}1\\0\end{pmatrix} + \sum_{j=1}^{N} \frac{c_j}{\alpha - \alpha_j}\chi^j, \\ \chi^j &= \varphi^2(\alpha_j)e^{i\alpha_j x}, \end{aligned} \tag{3.48}$$

where the vector functions χ^j satisfy the following system of linear equations ($1 \leqq m \leqq N$):

$$\begin{aligned} \chi^m + \sum_{k,j=1}^{N} \frac{\overline{c}_k c_j \exp(2i(\alpha_m - \overline{\alpha}_k)x)}{(\alpha_m - \overline{\alpha}_k)(\overline{\alpha}_k - \alpha_j)}\chi^j &= \\ = \begin{pmatrix}0\\1\end{pmatrix} \exp(2i\alpha_m x) + \begin{pmatrix}1\\0\end{pmatrix} \sum_{k=1}^{N} \frac{\overline{c}_k \exp(2i(\alpha_m - \overline{\alpha}_k)x)}{\overline{\alpha}_k - \alpha_m}&. \end{aligned} \tag{3.49}$$

Proof. The existence of expansion (3.48) for $\chi^j = b_j^{-1}\varphi^1(\alpha_j)\exp(i\alpha_j x)$ follows from the rationality of the function $\widehat{\varphi}^1$ (with poles of first order only at the points

$\{\overline{\alpha}_j\}$) and the equation $\widehat{\varphi}^1(\infty) = \widehat{e}^1_-(\infty) = {}^t(1,0)$ (Lemma2.1). The functions χ^j are expressed by formulae (3.39) in terms of $\varphi^2(\alpha_j)$. Set $a^{-1}\varphi^1(\alpha) = \varphi^{2*}(\alpha) = (\varphi^2(\overline{\alpha}))^*$ in (3.48) and apply the anti-involution $*$ (recall that $\varphi^{2**} = -\varphi^2$). Substituting consecutively $\alpha = \alpha_1, \dots, \alpha_N, \overline{\alpha}_1, \dots, \overline{\alpha}_N$ in (3.48) and (3.48)*, we obtain the relations

$$\exp(i\overline{\alpha}_k x)\varphi^2(\alpha_k)^* = \begin{pmatrix}1\\0\end{pmatrix} + \sum_{j=1}^N \frac{\exp(i\alpha_j x)c_j}{\overline{\alpha}_k - \alpha_j}\varphi^2(\alpha_j),$$

$$-\exp(-i\overline{\alpha}_k x)\varphi^2(\alpha_k) = \begin{pmatrix}0\\-1\end{pmatrix} + \sum_{j=1}^N \frac{\exp(-i\overline{\alpha}_j x)\overline{c}_j}{\alpha_k - \overline{\alpha}_j}\varphi^2(\alpha_j)^*.$$

Cancelling $\{\varphi^2(\alpha_j)^*\}$, we get (3.49). □

Corollary3.3. *The multi-soliton solution of equation (3.45) corresponding to the set $\{\alpha_j, b_j, 1 \leqq j \leqq N\}$ is given by the formula*

$$r = -2i\sum_{j=1}^N c_j\chi_2^j, \qquad \chi^j = \begin{pmatrix}\chi_1^j\\ \chi_2^j\end{pmatrix}, \tag{3.50}$$

and satisfies the relation

$$\int_{-\infty}^x r\overline{r}(x')\,dx' = 2i\sum_{j=1}^N c_j\chi_1^j.$$

In particular, for $N = 1$, $\lambda_1 = \xi + i\eta$, $b_1^0 = \exp(2\eta x_0 + i\varphi_0)$, $x_0, \varphi_0 \in \mathbb{R}$ and $\eta > 0$,

$$r(x,t) = 2\eta\frac{\exp(2i\xi x + 4i(\xi^2 - \eta^2)t + i\varphi_0)}{\cosh(2\eta(x - x_0 + 4\xi t))}, \tag{3.51}$$

$$\int_{-\infty}^x r\overline{r}(x')\,dx' = 4\eta\frac{1}{1 + e^{4\eta(x_0 - x)}}. \tag{3.51'}$$

Proof. Formula (3.50) and the relation for $r\overline{r}$ are derived from (3.42) for $\mu = -2i$ and the limit of expression (3.48) for $\alpha \to \infty$. We also use the expansion

$$a(\alpha) = 1 + \alpha^{-1}\sum_{j=1}^N(\alpha_j - \overline{\alpha}_j) + o(\alpha^{-1}).$$

For the one-soliton solution r we have (see (3.49)):

$$\chi_2^1 = \exp(2i\alpha_1 x)\frac{1}{1 + c_1\bar{c}_1 \exp(-4\eta x)(4\eta^2)^{-1}},$$
$$\chi_1^1 = i\bar{c}_1 \exp(-4\eta x)(2\eta)^{-1}\frac{1}{1 + c_1\bar{c}_1 \exp(-4\eta x)(4\eta^2)^{-1}}.$$

Substituting $c_1 = 2i\eta b_1$ ($a' = \partial a/\partial\alpha = 2i\eta(\alpha - \overline{\alpha}_1)^2$), we get the formulae

$$\chi_2^1 = \frac{e^{2i\lambda_1 x}}{1 + e^{4\eta(x_0 - x)}}, \qquad \chi_1^1 = \frac{e^{-i\varphi_0 + 2\eta x_0 - 4\eta x}}{1 + e^{4\eta(x_0 - x)}}. \tag{3.52}$$

As $r = -2ic_1\chi_2^1$,

$$\int_{-\infty}^{x} r\overline{r}(x')\,dx' = 4\eta + 2ic_1\chi_1^1 = 4\eta\left(1 - \frac{e^{4\eta(x_0-x)}}{1 + e^{4\eta(x_0-x)}}\right).$$

□

It is easy to check that formula (3.51) coincides with formula (3.44) which was obtained using Bäcklund-Darboux transformations, when $\varphi_1^2 = \varphi_2^1 = 0$, $\varphi_1^1 = \exp(-i\alpha x)$ and $\varphi_2^2 = \exp(i\alpha x)$. Note also that $\int_{-\infty}^{\infty} r\overline{r}(x)\,dx = 4\eta$ (see (3.51')). This follows from the identity

$$-\frac{2}{\pi}\int_{-\infty}^{\infty} \log|a(\alpha)|\,d\alpha + \sum_{j=1}^{N} \operatorname{Im}\alpha_j = \omega\int_{-\infty}^{\infty} r\overline{r}(x)\,dx,$$

which is obtained from (3.21) with $n = 2$, $q_2^1 = -r$ and $\mu_{1,2} = \mp i$. Compare (3.50) and the formula of Exercise1.15 as an exercise.

Let $r^0(x)$ be a function which is absolutely integrable on the real axis and has the property that its derivative r_x^0 is absolutely integrable on $\mathbb{R}$ also. For a pair of points $-\infty \leqq y < z \leqq \infty$ we introduce a truncated function

$$r_y^{z0}(x) = \Theta(x - y)\Theta(z - x)r^0(x),$$

where $\Theta(x) = 0$ for $x < 0$ and $\Theta(x) = 1$ for $x \geqq 0$. Using r_y^{z0}, we will construct the monodromy matrix T_y^{z0} of equation (3.43), the corresponding coefficients a_y^z and b_y^{z0} (see (3.37)) and vector functions $\varphi^{10} = e_-^{10}$ and $\varphi^{20} = e_+^{20}$ (the index 0 indicates independence of the parameter t). By the definitions of T and $\varphi^{1,2}$ we obtain the relations

$$\begin{aligned} \varphi_1^{10}(x;\alpha) &= a_-^x(\alpha)e^{-i\alpha x}, & \varphi_2^{10}(x;\alpha) &= b_-^{x0}(x;\alpha)e^{i\alpha x}, \\ \varphi_2^{20}(x;\alpha) &= a_x^+(\alpha)e^{i\alpha x}, & \varphi_1^{20}(x;\alpha) &= \omega\overline{b}_x^{+0}(x;\alpha)e^{-i\alpha x}. \end{aligned} \tag{3.53}$$

From these relations we can write a_y^z and b_y^z in terms of φ^1 and φ^2 for any y and z. Here and below indices $\pm$ attached to r, a or b denote $\pm\infty$. It is interesting to examine the zeros of a_y^z in the upper half-plane, which determine the asymptotics for large t of the solution $r(x,t)$ with the initial condition $r(x,0) = r_y^{z0}(x)$. In order to recover $r(x,t)$ from a_y^z and b_y^{z0} we need to solve the corresponding Riemann problem (3.41).

Let us illustrate the general approach above for the N-soliton function (see (3.17)). Any such function $r^0 = r(x,0)$ is given by formula (3.50) at $t = 0$. The relations

$$a_-^x(\alpha) = a(\alpha) + \sum_{j=1}^N c_j \frac{a(\alpha)}{\alpha - \alpha_j} \chi_1^j, \tag{3.54a}$$

$$b_-^x(\alpha) = e^{-2i\alpha x} \sum_{j=1}^N c_j \frac{a(\alpha)}{\alpha - \alpha_j} \chi_2^j \tag{3.54b}$$

hold by (3.53) and (3.48). Hence for the calculation of a_y^z and b_y^z, one needs only the following property of T:

$$T_\zeta^z(\alpha) T_y^\zeta(\alpha) = T_y^z(\alpha), \qquad y < \zeta < z,$$

where

$$T_y^z = \begin{pmatrix} a_y^z & -\overline{b}_y^z \\ b_y^z & \overline{a}_y^z \end{pmatrix}.$$

In the special case $N = 1$ we obtain:

Proposition3.6. *Let $N = 1$, $\lambda_1 = \xi + i\eta$, $b_1^0 = e^{2\eta x_0 + i\varphi_0}$, $l_x = \exp(4(x_0 - x)\eta)$. Then for the function r^0 defined by (3.51) at $t = 0$, we get the formulae*

$$\begin{aligned}
a_-^x &= 1 - \frac{2i\eta}{\alpha - \overline{\alpha}_1}(1 + l_x)^{-1} &&= \frac{\alpha - \xi - i\eta \tanh(2\eta(x - x_0))}{\alpha - \overline{\alpha}_1}, \\
b_-^x &= \frac{2i\eta e^{2i(\alpha_1 - \alpha)x} b_1^0}{\alpha - \overline{\alpha}_1}(1 + l_x)^{-1}, \\
a_x^+ &= 1 - \frac{2i\eta}{\alpha - \overline{\alpha}_1}(1 + l_x^{-1})^{-1} &&= \frac{\alpha - \xi + i\eta \tanh(2\eta(x - x_0))}{\alpha - \overline{\alpha}_1}, \\
b_x^{+0} &= -\frac{2i\eta e^{2i(\alpha_1 - \alpha)x} b_1^0}{\alpha - \alpha_1}(1 + l_x)^{-1},
\end{aligned}$$

$$a_y^z = 1 + \frac{2i\eta}{(1+l_y)(1+l_z)} \frac{l_y}{(\alpha - \overline{\alpha}_1)} (e^{2i(\alpha-\overline{\alpha}_1)(z-y)} - 1) - \frac{2i\eta}{(1+l_y)(1+l_z)} \frac{l_z}{(\alpha - \alpha_1)} (e^{2i(\alpha-\alpha_1)(z-y)} - 1),$$

$$b_y^z = \frac{2i\eta b_1^0}{(\alpha-\alpha_1)(1+l_y)(1+l_z)} e^{2i(\alpha_1-\alpha)z} \left(1 + l_y \frac{\alpha-\alpha_1}{\alpha-\overline{\alpha}_1}\right) - \frac{2i\eta b_1^0}{(\alpha-\alpha_1)(1+l_y)(1+l_z)} e^{2i(\alpha_1-\alpha)y} \left(1 + l_z \frac{\alpha-\alpha_1}{\alpha-\overline{\alpha}_1}\right).$$

Proof. We apply formulae (3.54) for the χ_2^1 and χ_1^1 from (3.52) and the relations for T. For example, the identity $T_x^{+\infty} T_{-\infty}^x = \text{diag}(a, \overline{a})$ can be used to calculate a_x^+ and b_x^+ in terms of a_-^x and b_-^x. □

It is easy to see that a_-^x has a zero in the upper half-plane $\operatorname{Im} \alpha \geqq 0$ only when $x > x_0$. Analogously, a_x^+ has a zero in the upper half-plane only when $x < x_0$. This zero is $\xi \pm i\eta \tanh(2\eta(x - x_0))$ respectively. The value $x = x_0$ is called the "center" of the soliton $r(x, 0)$. The discrete spectrum vanishes when we cut off more than one half of the one-soliton.

Exercise3.5. *Show that the coefficient a_y^z calculated above for $N = 1$ has only zeros of the form $\xi + i\rho\eta$, $\rho \in \mathbb{R}$, if any. For $z - x_0 = k = x_0 - y$ (i.e. for the symmetric truncation), ρ satisfies the equation*

$$(\kappa + \kappa^{-1} - 2)h^\rho = 2 + h^{-1}\kappa + h\kappa^{-1},$$

where $\kappa = (\rho - 1)/(\rho + 1)$ and $h = \exp(-2\eta k)$. In particular, a_y^z has a zero in the upper half-plane only when $e^{4\eta k} > 3 + \sqrt{8}$. □

Superposition of "semi-solitons". Interesting effects arise for the sum of two sequential (nonintersecting) semi-soliton functions r. Let $R^0(x) = \widetilde{r}_-^{x'0}(x) + r_{x'}^{+0}(x)$ for a certain fixed x', where r^0 and $\widetilde{r}^0$ are defined by formula (3.31) at $t = 0$ with the same value of λ_1, but with different coefficients $b_1^0 = \exp(2\eta x_0 + i\varphi_0)$ and $\widetilde{b}_1^0 = \exp(2\eta\widetilde{x}_0 + i\widetilde{\varphi}_0)$. Assume further that $\widetilde{x}_0 \leqq x' \leqq x_0$. We set:

$$l = \exp(4(\eta(x_0 - x')), \qquad \widetilde{l} = \exp(4\eta(\widetilde{x}_0 - x))$$

$$(l\widetilde{l})^{1/2} = \exp(2\eta(x_0 + \widetilde{x}_0 - 2x')), \qquad u = \frac{2i\eta}{\overline{\alpha}_1 - \alpha}.$$

The coefficients A and B^0 of the matrix T corresponding to R_0 are written as follows:

$$A = 1 + \frac{1}{(1+l)(1+\widetilde{l})}\left(u(1+2l+l\widetilde{l}) + u^2(l - (l\widetilde{l})^{1/2}e^{i(\widetilde{\varphi}_0-\varphi_0)})\right),$$

$$B^0 = -\frac{\exp(2i(\alpha_1-\alpha)x'}{(1+l)(1+\widetilde{l})}\left(u(\widetilde{b}_1^0 - b_1^0) + \overline{u}(b_1^0\widetilde{l} - \widetilde{b}_1^0 l)\right),$$

where $\overline{u} = 2i\eta/(\alpha-\alpha_1)$. For the zeros of A we get a quadratic equation with respect to u. For simplicity, let us assume that $x' - \widetilde{x}_0 = k = x_0 - x'$ and set $\sigma = \exp(2\eta k)$ and $\varepsilon = e^{i(\widetilde{\varphi}_0-\varphi_0)}$. Then the formula for A takes the form

$$A = 1 + 2u(1+\sigma^{-1})^{-1} + u^2\frac{\sigma-\varepsilon}{(1+\sigma)(1+\sigma^{-1})}.$$

As a first illustration, let us consider the case when $\varepsilon = -1$, i.e., assume that initial one-soliton functions are in opposite phase. We can find two zeros of A by simple calculation:

$$\alpha_{1,2} = (\xi \pm \eta\cosh^{-1}(\eta k)) + i\eta\tanh(\eta k).$$

The multi-soliton constituent of R^0 (i.e., a solution of the NS with the same discrete scattering data as R^0, but without continuous scattering data) is the superposition of two one-solitons with different velocities $v_1 = -4\xi - 4\eta/\cosh(\eta k)$ and $v_2 = -4\xi + 4\eta/\cosh(\eta k)$, and with the same amplitude. This means, qualitatively, that we obtain a pair of one-solitons of identical shape traveling apart asymptotically.

We will study the case of identical phase in more detail. Let $\varepsilon = 1$. Then the zeros $\alpha_{1,2}$ are written in the form:

$$\alpha_{1,2} = \xi + i\lambda_{1,2}\eta, \qquad \lambda_{1,2} = \frac{\sinh(2\eta k) \pm 1}{\cosh(2\eta k)}.$$

For $e^{2\eta k} > 1 + \sqrt{2}$, the coefficent A has two zeros in the upper half-plane. For $e^{2\eta k} \leqq 1 + \sqrt{2}$ there is only one zero, α_1. Note that the velocities of the resulting one-solitons (from R^0) are equal to the velocity $v = -4\xi$ of the initial solitons, but they have different amplitudes.

If we formally set $\overline{u} = 2i\eta(\alpha - \alpha_1)$ for $\alpha \in \mathbb{C}$, then we can consider the above formula for B^0 also for $\operatorname{Im}\alpha > 0$. Let us substitute $\alpha = \alpha_{1,2}$ in this formula. We remind that the coefficient b is not analytic in the upper half-plane, generally speaking. However in the present case we can show that $B^0(\alpha_{1,2}) = \widetilde{b}_-^{x'0}/a_{x'}^+(\alpha_{1,2}) = B_{1,2}^0$. Here $B_{1,2}^0$ are b-coefficients of R^0 defined at the zeros $\alpha_{1,2}$ of the coefficient A by the procedure of the previous section. This follows immediately from formulae (3.39) and (3.53) (substitute $x = x'$). By a simple calculation we get:

$$B_{1,2}^0 = \pm\exp(i\varphi_0 + 2\lambda_{1,2}\eta x'),$$

where the signs "+" and "−" correspond to α_1 and α_2 respectively (if $\operatorname{Im}\alpha_2 > 0$).

Exercise3.6. *Show that*

$$C^0_{1,2} \stackrel{\text{def}}{=} \frac{B^0_{1,2}}{A'(\alpha_{1,2})} = i\eta \exp(i\varphi_0 + 2\lambda_{1,2}\eta x')\frac{(1 \pm e^{2\eta k})^2}{2\cosh(2\eta k)}.$$

Construct the two-soliton solution $R_s(x,t)$ from the scattering data $\{\alpha_{1,2}, C^0_{1,2}\}$ for $\operatorname{Im}\alpha_2 > 0$, using formula (3.50). Check the relation

$$\int_{-\infty}^{\infty} |R|^2 - |R_s|^2 dx = 4\eta(1 - \tanh(2\eta k)),$$

where $R(x,t)$ is a solution of (3.45) for which $R(x,0) = R^0(x)$. Show that when $\xi = 0$ the function $|R_s(x,t)|^2$ is periodic with respect to t with the period $\tau_k = \frac{\pi}{2}(\alpha_2^2 - \alpha_1^2) = \frac{\pi}{8\eta^2}\cosh^2(2\eta k)/\sinh(2\eta k)$. □

Variations of the discrete spectral data ($N = 1$). Let $r(x,t)$ be the one-soliton solution given by formula (3.51). We compute the variations $\delta\alpha_1 = \widetilde{\alpha}_1 - \alpha_1$ and $\delta b^0_1 = \widetilde{b}^0_1 - b^0_1$ in the case of an arbitrary (small) variation $\delta r^0 = \delta r^0(x)$ of the initial value $r^0 = r(x,0)$. Here $\widetilde{\alpha}_1$ is the zero of the coefficient $\widetilde{a}$ of the function $\widetilde{r}^0 = r^0 + \delta r^0$ and $\widetilde{b}^0_1$ is given by formula (3.39) for $\widetilde{r}^0$ ($\alpha = \widetilde{\alpha}_1$, $t = 0$). We assume that $\widetilde{r}^0$ tends rapidly to zero when $x \to \pm\infty$ as before, but maybe not to a pure one-soliton function.

Combining the results of Proposition2.4 with formulae (3.53), we get the following general identities:

$$\frac{\delta a(\alpha)}{\delta r^0(x)} = -a^x_-(\alpha)\overline{b}^{+0}_x(\alpha)\exp(-2i\alpha x), \tag{3.55a}$$

$$\frac{\delta a(\alpha)}{\delta \overline{r^0}(x)} = -b^{x0}_-(\alpha)a^+_x(\alpha)\exp(2i\alpha x), \tag{3.55b}$$

which hold for $\operatorname{Im}\alpha \geqq 0$ ($\overline{b}^{+0}_x(\alpha) \stackrel{\text{def}}{=} \overline{b^{+0}_x(\overline{\alpha})}$). We also take the relation

$$\delta a(\alpha_1) + \partial a/\partial\alpha(\alpha_1)\delta\alpha_1 = 0$$

into account. Replacing a^x_-, $\overline{b}^{+0}_x$, a^+_x and b^{x0}_- by their expressions for the one-soliton function r^0 (Proposition3.6), we obtain the formula

$$\begin{aligned}\delta\alpha_1 = {} & i\eta\int_{-\infty}^{\infty} \frac{\exp(-i\varphi_0 - 2i\xi x - 2\eta(x - x_0))}{2\cosh^2(2\eta(x - x_0))}\delta r^0(x)\, dx \\ & + i\eta\int_{-\infty}^{\infty} \frac{\exp(i\varphi_0 + 2i\xi x + 2\eta(x - x_0))}{2\cosh^2(2\eta(x - x_0))}\overline{\delta r^0(x)}\, dx.\end{aligned} \tag{3.56}$$

The value $\alpha_1 + \delta\alpha_1$ is consistent with exact relations for the zero of the coefficient a^z_y in the limit when $y = -k$ and $z = k$ as $k \to \infty$.

Analogously, but after a longer computation we can establish the following formulae:

(3.57a)

$$\delta x_0 = \int_{-\infty}^{\infty} \frac{x - x_0}{2\cosh(2\eta(x - x_0))} \Delta^+_r \, dx,$$

(3.57b)

$$\delta\varphi_0 = -i \int_{-\infty}^{\infty} \left(\frac{1}{2\cosh(2\eta(x - x_0))} - \eta x \frac{\sinh(2\eta(x - x_0))}{\cosh^2(2\eta(x - x_0))} \right) \Delta^-_r \, dx,$$

$$\Delta^{\pm}_r \stackrel{\text{def}}{=} \exp(-2i\xi x - i\varphi_0)\delta r^0(x) \pm \exp(2i\xi x + i\varphi_0)\delta \overline{r^0(x)}.$$

Here $\delta x_0 = \widetilde{x}_0 - x_0$, $\delta\varphi_0 = \widetilde{\varphi}_0 - \varphi_0$, $\widetilde{x}_0$ and $\widetilde{\varphi}_0$ are not derived from $\widetilde{b}^0_1$, but from the coefficient $\widetilde{c}^0_1$, since c appears in formulae (3.50) and (3.51). More precisely, it is necessary to set $\widetilde{c}^0_1 = 2i\widetilde{\eta}\exp(2\widetilde{\eta}\widetilde{x}_0 + i\widetilde{\varphi}_0)$, where $\widetilde{\eta} = \operatorname{Im}\widetilde{\alpha}_1$. The coefficient $\widetilde{c}^0_1$ is defined by (3.39) for $\widetilde{r}^0$, as usual. It is convenient to use the general formula

$$\widetilde{c}^0_1 = \frac{\widetilde{b}^{\zeta 0}_-(\widetilde{\alpha}_1)}{\widetilde{a}^+_\zeta(\widetilde{\alpha}_1)\widetilde{a}'(\widetilde{\alpha}_1)}, \qquad \widetilde{a}' = \frac{\partial \widetilde{a}}{\partial \alpha},$$

where the value of $\widetilde{\alpha}_1 = \alpha_1 + \delta\alpha_1$ computed above, the derivative of $\widetilde{a}$ at that point, $\widetilde{b}^{\zeta 0}_0$ and $\widetilde{a}^+_\zeta$ (defined by the suitable analogues of formulae (3.55)) are substituted. The point $\zeta \in \mathbb{R}$ is chosen arbitrarily. Here we apply Proposition3.6 at full potential. Indeed, to compute the variation of a^+_ζ in (3.55), it is necessary to replace the index "$-$" by ζ (analogously for $b^{\zeta 0}_-$). Therefore we must use the formula for a^z_y and b^z_y.

Exercise3.7. *Deduce relations (3.57) by the above method. In particular, check the following identities resulting from Proposition2.5:*

$$\frac{\delta b^0(\alpha)}{\delta r(x)} = a^x_-(\alpha)\overline{a^+_x}(\alpha)\exp(-2i\alpha x),$$

$$\frac{\delta b^0(\alpha)}{\delta \overline{r}(x)} = -b^x_-(\alpha)b^+_x(\alpha)\exp(2i\alpha x),$$

and similar formulae for b^{z0}_y in which the indices "$-$" and "$+$" are changed by y and z respectively. (Use the fact that the expression for $\widetilde{c}^0_1$ is independent of ζ and let ζ tend to ∞ in the final formula for δc^0_1.) □

3.5. Application: DNS equation.

Consider the following generalization of the NS equation which is of practical importance:

$$ir_t = r_{xx} - i\omega(\bar{r}r^2)_x, \qquad \omega = \pm 1, \tag{3.58}$$

called the *derivative nonlinear Schrödinger equation* (DNS). This is also a soliton equation and the associated linear problem is close to (3.43). However it has new features and will illustrate better some of the general results of §2. In this section we apply the inverse problem method to (3.58). Before this, let us show how the NS arises as a limit of the DNS.

Exercise3.8. *Let r be a solution of (3.58). Then for $\beta > 0$ the function*

$$r'(x,t) = \frac{1}{\sqrt{2\beta}} e^{i\beta^{-2}t - i\beta^{-1}x} r(x - 2\beta^{-1}t, t)$$

satisfies the equation

$$ir'_t = r'_{xx} + 2\omega\bar{r}'(r')^2 - 2i\omega(\bar{r}'(r')^2),$$

which degenerates to the NS when $\beta \to \infty$.

Equation (3.58) is the compatibility condition of the following two equations for the invertible matrix function $\widetilde{\Phi}$.

$$\widetilde{\Phi}_x = \begin{pmatrix} -i\zeta^2 & -\omega\bar{r}\zeta \\ r\zeta & i\zeta^2 \end{pmatrix} \widetilde{\Phi}, \tag{3.59a}$$

(3.59b)

$$i\widetilde{\Phi}_t\widetilde{\Phi}^{-1} = \begin{pmatrix} 2\zeta^4 - \omega\zeta^2 r\bar{r} & -2i\omega\zeta^3\bar{r} + \omega\zeta\bar{r}_x + i\zeta r\bar{r}^2 \\ 2i\zeta^3 r + \zeta r_x - i\omega\zeta r^2\bar{r} & -2\zeta^4 + \omega\zeta^2 r\bar{r} \end{pmatrix}.$$

Though the dependence on the "spectral parameter" ζ is complicated, this greatly resembles the zero curvature representation (1.28) from §1 for the NS equation. In order to apply the results we already know (without developing the scattering theory with quadratic dependence on the parameter from scratch), we will use the parameter $\alpha = \zeta^2$ instead of ζ. Let us fix notation:

$$\mu_-(x) = -\frac{\omega}{2}\int_{-\infty}^{x} r\bar{r}(x')\,dx', \qquad \mu_+(x) = -\frac{\omega}{2}\int_{x}^{\infty} r\bar{r}(x')\,dx', \qquad \mu = \mu_- + \mu_+.$$

Note that μ does not depend on t.

Lemma3.4. *Set* $\Phi = \begin{pmatrix} 1 & 0 \\ 0 & \zeta^{-1} \end{pmatrix} \widetilde{\Phi} \begin{pmatrix} 1 & 0 \\ 0 & \zeta \end{pmatrix}$, *where* $\widetilde{\Phi}$ *is defined by (3.59). Then*

$$\Phi_x \Phi^{-1} = \begin{pmatrix} -i\alpha & -\omega\overline{r}\alpha \\ r & i\alpha \end{pmatrix}. \tag{3.60}$$

Analogously, only α *appears in the expression for* $\Phi_t\Phi^{-1}$. *For this* Φ, *if we introduce* Φ^{red} *from the relation*

$$\Phi(\Phi^{\mathrm{red}})^{-1} = \begin{pmatrix} 1 & \frac{i\omega}{2}\overline{r} \\ 0 & 1 \end{pmatrix} \begin{pmatrix} \exp(i\mu_-(x)) & 0 \\ 0 & \exp(-i\mu_-(x)) \end{pmatrix}, \tag{3.60a}$$

then

$$\Phi^{\mathrm{red}}_x(\Phi^{\mathrm{red}})^{-1} = \begin{pmatrix} -i\alpha & \exp(-2i\mu_-)(\overline{r}^2 r/4 - i\omega\overline{r}_x/2) \\ r\exp(2i\mu_-) & i\alpha \end{pmatrix}. \tag{3.60b}$$

□

Following the procedure given in §2, we can construct solutions $\widetilde{E}_\pm(x;\zeta)$ of equation (3.59a) for $\alpha = \zeta^2 \in \mathbb{R}$ normalized as follows:

$$\widetilde{E}_\pm(x;\zeta) \to \exp\begin{pmatrix} -i\alpha x & 0 \\ 0 & i\alpha x \end{pmatrix} \qquad \text{when } x \to \pm\infty.$$

We assume that the functions r and r_x are absolutely integrable with respect to x on the real axis. Let us introduce the matrix $\widetilde{T}(\zeta)$ by the usual relation $\widetilde{E}_- = \widetilde{E}_+\widetilde{T}$. It is more convenient to use $E_\pm = \begin{pmatrix} 1 & 0 \\ 0 & \zeta^{-1} \end{pmatrix} \widetilde{E}_\pm \begin{pmatrix} 1 & 0 \\ 0 & \zeta \end{pmatrix}$ and $T = \begin{pmatrix} 1 & 0 \\ 0 & \zeta^{-1} \end{pmatrix} \widetilde{T} \begin{pmatrix} 1 & 0 \\ 0 & \zeta \end{pmatrix}$ rather than $\widetilde{E}_\pm$ and $\widetilde{T}$ (see Lemma3.4). These matrices $E_\pm$ and T are the $E_\pm$-functions and the monodromy matrix for equation (3.60). These functions depend directly on $\alpha \in \mathbb{R}$.

The matrix $\widetilde{T}$ has the form $\widetilde{T} = \begin{pmatrix} \widetilde{a} & -\omega\overline{\widetilde{b}} \\ \widetilde{b} & \overline{\widetilde{a}} \end{pmatrix}$ for suitable functions $\widetilde{a}(\alpha)$ and $\widetilde{b}(\zeta)$ ($\alpha \in \mathbb{R}$). Indeed, $\widetilde{T}$ is $*$-unitary and $\det \widetilde{T} = 1$, since $\widetilde{\Phi}_x\widetilde{\Phi}^{-1}$ of (3.59a) is $*$-anti-hermitian and $\mathrm{Sp}\,\widetilde{\Phi}_x\widetilde{\Phi}^{-1} = 0$. Let us recall that the anti-involution $*$ is given by the formula $A^* = \Omega A^+\Omega$, $\Omega = \mathrm{diag}(1,\omega)$. Set

$$T = \begin{pmatrix} a & -\omega\overline{b}\alpha \\ b & \overline{a} \end{pmatrix}, \qquad \varphi^1 = e^1_-, \qquad \varphi^2 = e^2_+, \qquad \Phi = (\varphi^1, \varphi^2), \tag{3.61}$$

Here $a = \widetilde{a}$ and $b = \zeta^{-1}\widetilde{b}$ by the definition of T and $\widetilde{T}$. The functions a, b and Φ depend on $\alpha \in \mathbb{R}$. We have the relations

$$(3.61') \qquad a\bar{a} + \omega\alpha b\bar{b} = 1, \qquad \varphi^1 = ae^1_+ + be^2_+, \qquad \varphi^2 = \alpha\omega\bar{b}e^1_- + ae^2_-.$$

Let

$$\varphi^*(\alpha) = {}^t(\overline{\varphi_2(\overline{\alpha})}, -\omega\overline{\varphi_1(\overline{\alpha})}) \qquad \text{for } \varphi = {}^t(\varphi_1, \varphi_2)$$

as above.

Proposition3.7. *a) The functions a, φ^1 and φ^2 are continuous and analytically continued with respect to α in the upper half-plane* $\operatorname{Im}\alpha > 0$. *When* $|\alpha| \to \infty$ *and* $\operatorname{Im}\alpha \geqq 0$,

$$(3.62) \quad a(\alpha) \to e^{i\mu}, \qquad \Phi \exp\begin{pmatrix} i\alpha x & 0 \\ 0 & -i\alpha x \end{pmatrix} \to \begin{pmatrix} \exp(i\mu_-) & i\omega\overline{r}\exp(i\mu_+)/2 \\ 0 & \exp(i\mu_+) \end{pmatrix}.$$

b) For $\alpha \in \mathbb{R}$ the functions φ^1 and φ^2 are related by:

$$(3.63) \qquad \frac{1}{a}(\varphi^1, \varphi^2)\begin{pmatrix} 1 & \bar{b} \\ -b & -\omega\alpha^{-1} \end{pmatrix} = \begin{pmatrix} 1 & 0 \\ 0 & \alpha^{-1} \end{pmatrix}(\varphi^{2*}, \varphi^{1*}).$$

Proof. By considering Φ^{red}, we establish the analyticity and asymptotic behaviour of the functions a and φ^1. Indeed, the functions

$$(E^{\text{red}})_- \overset{\text{def}}{=} (E_-)^{\text{red}},$$
$$(E^{\text{red}})_+ \overset{\text{def}}{=} (E_+)^{\text{red}}\operatorname{diag}(e^{i\mu}, e^{-i\mu})$$

have the same normalizations as $E_\pm$, and satisfy (3.60b). Correspondingly, we have $a = a^{\text{red}}\exp(i\mu)$ and $b = b^{\text{red}}\exp(-i\mu)$. We then apply theorems from §2 to equation (3.60b) to prove statement a). Furthermore, the functions $\widetilde{\varphi}^1 = e^1_-$ and $\widetilde{\varphi}^2 = \widetilde{e}^2_+$ satisfy relation (3.41), in which b is replaced by $\widetilde{b} = b\zeta$ and we substitute $\overline{\widetilde{b}} = \zeta\bar{b}$ for $\bar{b}$. Thus in order to prove relation (3.63), it suffices to conjugate by $\begin{pmatrix} 1 & 0 \\ 0 & \zeta \end{pmatrix}$. □

Analogous to (3.39), we set

$$(3.64) \qquad \varphi^1(\alpha_j) = b_j\varphi^2(\alpha_j), \qquad b_j \in \mathbb{C}^*, \qquad 1 \leqq j \leqq N$$

at the zeros $\alpha_1, \dots, \alpha_N$ of the coefficient a in the upper half-plane (we assume that $a \neq 0$ on the real axis). Then *the function r is uniquely determined by the scattering data $b(\alpha)$, $\alpha \in \mathbb{R}$, $\{\alpha_j, b_j\}$*. Let us discuss this in detail.

The right hand side of (3.59a) vanishes when $\zeta = 0$. Consequently, $\widetilde{E}_\pm(\zeta = 0) = \boldsymbol{I}$ and

$$a(0) = 1, \qquad \Phi(\alpha = 0) = \begin{pmatrix} 1 & 0 \\ (\cdot) & 1 \end{pmatrix}, \tag{3.65}$$

where $(\cdot)$ denotes a certain function of x. The asymptotics

$$\Phi \exp\begin{pmatrix} i\alpha x & 0 \\ 0 & -i\alpha x \end{pmatrix} \to \begin{pmatrix} (\cdot) & (\cdot) \\ 0 & (\cdot) \end{pmatrix}$$

when $|\alpha| \to \infty$ (see (3.62)), the normalization condition (3.65), and relation (3.64) make the Riemann problem (3.63) uniquely solvable for known a, b and $\{\alpha_j, \beta_j\}$. However, the coefficient a can be recovered from its zeros and the value of the function $|a|^2 = 1 - \omega\alpha|b|^2$ on the real axis because of the normalization $a(0) = 1$. Therefore a is uniquely determined by b and $\{\alpha_j\}$.

The function r can be found using Φ from the corresponding differential equation. In the case of the DNS (as in the case of the NS equation), it is convenient to use (3.60) in a neighbourhood of $\alpha = \infty$. When $|\alpha| \to \infty$ and $\operatorname{Im}\alpha \geqq 0$, we have the following relations:

$$\varphi_1^2 \exp(-i\alpha x) \to \frac{i}{2}\omega\bar{r}e^{i\mu_+}, \tag{3.66a}$$

$$\alpha\varphi_2^1 \exp(i\alpha x) \to \frac{i}{2} r e^{i\mu_-}. \tag{3.66b}$$

The first of these is included in formula (3.62), while the second is obtained by substituting (a part of) the asymptotic expansion of $\varphi^1 \exp(i\alpha x)$ for $|\alpha| \to \infty$ into (3.60). We can also derive b) from a), using the equality $\det \Phi(\alpha = \infty) = e^{i\mu}$ and the fact that $\widetilde{\Phi}$ is $*$-unitary in a neighbourhood of $\alpha = \infty$.

Note that for the first time in this chapter, it turned out that it is useful to normalize Φ at the two points $\alpha = 0, \infty$. The DNS equation can be also regarded as a special case of soliton equations for which the associated linear problem depends on the spectral parameter polynomially (go back to $\widetilde{\Phi}$, ζ from Φ, α). The Φ-functions of such equations are analytic in sectors of the complex plane (not necessarily in half-planes) and have more complicated asymptotic expansions for $|\alpha| \to \infty$ (see (3.62)).

The dependence of the scattering data on t is determined in the standard way. It suffices to study the behaviour of $\Phi_t\Phi^{-1}$ at $x \to \pm\infty$. Let us suppose that $r(x, t)$

satisfies (3.56) and that r, r_x and r_{xx} are absolutely integrable with respect to x on the real axis. Then $a_t = 0$, $(\alpha_j)_t = 0$ and

$$b(t;\alpha) = \exp(4i\alpha^2 t)b^0(\alpha), \qquad b_j(t) = \exp(4i\alpha_j^2 t)b_j^0, \qquad 1 \leqq j \leqq N, \tag{3.67}$$

i.e., they satisfy relation (3.46) as they do in the case of the NS. Analogously

$$c_j(t) \overset{\text{def}}{=} b_j a'(\alpha_j)^{-1} = \exp(4i\alpha_j^2 t)c_j^0,$$

where $a' = \partial a/\partial\alpha$.

N-soliton solutions. Let us construct solutions of (3.58) for which

$$b(\alpha) \equiv 0 \Leftrightarrow a(\alpha) = \exp(i\mu)\prod_{j=1}^{N}\frac{\alpha-\alpha_j}{\alpha-\overline{\alpha}_j},$$

$\operatorname{Im}\alpha_j > 0$ (cf. (3.47)). As in the previous section, $\widehat{\varphi}^1 = \varphi^1\exp(i\alpha x)$ and $\widehat{\varphi}^2 = \varphi^2\exp(-i\alpha x)$ are rational with respect to α and formula (3.63) holds on the entire complex plane.

Proposition3.8. *The N-soliton function φ^1 has a representation of the form:*

$$a^{-1}\varphi^1(\alpha)e^{i\alpha x} = \begin{pmatrix}1\\0\end{pmatrix}e^{-i\mu_+} + \sum_{j=1}^{N}\frac{c_j}{\alpha-\alpha_j}\widehat{\chi}^j, \qquad \widehat{\chi}^j \overset{\text{def}}{=} \varphi^2(\alpha_j)e^{i\alpha_j x}, \tag{3.68}$$

where the vector functions $\chi^j = \begin{pmatrix}\chi_1^j\\ \chi_2^j\end{pmatrix} \overset{\text{def}}{=} \operatorname{diag}(u, u^{-1})\widehat{\chi}^j$ *and* $u \overset{\text{def}}{=} e^{i\mu_+}$ *are defined by the following system of equations:*

$$\omega\chi_1^m + \sum_{k,j=1}^{N}\alpha_m C_{k,j}^m\chi_1^j = \alpha_m\sum_{k=1}^{N}\frac{\overline{c}_k\exp(2i(\alpha_m-\overline{\alpha}_k)x)}{\overline{\alpha}_k-\alpha_m}, \tag{3.69a}$$

$$\chi_2^m + \omega\sum_{k,j=1}^{N}\overline{\alpha}_k C_{k,j}^m\chi_2^j = \exp(2i\alpha_m x), \tag{3.69b}$$

where

$$C_{k,j}^m \overset{\text{def}}{=} \frac{\overline{c}_k c_j\exp(2i(\alpha_m-\overline{\alpha}_k)x)}{(\alpha_m-\overline{\alpha}_k)(\overline{\alpha}_k-\alpha_j)},$$

(cf. (3.49)), and

$$u = e^{i\mu_+} = 1 - \sum_{j=1}^{N} c_j \alpha_j^{-1} \chi_1^j.$$

The corresponding solution r of the DNS is given by the formula (cf. (3.50))

$$r = -2i \left(\sum_{j=1}^{N} c_j \chi_2^j \right) u^2. \tag{3.70}$$

Proof. This is similar to the proofs of Proposition3.5 and Corollary3.3. Using the expansion (3.68) and conjugating by $*$ at the points $\overline{\alpha}_1, \dots, \overline{\alpha}_N, \alpha_1, \dots, \alpha_N$, we get system (3.69). We use χ^j instead of $\widehat{\chi}^j$ to eliminate $e^{i\mu_+}$, that is also an unknown function and is to be computed via $\{\alpha_j, c_j\}$. The expression for u in terms of $\{\chi_1^j\}$ can be found by substituting the value $\alpha = 0$ in (3.68) and using formula (3.65). Finally, (3.70) follows immediately from relation (3.66b) (recall that $a(\infty) = e^{i\mu}$, $\mu = \mu_- + \mu_+$). □

Now we write down the formula for the one-soliton solution of the DNS. Let $N = 1$ and $\alpha_1 = \xi + i\eta$. We set hereafter

$$\eta = \Delta^2 \sin\gamma, \qquad \xi = \omega\Delta^2 \cos\gamma, \qquad c_1^0 = 2\eta\Delta^{-1}\exp(2\eta x_0 + 2i\sigma_0),$$

where $\Delta \in \mathbb{R}_+$, and the angle γ is chosen such that $0 < \gamma < \pi$. Note that $\alpha_1 = i\Delta^2 \exp(i\omega(\gamma - \pi/2)) = \omega\Delta^2 e^{i\omega\gamma}$. As always in this section, $\omega = \pm 1$ is from equation (3.58). Since $a(\alpha) = \dfrac{\alpha - \alpha_1}{\alpha - \overline{\alpha}_1} e^{i\mu}$ and $a(0) = 1$, we have $e^{i\mu} = \overline{\alpha}_1/\alpha_1 = e^{-2i\omega\gamma}$. Let

$$\theta = \eta(x - x_0 + 4\omega t\Delta^2 \cos\gamma), \qquad \sigma = \xi x + \sigma_0 + 2t\Delta^4 \cos(2\gamma).$$

Using Proposition3.8, we get

$$\chi_2^1 = \frac{4\eta^2 e^{2i\alpha_1 x}}{4\eta^2 + \omega\overline{\alpha}_1 c_1 \overline{c}_1 e^{-4\eta x}}, \qquad \chi_1^1 = \frac{2i\eta\alpha_1 \overline{c}_1 e^{-4\eta x}}{\omega 4\eta^2 + \alpha_1 c_1 \overline{c}_1 e^{-4\eta x}}, \tag{3.71}$$

$$u = e^{i\mu_+} = \frac{\omega + \overline{\alpha}_1 e^{4\eta(x_0 - x)}}{\omega + \alpha_1 e^{4\eta(x_0 - x)}} = \frac{e^{4\theta} + e^{-i\omega\gamma}}{e^{4\theta} + e^{i\omega\gamma}}. \tag{3.72}$$

Substituting χ_2^1, u^2 and the values of all the parameters into (3.70) and taking the evolution relation (3.67) into account, we get the formula:

$$r(x,t) = -4i\Delta \sin\gamma \frac{e^{2\theta + 2i\sigma}}{e^{4\theta} + e^{i\omega\gamma}} \frac{e^{4\theta} + e^{-i\omega\gamma}}{e^{4\theta} + e^{i\omega\gamma}}. \tag{3.73}$$

We note that by (3.72), we have the relation

$$\frac{1}{2}\int_{-\infty}^{\infty} r\overline{r}(x)\,dx = -\omega\mu = 2\gamma, \tag{3.74}$$

resulting in

$$\frac{1}{2}\int_{-\infty}^{\infty} r\overline{r}(x)\,dx = 2\sum_{j=1}^{N}\gamma_j \tag{3.74a}$$

for an arbitrary N-soliton solution r given by formula (3.70). Here $\gamma_j = \arg\alpha_j$ for $\omega = 1$ and $\gamma_j = \pi - \arg\alpha_j$ when $\omega = -1$ ($0 < \arg\alpha_j < \pi$). For λ_j with pairwise distinct $\xi_j = \operatorname{Re}\alpha_j$, the last formula follows directly from (3.74). Indeed, asymptotically for large t, the function r is a sum of one-soliton functions r_j which correspond to each zero α_j for $1 \leqq j \leqq N$. The distances between any two centers of r_j can be arbitrarily large. The point $x_0 = 4\xi t$ is called the center of the one-soliton function $r(x,t)$ in the form of (3.73). Using (3.74) and the fact that $\int_{-\infty}^{\infty} r\overline{r}(x)\,dx$ is independent of t, we get the desired identity. By continuity, we can check it for any $\{\xi_j\}$. More rigorously, we can use the relation $a(0) = e^{i\mu}\prod_{j=1}^{N}(\alpha_j/\overline{\alpha}_j) = 1$, which holds for any N-soliton r. This implies (3.74a) modulo 2π.

It follows from identity (3.74) that the "energy" $\frac{1}{2}\int_{-\infty}^{\infty} r\overline{r}(x)\,dx$ of a one-soliton solution of the DNS does not exceed 2π. This is not the case for the NS, as one-soliton solutions of this equation may have any energy by the formula

$$\frac{1}{2}\int_{-\infty}^{\infty} r\overline{r}(x)\,dx = 2\eta$$

(see Corollary3.3). We can explain this qualitatively as follows. As the DNS equation has a higher order of non-linearity, its solutions which have a stable shape must be relatively narrower and steeper than such solutions of the NS equation. The other important difference between multi-soliton solutions of the DNS and those of the NS is the appearance of much more complicated phase modulations. The next exercise reflects this.

Exercise3.9. a) *Let $r(x) = \Theta(x-y)\Theta(z-x)R$, where $R \in \mathbb{C}^*$ is a certain constant, $\Theta(x) = 1$ for $x \geqq 0$ and $\Theta(x) = 0$ for $x < 0$. By integrating equation (3.60b), show that for such a function r and $M \stackrel{\text{def}}{=} \omega R\overline{R}$ the coefficients a and b have the form*

$$a = e^{i\alpha(z-y)}\left(\cos(\alpha(z-y)\sqrt{1+M/\alpha}) - \frac{\sin(\alpha(z-y)\sqrt{1+M/\alpha})}{\sqrt{1+M/\alpha}}\right),$$

$$b = \frac{Re^{-i\alpha(z+y)}}{i\alpha\sqrt{1+M/\alpha}}\sin(\alpha(z-y)\sqrt{1+M/\alpha}).$$

b) For the same "step-like" r, but for the coefficients a and b given in §3.4 for the case of the NS, check the formulae

$$a = e^{i\alpha(z-y)}\left(\cos((z-y)\sqrt{\alpha^2+M}) - \frac{\alpha\sin((z-y)\sqrt{\alpha^2+M})}{\sqrt{\alpha^2+M}}\right),$$

$$b = \frac{Re^{-i\alpha(z+y)}}{i\sqrt{\alpha^2+M}}\sin((z-y)\sqrt{\alpha^2+M}).$$

Show that in the upper half-plane, the number of zeros and the imaginary parts of the zeros of a for the NS ($\omega = 1$) for rather small M grow almost uniformly relative to M. This is not true for the coefficient a of the step-like function in the case of the DNS (see a)). Check this by numerical experiments. □

Computation of $\delta\alpha_1$ for one-soliton. Using the method developed in the previous section let us find the variation $\delta\alpha_1$ for the one-soliton function $r^0(x)$ of the DNS given by formula (3.74) at $t = 0$. We have the relations:

$$\frac{\delta a(\alpha)}{\delta r(x)} = -\omega\alpha a_-^x(\alpha)\overline{b}_x^{+0}(\alpha)\exp(-2i\alpha x), \tag{3.75a}$$

$$\frac{\delta a(\alpha)}{\delta \overline{r}(x)} = -\omega\alpha b_-^{x0}(\alpha)a_x^+(\alpha)\exp(2i\alpha x), \tag{3.75b}$$

Here, the coefficients a_y^z and b_y^{z0} are defined as in the previous section for

$$r_y^{z0} = \Theta(z-x)\Theta(x-y)r^0(x).$$

These identities follow from (3.55). Indeed, the linear problem (3.59a) is obtained from (3.43) for $\mu_{1,2} = \mp i$ by replacing r and $\overline{r}$ by $r\zeta$ and $\overline{r}\zeta$ respectively. Consequently, for the DNS the right hand sides of (3.75) coincide with the right hand sides of formulae (3.55) after multiplying by $\omega\zeta$ and substituting $\widetilde{b} = \zeta b$ for b. The appearance of the coefficient ω is connected with the fact that identities (3.55) were written under the assumption that $\omega = 1$.

The coefficients a_-^x and b_-^{x0} are connected with φ^1 by the usual relations (cf. (3.53))

$$a_-^x\exp(-i\alpha x) = \varphi_1^1, \qquad b_-^{x0}\exp(i\alpha x) = \varphi_2^1.$$

Using (3.68), (3.71), and (3.72) we can determine φ^1 and, consequently, a_-^x and b_-^{x0} for the one-soliton function r^0. We have (in the notation of (3.73)):

$$a_-^x = e^{i\mu - i\mu_+}\left(1 - \frac{2i\eta}{(\alpha-\overline{\alpha}_1)}\frac{\Delta^2}{\Delta^2+\omega\alpha_1 e^{-4\theta}}\right), \tag{3.76a}$$

$$b_-^x = \frac{e^{i\mu+i\mu_+-2i\alpha x}}{(\alpha-\overline{\alpha}_1)}\frac{2\Delta\eta e^{2i\sigma-2\theta}}{\Delta^2+\omega\overline{\alpha}_1 e^{-4\theta}}. \tag{3.76b}$$

Using the multiplicative property of the T, we can derive the following formulae from (3.76):

$$a_x^+ = e^{i\mu_+}\left(1 - \frac{2i\eta}{(\alpha - \overline{\alpha}_1)} \frac{\overline{\alpha}_1 e^{-4\theta}}{\omega\Delta^2 + \overline{\alpha}_1 e^{-4\theta}}\right), \tag{3.77a}$$

$$b_x^+ = \frac{-e^{i\mu_+ - 2i\alpha x}}{(\alpha - \alpha_1)} \frac{2\Delta\eta e^{2i\sigma - 2\theta}}{\Delta^2 + \omega\overline{\alpha}_1 e^{-4\theta}}. \tag{3.77b}$$

Substituting (3.76) and (3.77) into (3.75), and then into the relation

$$\delta\alpha_1 = -\frac{2i\Delta^2 \sin\gamma}{\exp(-2i\omega\gamma)}\delta a(\alpha_1),$$

we get the identities

$$\frac{\delta\alpha_1}{\delta r(x)} = -2\frac{\Delta^3 \sin\gamma}{e^{-2i\omega\gamma}} \frac{e^{-2(\theta+i\sigma)}}{(e^{2\theta} + e^{-2\theta - i\omega\gamma})^2},$$

$$\frac{\delta\alpha_1}{\delta \overline{r}(x)} = 2\frac{\Delta^3 \sin\gamma}{e^{-2i\omega\gamma}} \frac{e^{2(\theta+i\sigma)}}{(e^{2\theta} + e^{-2\theta + i\omega\gamma})^2}.$$

The following form is more convenient

$$\frac{\delta\Delta^2}{\delta r(x)} = -\frac{\omega\Delta^2 \sin\gamma(e^{-i\omega\gamma + 2(\theta - i\sigma)} - e^{i\omega\gamma - 2(\theta + i\sigma)})}{(e^{2\theta} + e^{-2\theta}e^{-i\omega\gamma})^2},$$

$$\frac{\delta\gamma}{\delta r(x)} = -i\Delta \sin\gamma e^{-2i\sigma} \frac{e^{2\theta - i\omega\gamma} + e^{-2\theta + i\omega\gamma}}{(e^{2\theta} + e^{-2\theta}e^{-i\omega\gamma})^2},$$

$$\frac{\delta\Delta^2}{\delta\overline{r}(x)} = \overline{\left(\frac{\delta\Delta^2}{\delta r(x)}\right)}, \qquad \frac{\delta\gamma}{\delta\overline{r}(x)} = \overline{\left(\frac{\delta\gamma}{\delta r(x)}\right)}.$$

Exercise3.10. a) *Show that the zeros of the coefficients a_-^x and a_x^+ (from formulae (3.76a) and (3.77a)) are equal to $\xi + i\rho\eta$ and $\xi - i\overline{\rho}\eta$ respectively, where $\rho = (\Delta^2 - \omega\alpha_1 e^{-4\theta})/(\Delta^2 + \omega\alpha_1 e^{-4\theta})$. Check that, as in the case of the NS, a_-^x (or a_x^+) has a zero in the upper half-plane, if and only if $x > x_0$ (or $x < x_0$ for a_x^+).*

b) *Derive formulae for the coefficients a_y^z and b_y^{z0} from (3.76) and (3.77) (cf. Proposition3.6) and formulae for the coefficients a and b in the case of the "superposition" of nonintersecting semi-solitons (see §3.4).* □

The trace formulae. Let us briefly describe the relationship between local integrals of motion (3.58) and the scattering data. We assume that all derivatives of r with respect to x are absolutely integrable on the real axis. Following the methods of §1, Ch.I, we construct a formal invertible solution, $\Psi = \widehat{\Psi} \exp\left(\alpha \begin{pmatrix} -i & 0 \\ 0 & i \end{pmatrix} x\right)$, of equation (3.60a), where $\widehat{\Psi}$ is a formal matrix power series in α^{-1}. Set

$$\zeta = (\log \widehat{\psi}_1^1)_x = \sum_{j=1}^{N} \zeta_j \alpha^{-j}.$$

Then the series ζ does not depend on the choice of Ψ and its coefficients are expressed as differential polynomials of r and $\overline{r}$.

Let us compute ζ by the method of Proposition1.3 of Ch.I (§1.4). Here $n = 2$. We set

$$q_2^1 = -r \exp(2i\mu_-), \qquad q_1^2 = -\exp(-2i\mu_-)(\overline{r}^2 r/4 - i\omega \overline{r}_x/2).$$

In particular, by formula (1.18),

$$\zeta_1 = \frac{q_1^2 q_2^1}{\mu_1 - \mu_2} = \frac{i}{2}\frac{(r\overline{r})^2}{4} + \frac{\omega}{2} r\overline{r}_x$$

(recall that $\mu_{1,2} = \pm i$ for the DNS).

The right hand side of equation (3.60b) is not $*$-anti-hermitian. However, the results of §3.3 (especially Lemma3.3 and Propsosition3.3) can be easily proved in this more general situation. For Lemma3.3, it is almost obvious. For the proof of Proposition3.3, we must use the relation

$$|a^{\text{red}}|^2 + \alpha\omega |b^{\text{red}}|^2 = 1$$

and the analogous property of the matrix $(E^{\text{red}})_\pm$ instead of the fact that $E_\pm$ and T are $*$-unitary (see the proof of Proposition3.7). These properties hold since $\widetilde{E}_\pm$ are $*$-unitary. We remind that $a^{\text{red}} = a \exp(-i\mu)$ and $b^{\text{red}} = b \exp(i\mu)$. In particular, we obtain that $\log |a|$ decreases more rapidly than any power of $|\alpha|$ when $|\alpha| \to \infty$, and the vector function $\widehat{\varphi}^{1\text{red}} = e_-^{1\text{red}}(x; \alpha) \exp(i\alpha x)$ (where $e_-^{1\text{red}}$ is the first column of $(E^{\text{red}})_-$) has an asymptotic expansion, i.e. is a power series of α^{-1} when $|\alpha| \to \infty$. It follows from this, as in §3.3, that ζ coincides with the asymptotic series for $(\log \widehat{\varphi}_1^{1\text{red}})_x$.

Using Lemma2.2 (§2), we can compute the limits of $\log \widehat{\varphi}_1^{1\mathrm{red}}$ for $x = \pm\infty$. Expressing a^{red} in terms of a, we get the following variant of Proposition3.4:

$$\log a = i\mu + \sum_{k=1}^{\infty} c_k \alpha^{-k}, \qquad (|\alpha| \to \infty, \operatorname{Im} \alpha \geqq 0)$$

$$c_k = \frac{i}{\pi} \int_{-\infty}^{\infty} \alpha^{k-1} \log |a(\alpha)|\, d\alpha - \frac{1}{k} \sum_{j=1}^{N} (\alpha_j^k - \overline{\alpha}_j^k),$$

where the α_j are the zeros of a in the upper half-plane counted with multiplicities (cf. (3.18) for $t_p = t_1 = a$), and

$$c_k = \int_{-\infty}^{\infty} \zeta_k dx, \qquad k \geqq 1.$$

We assume here that a does not have any zeros on the real line. As in Corollary3.2, all the c_k are integrals of motion of equation (3.58), i.e. they do not depend on t. The simplest integral of (3.58) is $\mu = \frac{1}{2} \int_{-\infty}^{\infty} r\overline{r}(x)\, dx$. Using the formula for ζ above, we obtain the next integral and the first of the trace formulae for the DNS:

$$4 \sum_{j=1}^{N} \operatorname{Im} \alpha_j - \frac{2}{\pi} \int_{-\infty}^{\infty} \log |a|\, d\alpha = \int_{-\infty}^{\infty} \left(\frac{\omega}{2} \operatorname{Im}(r_x \overline{r}) - \frac{(r\overline{r})^2}{4} \right) dx.$$

Here we use the relation $\int_{-\infty}^{\infty} (r\overline{r})_x dx = 0$.

In conclusion, we present some results on the limiting passage from the DNS to the NS equation.

Exercise3.11. a) *Let $r'(x) = \Theta(x-y)\Theta(z-x)R$ for $R \in \mathbb{C}^*$, $\sqrt{\beta} > 0$ and $r = \sqrt{2\beta} e^{i\beta^{-1}x} r'(x)$ (r' is defined by the formula in Exercise3.8 in terms of r at $t = 0$). Show that for such a function r, $M = R\overline{R}$ and $\omega = 1$, the coefficients a and b are given by the formulae*

$$a(\alpha + \beta^{-1}/2) = e^{i\alpha(z-y)} (\cos((z-y)S) - \alpha \sin((z-y)S)/S),$$
$$b(\alpha + \beta^{-1}/2) = e^{-i\alpha(z-y)} \sin((z-y)S) \sqrt{2\beta} R/(iS),$$

where $S = \sqrt{\alpha^2 + M + 2\alpha\beta M}$. (This is a generalizes of the result of Exercise3.9a.

b) *Check that when $\beta \to 0$, the function $a(\alpha + \beta^{-1}/2)$ defined above uniformly converges for $\alpha \in \mathbb{R}$ to the coefficient $a(\alpha)$ for the step function r' (see a) in the case of the NS, calculated in Exercise3.9b.* □

Exercise3.12. *In formula (3.73) for the one-soliton solution $r(x,t)$ of the DNS equation, let $\alpha_1 = \alpha_1^0 + \omega\beta^{-1}/2$ (i.e., $\eta = \eta^0$, $\xi = \xi^0 + \omega\beta^{-1}/2$) and $2\sigma_0 = \pi/2 + \varphi_0$ without changing x_0. Then, when $\beta \to \infty$, the function r' written in terms of r by the formula in Exercise3.8 uniformly converges to the one-soliton solution of the NS equation of (3.51) with parameters η^0, ξ^0, φ_0 and x_0.* □

3.6. Comments. The study of concrete integrable equations using the inverse scattering problem method is a very important part of the soliton theory. The examples of the equations for which we apply this method are maily contained in the books [ZMaNP] and [TF2], and in [AS], [CD], [Lam]. There are quite a few papers on 2×2-matrix equations such as the NS, the Heisenberg magnet, and the Sin-Gordon equations. We are not attempting to review all of the literature on the subject.

V. E. Zakharov and A. V. Mikhailov discovered that the inverse problem method could be applied to the PCF equation (see [ZMaNP], [ZMi1,2]). They obtained formulas for several multi-soliton solutions and considered the basic reductions of the PCF, including the reduction to the S^{n-1} fields (see [ZMaNP], [ZMi3]). Here, we follow [Ch8,2], where the stabilizing case was considered in detail. In particular, we give a complete proof of the evolution relation for T, S and the discrete scattering data (see §3.1) and verify the existence of the asymptotic expansions for the Φ-functions. This differs from [ZMaNP] which suggested only a general approach to the study of the PCF based on the Riemann problem. We also analyze the symmetries of the scattering data for the O_n and S^{n-1}-fields, following [Ch2]. The computation of local integrals of motion of the PCF in the stabilizing case via the scattering data was done for the first time in [Ch8]. Our study of the trace formulae and the integrals of motion is similar in many ways to the study of the n-wave problem by V. E. Zakharov and S. V. Manakov ([ZMa], [Mana1]). Let us note that as in §2 (see §2.6), we do not assume the numbers $\{\mu_j\}$ to be pairwise distinct.

We also mention the contribution made by A. B. Shabat to the matrix inverse problem method (see [Sh1]) and the results from [TF2] related to chiral fields and other matrix equations. I. M. Krichever [Kri5] applied the Riemann problem to the construction of general local solutions of the PCF equation and the Sin-Gordon equation (actually, his approach was close to that from [ZMaNP]). We note that a system which is equivalent to the PCF system was studied in [BT] using the inverse problem method.

Regarding the trace formulae, the results by L. D. Faddeev (see, for example, [F2], [TF2]) were very important. The involutivity of integrals of motion (the end of §3.2) is an analytic form of the results by I. M. Gelfand and L. A. Dickey in the special case of first-order matrix differential operators (see [GD3,4]). A more detailed discussion of the involutivity can be found in [TF2]. We mention that M. Lüscher and K. Pohlmeyer suggested a construction of certain non-local integrals of motion for the PCF. Their construction is connected with the construction of T given in this chapter. Many studies are related to local and non-local conservation laws for chiral fields. Among others, we refer the reader to the works by A. M. Polyakov, A. B. Zamolodchikov, K. Pohlmeyer, G. Eichenherr, and M. Forger which are interesting from the viewpoint of both physics and mathematics.

In the special case of $n = 2$, the basic results of this section are known. More information related to this case can be found in [ZMaNP], [TF2], and [Ch5]. In particular, [TF2] contains a complete mathmatical foundation of the inverse problem method for $n = 2$. We followed [ZMaNP] in deriving the formula for the N-soliton solutions of the NS equation in §3.4. The scattering data for the "truncated" N-soliton solutions for the NS were calculated in [VCh1]. The formulae for the variations of the discrete spectral data in the one-soliton case are also contained there. The formula analogous to (3.56) for $\delta\alpha_1$ can be found in the works on the perturbations of one-soliton solutions, say, in the paper by V. I. Karpman and E. M. Maslov (1977). These formulae were applied to the problem of recovering the envelopes of optical impulses in optical fibres in [VCh2]. The integrability of the NS equation by the inverse problem method is due to V. E. Zakharov and A. B. Shabat. We note also [ZT] on the connection of the NS equation and the HM. The reader can find a more complete list of references on the NS and HM equations in [TF2].

In this section we did not consider the inverse problem method for the Sin-Gordon equation since this is well-studied. See [Takh], [AKNS1], [ZTF], [TF1] and [AKNS2] in addition to the references cited above. There are other soliton equations similar to the equations in this book, for example, the Pohlymeyer-Getmanov-Lund-Regge system on the chiral invariants of SU_2-fields (see [G], [LuR], [Poh], and [ZMaNP]), the massive Thirring model (E. A. Kuznetsov, A. V. Mikhailov and others). The technique of this chapter can be easily applied to these equations.

Our consideration of the DNS equation is based on [KaN]. We refine and systematize the computations from this paper. T. Kavata and H. Inoye also studied the inverse problem for this equation. Similar aspects of the DNS were studied by M. Wadati, H. Sanuki, K. Konno, I. Ichikawa and others. In collaboration with V. A. Vysloukh, the author obtained the formula for $\delta\alpha_1$ for the DNS, as well as the the formulae for the coefficients a and b in the case of a "step-function" (analogous formulae for the NS were known), and the analysis of the limiting procedure from the DNS to the NS.

REFERENCES

The information about English translations of papers/books in Russian is given after ";" or in { } (coinciding data are omitted).

[AKNS1] Ablowitz M.J., Kaup D. J., Newell A.C., Segur H.: Method for solving the Sine-Gordon eguation. Phys. Rev. Lett. **30**:25, 1262-1264 (1973).

[AKNS2] —,—,—,—: The inverse scattering transform – Fourier analisys for nonlinear problems. Stud. Appl. Math. **53**:4, 249- 315 (1974).

[AS] Ablowitz M.J., Segur H.: Solitons and the inverse scattering transform. SIAM, Philadelphia, 1981.

[A1] Arnold V.I.: Mathematical Methods of Classical Mechanics. Moscow, Nauka, 1974, Graduate Texts in Mathematics 60, New-York- Berlin- Heidelberg, Springer, 1978.

[A2] —: Ordinary differential equations. Moscow, Nauka, 1971.

[ADHM] Atiyah M.F., Drinfeld V.G., Hitchin N.J., Manin Yu. I.: Construction of instantons. Phys. Letters **A65**:3, 185-187 (1978).

[BBM] Babich M.V., Bobenko A.I., Matveev V.B.: Reductions of Riemann theta-functions to theta-functions of smaller genus and symmetries of algebraic curves. Dokl. Akad. Nauk SSSR **272**:1, 13-17 (1983).

[Ba] Baker H.F.: Note on the foregoing paper "Commutative ordinary differential operators" by J.L.Burchnall and T.W. Chaundy. Proc. Royal Soc. London **A118**, 584-593 (1928).

[Bä] Bäcklund A,V.: Zur Theorie der Partiellen Differential-gleichungen erster Ordnung. Mathematische Annalen **18**, 285-328 (1880).

[BC1] Beals R., Coifman R.: Inverse scattering and evolution equations. Comm. Pure Appl. Math. **37**, 29-42 (1985).

[BC2] —,—: Scattering and inverse scattering for first order systems, II. Inverse Problems **3**, 577-593 (1987).

[BDZ] Beals R., Deift P., Zhou X.: The inverse scattering transform on the line. In "Important Developments in Soliton Theory", A. S. Fokas and V. E. Zakharov, eds., Springer, Berlin, 1993, 7-32.

[BD] Belavin A.A., Drinfeld V.G.: Solutions of the classical Yang - Baxter equation for semisimple Lie algebras. Funct. Anal. and Appl. **16**:3, 1-29 {159-180} (1982).

[BZ] Belavin A.A., Zakharov V.E.: A multidimentional method of the inverse scattering problem and the duality equation for the Yang- Mills field. Pis'ma Zh. Exp. Teor. Fiz. **25**:12, 603-607 (1977).

[BE] Belokolos E.D., Enolsky V.Z.: Generalized Lamb ansatz. Theor. Math. Phys. **53**:2, 271-282 (1982).

[Bo] Bogoyavlensky O.I.: The integrals of higher stationary KdV equations and the eigenvalues of the Hill operator. Funct. Anal. and Appl. **10**:2, 9-12 {92-95} (1976).

[BN] Bogoyavlensky O.I., Novikov S.P.: The connection between the Hamiltonian formalisms of stationary and non-stationary problems. Funct. Anal. and Appl. **10**:1, 9-13 {8-11} (1976).

[Bor] Borovik A.E.: N-soliton solutions of the nonlinear Landau-Lifschiz equations. Pis'ma Zh. Exp. Teor. Fiz. **28**:10, 629-632 (1978).

[BC] Burchnall J.L., Chaundy T.W.: Commutative ordinary differential operators. Proc. Roy. Soc. London **A118** , 557-583 (1928).

[BT] Budagov A.S., Takhtajan L.A.: A nonlinear one-dimentional model of classical field theory with internal degrees of freedom. Dokl. Akad. Nauk SSSR **235**:4, 805-808 (1977).

[CD] Calogero F. Dagasperis A.: Spectral transform and solitons. I. Amsterdam, North-Holland, 1982.

[Ca] H. Cartan.: Théorie élémentaire des fonctions analytiques d'une ou plusieurs variables complexes. Hermann, Paris, 1961.

[CGW] Chau L. L., Wu Y.S.: Kac-Moody algebra in the self-dual Yang-Mills equations. Phys. Rev. **D25**, 1086-1094 (1982).

[Ch1] Cherednik I.V.: Algebraic aspects of two-dimensional chiral fields I. Itogi Nauki (Modern Problems in Math.), VINITI Akad. Nauk SSSR, Moscow **17**, 175-218 (1980); J. Sov. Math. (Plenum P.C.) **21**, 601-636 (1983).

[Ch2] —: Algebraic aspects of two-dimensional chiral fields II. Itogi Nauki (Algebra, Geometry, Topology), VINITI Akad. Nauk SSSR, Moscow **18**, 73-150 (1980); J. Sov. Math. (Plenum P.C.) **18**, 211-254 (1982).

[Ch3] —: Integrable differential equations and coverings of elliptic curves. Izv. Akad. Nauk. SSSR, Ser. Math. **47**:2, 384-406 (1983); **22**,357-377 (1983).

[Ch4] —: On a generalization of the differential equations KdV and Sin-Gordon. Funct. Anal. and Appl. **13**:1, 81-82 {68-69} (1979).

[Ch5] —: Differential equations for Baker-Akhiezer functions of algebraic curves. Funct. Anal. and Appl. **12**:3, 45-54 {195-203} (1978).

[Ch6] —: Local conservation laws for principal chiral fields (d=1). Theor. Math. Phys. **38**:2, 179-185 {120-124} (1979).

[Ch7] —: On finite-zoned solutions of the duality equation over S^4 and two-dimensional relativistically- invariant systems. Dokl. Akad. Nauk SSSR **246**:3, 575-578 (1979).

[Ch8] —: Conservation laws and elements of the scattering theory for principal chiral fields (d=1). Theor. Math. Phys. **41**:2, 236-244 (1979) {997-1002 (1980)} .

[Ch9] —: Quantum and classical chains for principal chiral fields. Funct. Anal. and Appl. **16**:3, 89-90 {75-76} (1982).

[Ch10] —: On regularity of "finite-zoned" solutions of matrix integrable equations. Dokl. Akad. Nauk SSSR **266**:3, 593-597 (1982); Soviet Phys. Doklady **27**.

[Ch11] —: Loop Groups in Soliton Theory. Chapters III-IV of "Algebraic Methods of Soliton Theory"(in Russian) (1994).

[Ch12] —: On an integrability of the equation of two-dimensional asymmetric O_3-fields and its quantum counterpart. Yadernaya Phys. {Soviet Journal of Nuclear Phys.} **33**:1, 278-282 {144-145} (1981).

[Ch13] —: On a definition of the τ-function for generalized affine Lie algebras. Funct. Anal. and Appl. **17**:3, 93-95 {243-245} (1983).

[C] Corones J.P.: Solitons and simple pseudopotentials. J. Math. Phys. **17**:5, 1867-1872 (1976).

[Dar] Darboux G.: Sur une proposition relative aux equations lineaires. Compt. Rend. **94**, 1456-1469 (1882).

[Dat1] Date E.: On a construction of multi-soliton solutions of the Pohlmeyer-Lund- Regge system and the Classical Massive Thirring model. Proc. Japan. Acad. **A55**:8, 278-281 (1979).

[Dat2] —: Multi-soliton solutions and quasi-periodic solutions of nonlinear equations of sine- Gordon type. Osaka J. Math. **19**:1, 125-158 (1982).

[DKM] Date E., Kashiwara M., Miwa T.: Vertex operators and τ-functions. Transformation groups for soliton equation. II. Proc. Japan. Acad. **A57**:8, 387-392 (1981).

[DJM] Date E., Jimbo M., Miwa T.: Method for generating discrete soliton equations. IV. Preprint RIMS-414, Kyoto Univ., 1982.

[DT] Deift P., Trubowitz E.: Inverse scattering on the line. Comm. Pure Appl. Math. **32**, 121-251 (1979).

[DS] Devinatz A., Shinbrot M.: General Wiener-Hopf operators. Trans. Amer. Math. Soc. **145**, 467-494 (1969).

[DB] Dodd R.K., Bullough R.K.: Polynomial conserved densities for the Sine-Gordon equations. Proc. Roy. Soc. London. **A352**:1671, 481-503 (1977).

[Dr] Drinfeld V.G.: On commutative subrings of certain non-commutative rings. Funct. Anal. and Appl. **11**:1, 11-14 (1977).

[Dr1] —: Quantum groups. In: Differential goemetry, Lie groups and mechanics.V. Zap. Nauchn. Semin. LOMI **155**, 18-49 (1986).

[DrM] Drinfeld V. G., Manin Yu.I.: The instantons and sheaves on CP^3. Funct. Anal. and Appl. **13**:2, 59-74 (1979).

[DrS] Drinfeld V. G., Sokolov V.V.: Equations of Korteweg-de Vries type and simple Lie algebras. Dokl. Akad. Nauk SSSR **258**:1, 11-16 (1981); Sov. Math. Doklady **23**, 457-461.

[DrS1] —, —: Lie algebras and and equations of Korteweg - de Vries type. Itogi Nauki (Modern Problems in Math.), VINITI Acad. Nauk SSSR, Moscow **24**, 81-180 (1984).

[Du1] Dubrovin B.A.: The periodic problem for the Korteweg-de Vries equation in the class of finite-zoned potentials. Funct. Anal. and Appl. **9**:3, 41-51 (1975).

[Du2] —: Theta functions and nonlinear equations. Uspekhi Mat. Nauk **36**:2, 11-80 (1981); Russian Math. Surveys, 11-92 (1982).

[Du3] —: Matrix finite-gap operators. Itogi Nauki (Modern Probl. in Math.) VINITI Acad. Nauk SSSR, Moscow **23**, 33-78 (1983).

[DuK] Dubrovin B.A.,Krichever I.M., Novikov S.P.: The Schrödinger equation in periodic field and Riemann surfaces. Dokl. Akad. Nauk SSSR **229**:1, 15-18 (1976).

[DuMN] Dubrovin B.A., Matveev V.B., Novikov S. P.: Nonlinear equations of Korteweg- de Vries type, finite-zone linear operators and abelian varieties. Uspekhi Mat. Nauk **31**:1, 55-136 (1976); Russian Math. Surveys, 59-146 (1976).

[DuNa] Dubrovin B.A., Natanson S.M.: Real two-gap solutions of the sine-Gordon equations. Funct. Anal. and Appl. **16**:1, 27-43 {21-33} (1982).

[DuNo] Dubrovin B.A., Novikov S.P.: Algebro-geometrical Poisson brackets for real finite-gap solutions of the sine-Gordon and nonlinear Schrödinger equations. Dokl. Akad. Nauk SSSR **267**:6, 1295-1300 (1982); Sov. Math. Doklady **26**, 760-765 (1983).

[DuNoF] Dubrovin B.A., Novikov S.P., Fomenko A.T.: Modern geometry. Methods and Applications. Moscow, Nauka (1979); Graduate Texts in Math. **93**, **104**, New York - Berlin -Heidelberg -Tokyo, Springer, 1984-85.

[Ei] Eisenhart L.P.: A treatise on the differential geometry of curves and surfaces. New York, N.Y., 1960.

[EsW] Estabrook F.B., Walhquist H.D.: Prolongation structures of nonlinear evolution equations. II. J.Math.Phys. **17**:7 1293-1297 (1976).

[F1] Faddeev L.D.: Properties of the S-matrix of the one-dimentional Schrödinger equation. Trudy Mat. Inst. Steklov **73**, 314-336 (1964).

[F2] —: The inverse problem of quantion scattering theory.II. Modern. Probl. in Math., VINITI Akad. Nauk SSSR, Moscow **3**, 93-181 (1974); Soviet J. Math. (Plenum P.C.) **5**, (1976).

[F3] —: A Hamiltonian interpretation of the inverse scattering method. In. Solutions, edited by Bullough R.K., Caudrey P.J. Berlin, New York, Springer, 339-354 (1980).

[F4] —: Quantum completely integrable models in field theory. Mathematical Physics Review. Sect. C. **1**. 107-155 (1980).

[Fay] Fay J.: Theta-functions on Riemann surfaces. Lect. Notes in math., Springer **352** (1973).

[Fl] Flashka H.: Construction of conservation laws for Lax equations: comments on a paper by G.Wilson. Quart. J. Math **34**:133, 61-65 (1983).

[FNR] Flashka H., Newell A. C., Ratiu T.: Kac-Moody Lie algebras and solution equations. II-IV. Physica D **9**:2, 303-345 (1983).

[GGKM] Gardner C.S., Greene J.M., Kruskal M.D., Miura R.M.: Method for solving the Korteweg-de Vries equation. Phys. Rev. Lett. **19**:19, 1095-1097 (1967).

[GD1] Gelfand I.M., Dikey L.A.: Lie algebra structure in formal variational calculus. Funct. Anal. and Appl. **10**:1, 18-25 (1976).

[GD2] —, —: Asymptotic behaviour of the resolvent of Sturm-Liouville equations and the algebra of the Korteweg-de Vries equations. Usp. Mat. Nauk. **30**:5, 67-100 (1975); Russian Math. Surveys, 77-133.

[GD3] —, —: The resolvent and Hamiltonian systems. Funct. Anal. and Appl. **11**:2, 11-27 {93-104} (1977).

[GD4] —, —: Calculation of jets and nonlinear Hamiltonian equations. Funct. Anal. and Appl. **12**:2, 8-23 (1978).

[G] Getmanov B.S.: A new Lorentz-invariant system with exact multisoliton solutions. Pis'ma Zh. Exp. Teor. Fiz. **25**:2, 132-136 (1977).

[GK] Gohberg I.C., Krein M. G.: Systems of integral equations on the half-line with kernels depending on the difference of the arguments. Uspekhi Mat. Nauk **13**:2, 3-72 (1958).

[H] Hilbert D.: Über Flächen von Konstanter Gausscher Krümmung. Trans. Amer. Math. Soc. **2**, 88-99 (1901).

[I] Its A.R.: Inversion of hyperelliptic integrals and integration of nonlinear differential equations. Vestnik Leningrad Univ., Ser. mat.-mech.-astr. **7**:2, 39-46 (1976).

[IK] Its A.R., Kotlyarov V.P.: Explicit formulas for solutions of the nonlinear Schrödinger equation. Dokl. Akad. Nauk Ukr. SSR, **A11**, 965-968 (1976).

[IM] Its A.R.,Matveev V.B.: On Hill operators with a finite number of lacunae. Funct. Anal. and Appl. **9**:1, 69-70 (1975).

[Kac] Kac V.G.: Infinite dimensional Lie algebras. Birhäuser. Boston, Basel, Stuttgart, 1983.

[KaN] Kaup D.J., Newell A.C.: An exact solution for a derivative nonlinear Schrödinger equation. J. Math. Phys. **19**:4, 798-801 (1978).

[KoK] Kosel V.A., Kotlyarov V.P.: Almost-periodic solutions of the equations $u_{tt} - u_{xx} + \sin u = 0$. Dokl. Akad. Nauk Ukr. SSR **A-10**, 878-881 (1976).

[Kre] Krein M.G.: On the theory of accelerants and S-matrices of canonical differential systems. Dokl. Akad. Nauk SSSR **111**:6, 1167-1170 (1956).

[Kri1] Krichever I.M.: Methods of algebraic geometry in the theory of nonlinear equations. Uspekhi Mat. Nauk **32**:6, 183-208 {185-213} (1977).

[Kri2] —: Algebraic curves and nonlinear difference equations. Usp. Mat. Nauk **33**:4, 215-216 (1978).

[Kri3] —: Integration of nonlinear equations by methods of algebraic geometry. Funct. Anal. and Appl. **11**:1, 15-33 (1977).

[Kri4] —: Commutative rings of of linear ordinary differential operators. Funct. Anal. and Appl. **12**:3, 20-31 (1978).

[Kri5] —: Analogue of the D'Alembert formula for the equation of principal chiral field and the sine-Gordon equation. Dokl. Akad. Nauk SSSR **253**:2, 288-292 (1980); Sov. Math. Doklady **22**:2, 79-84 (1981).

[Kri6] —: Nonlinear equations and elliptic curves. Itogi Nauki (Modern Problems in Math.), VINITI Akad. Nauk SSSR **23**, 79-136 (1983).

[KN] Krichever I.M., Novikov S.P.: Holomorphic bundles over Riemann surfaces and the Kadomtsev- Petviashvili equation. Funct. Anal. and Appl. **12**:4, 41-52 {276-286} (1978).

[K1] Kulish P.P.: Conservation laws for the string in a static field. . Theor. Math. Phys. **33**:2, 272-275 (1977).

[K2] —: On action-angle variables for the multi-component nonlinear Schrödinger equation. In: Boundary-value problems of mathematical physics and related questions in function theory. 14. Zapiski Nauchn. Semin. LOMI **115**, 126-136 (1982).

[KW] Kupershmidt B.A., Wilson G.: Conservation laws and symmetry of generalized Sine-Gordon equations. Comm. Math. Phys. **81**, 189-202 (1981).

[Lam] Lamb G.L.,Jr.: Elements of Soliton Theory. New York: Wiley, 1980.

[La] Lang S.: Introduction to Algebraic and Abelian Functions. Mir. Moscow (1976).

[Lax1] Lax P.D.: Integral of nonlinear equations of evolution and solitary waves. **21**:5, 467-490 (1968).

[Lax2] —: Periodic solutions of the KdV equation. Comm.Pure Appl. Math. **28**:1, 141-188 (1975).

[LeS] Leznov A.N., Saveliev M.A.: Representation of zero curvature for the system of nonlinear partial differential equations $\chi_{\alpha,z\bar{z}} = (\exp K\chi)\alpha$ and its integrability. Lett. Math. Phys. **3**:5, 489-494 (1979).

[Lu] Lund F.: Classically solvable field theory model. Ann. of Phys. **115**, 251-268 (1978).

[LuR] Lund F., Regge T.: Unified approach to strings and vertices with solitons. Phys. Rev. **D14**, 1524-1535 (1976).

[Mana1] Manakov S. V.: Example of a completely integrable nonlinear wave field with nontrivial dynamics (Lee model). Theor.Math.Phys. **28**:2, 172-179 {709-714} (1976).

[Mana2] —: On the theory of two-dimentional stationary self- focusing of electromagnetic waves. Zh. Exp. Teor. Fiz. **65**:2, 505-516 (1973); Sov. Phys. JETP **38**, 248-253 (1974).

[Mana3] —: Note on the integration of Euler's equations of the dynamics of an n-dimentional rigid body. Funct. Anal. and Appl. **10**:4, 93-94 {328-329} (1976).

[Mani1] Manin Yu.I.: Algebraic aspects of nonlinear differential equations. Itogi Nauki (Modern Problems in Math.), VINITI Akad. Nauk SSSR, Moscow **11**, 5-152 (1978).

[Mani2] —: Matrix solitons and vector bundles over curves with singularities. Funct. Anal. and Appl. **12**:4, 53-64 (1978).

[Mar] Marchenko V.A.: Sturm-Liouville Operators and their Applications. Kiev, Naukova Dumka (1977).

[Mat1] Matveev V.B.: Some comments on the rational solutions of the Zakharov-Schabat equations. Lett. in Math. Phys. **3**, 503-512 (1979).

[Mat2] —: Darboux transformations and nonlinear equations. Compt. Rend. RCP **264**, 247-264 (1980).

[Mat3] —: Abelian functions and solitions. Preprint N 373, Inst. Theor. Phys. Univ., Wroclaw, 1979.

[McKvM] Mc Kean H.P., van Moerbeke P.: The spectrum of Hill's equation. Invent. Math. **30**:3, 217-274 (1975).

[MOP] Mikhailov A. V., Olshanetsky M. A., Perelomov A.M.: Two-dimentional generalized Toda lattice. Comm. Math. Phys. **79**:4, 473-488 (1981).

[Mi] Miura R.(editor): Bäcklund transformations. Lecture Notes in Math. Berlin: Springer, **515**, (1976).

[vMM] van Moerbeke P., Mumford D.: The spectrum of difference operators and algebraic curves. Acta Mathematica **143**, 93-154 (1979).

[Mu] Mumford D.: Tata lectures on theta. I,II. Progress in math. Boston, Basel, Stutgart: Burkhäuser Boston, (1983,1984).

[NP] Neveu A., Papanicolaou N.: Integrability of the classical $[\bar{\psi}_i\psi_i]_2^2$ and $[\bar{\psi}_i\psi_i]_2^2 - [\bar{\psi}_i\gamma_5\psi_i]_2^2$ interactions. Comm. Math. Phys. **58**:1,31-64 (1978).

[N1] Novikov S.P.: The periodic problem for the Korteweg-de-Vries equation. Funct. Anal. and Appl. **8**:3, 54-66 {236-246} (1974).

[N2] —: Commuting operators of rank $l > 1$ with periodic coefficients. Dokl. Akad. Nauk SSSR **263**:6, 1311-1314 (1982).

[N3] —: Two-dimentional Schrödinger operators in periodic fields. Itogi Nauki (Modern Problems in Math.), VINITI Akad. Nauk SSSR, Moscow **23**, 3-32 (1983).

[OPSW] Ogielski A.T., Prasad H.K., Sinha A., Wang L.L.C.: Bäcklund transformations and local conservation laws for principal chiral fields. Phys. Lett. **91B**:3-4, 387-391 (1980).

[Poh] Pohlmeyer K.: Integrable Hamiltonian systems and interaction through guadratic constraints. Comm. Math. Phys. **46**:3, 207-221 (1976).

[Pol] Polyakov A.M.: String representations and hidden symmetries for gauge fields. Phys. Lett. **82B**:2, 247-250 (1979).

[PS] Pressley A., Segal G.: Loop groups. Oxford University Press., 1988.

[RSTS] Reyman A.G., Semenov– Tian-Shansky M.A.: A family of Hamiltonian structures, a hierarchy of Hamiltonians and reduction for first order matrix differential operators. Funct. Anal. and Appl. **14**:2, 77-78 {146-148} (1980).

[Sa] Sall M.A.: Darboux transformations for non-abelian and non-local equations of Toda type. Theor. Math. Phys. **53**:2, 227-237 (1982).

[STSF] Semenov– Tian-Shansky M.A., Faddeev L. D.: On the theory of nonlinear chiral fields. Vestn. Leningr. Univ. **13**:3, 81-88 (1977).

[Se1] Serre J.-P.: Groupes algebriques et corps de classes. Paris Herman, 1959.

[Se2] —: Cohomologie Galoisienne. Berlin- Göttingen- Heidelberg- New York: Springer-Verlag, 1964.

[Sh1] Shabat A.B.: The inverse scattering problem for a system of differential equations. Funct. Anal. and Appl. **9**:3, 75-78 (1975).

[Sh2] —: An inverse scattering problem. Differencialnye Uravneniya **15**:10, 1824-1834 (1979) {1299-1307 (1980).

[Sk] Sklyanin E.K.; On complete integrability of the Landau-Lifshitz equations. LOMI preprint E-3-1979, Leningrad (1979).

[Sy] Sym A.: Soliton Surfaces. Lett. Nuovo Cimento **33**:12, 394-400, **36**:10, 307-312.

[Takh] Takhtajan L.A.: Exact theory of propagation of ultra-short optical pulses in two-level media. Zh. Exp. Teor. Fiz. **66**:2, 476-489 (1974).

[TF1] Takhtajan L.A., Faddeev L.D.: Essentially nonlinear one-dimentional model of classical field theory. Theor. Math. Phys. **21**:2, 160-174 (1974).

[TF2] —, —: A Hamiltonian approach in the soliton theory. Moscow, Nauka 1986.

[Taka] Takasaki K.: A new approach to the self-dual Yung-Millis equations. Comm. Math. Phys. **94**, 35-59 (1984).

[Tc] Tchebychef P.: Sur la coupe des vêtements. Assoc. Franc., Oeuvres II (1978).

[U] Ueno K.: Infinite dimentional Lie algebras acting on chiral field and the Riemann-Hilbert problem. Publ. RIMS, Kyoto Univ. **19**, 59-82 (1983).

[UN] Ueno K., Nakamura Y.: Transformation theory for anti-self-dual equations. Publ. RIMS, Kyoto Univ. **19**, 519-547 (1983).

[UT] Ueno K., Takasaki K.: Toda Lattice Hierarchy. Preprint RIMS-425, Kyoto Univ. (1983).

[VV] Vladimirov V.S., Volovich I.V.: Local and nonlocal currents for nonlinear equations. Theor. Math. Phys. **62**:1, 3-29 (1985).

[VCh1] Vysloukh V.A., Cherednik I.V.: On restricted N-soliton solutions of the nonlinear Schrödinger equation. Theor. Math. Phys. **71**:1, 13-20 (1987).

[VCh2] —, —: A model of reconstruction of the envelope of ultrashort optical pulses by the characteristics of their interaction with probe one-soliton pulses. Dokl. Akad. Nauk SSSR **299**:1, 283-287 (1988); Soviet Phys. Doklady **33**, 182-184.

[W1] Wilson G.: Commuting flows and conservation laws for Lax equations. Math.Proc. Camb.Phil. Soc. **86**:1, 131-143 (1979).

[W2] Wilson G.: On two constructions of conservation laws for Lax equations. Quart. J. Math. Oxford **32**:128, 491-512 (1981).

[ZJ] Zagrodzinski J., Javorski M.: Mixed solutions of the Sine-Gordon equation. Z. Phys. B. Condensed Matter **49**, 75-77 (1982).

[ZF] Zakharov V.E., Faddeev L. D.: Korteweg-de Vries equation, a completely integrable Hamiltonian system. Funct. Anal. and Appl. **5**:4, 18-27 {280-287} (1971).

[ZMa] Zakharov V.E., Manakov S.V.: The theory of resonant interaction of wave packets in nonlinear media. Zh. Eksp. Theor. Phys. **69**:5, 1654-1673 (1975); Sov. Phys. JETP **42**, 842-850 (1976).

[ZMaNP] Zakharov V.E.,Manakov S.V., Novikov S.P., Pitaievski L.P.: Theory of Solitons. The Inverse Problem Method. Moscow, Nauka, 1980; New York, Plenum, 1984.

[ZMi1] Zakharov V. E., Mikhailov A.V.: An examle of non-trivial scattering of solitons in two-dimensional classical field theory. Pis'ma v. JETP **27**:1, 47-51 (1978).

[ZMi2] —, —: Relativistically invariant two-dimentional models of field theory which are integrable by means of the inverse scattering problem method. Zh. Eksp. Theor. Phys. **74**:6, 1953-1973 (1978); Sov. Phys. JETP **47**, 1017-1027 (1978).

[ZMi3] —, —: On the integrability of classical spinor models in two-dim. space time. Comm. Math. Phys. **74**, 21-40 (1980).

[ZS1] Zakharov V. E., Shabat A.B.: Integration of nonlinear equations of mathematical physics by the method of the inverse scattering problem. I. Funct. Anal. and Appl. **8**:3, 43-53 {166-174} (1974).

[ZS2] —, —: Integration of the nonlinear equations of mathematical physics by the method of the inverse scattering problem. II. Funct. Anal. and Appl. **13**:3 13-22 {226-235} (1979).

[ZT] Zakharov V.E., Takhtajan L.A.: Equivalence of the nonlinear Schrödinger equation and the Heisenberg ferromagnet equation. Theor. Math. Phys. **38**:1, 26-35 {17-23} (1979).

[ZTF] Zakharov V.E., Takhtajan L.A., Faddeev L.D.: A complete description of solutions of the "sine-Gordon" equation. Dokl. Akad. Nauk SSSR **219**:6, 1334-1337 (1974); Sov. Phys. Doklady **19**, 824-826 (1975).

INDEX